TABLE

IB	IIB	IIIA	IVA	VA	VIA	VIIA	noble gases
							He 2
		B 5	C 6	N 7	O 8	F 9	Ne 10
		Al 13	Si 14	P 15	S 16	Cl 17	Ar 18
Cu 29	Zn 30	Ga 31	Ge 32	As 33	Se 34	Br 35	Kr 36
Ag 47	Cd 48	In 49	Sn 50	Sb 51	Te 52	I 53	Xe 54
Au 79	Hg 80	Tl 81	Pb 82	Bi 83	Po 84	At 85	Rn 86
111	112	113	114	115	116	117	118

Tb 65	Dy 66	Ho 67	Er 68	Tm 69	Yb 70	Lu 71
Bk 97	Cf 98	Es 99	Fm 100	Md 101	No 102	Lw 103

UNDERSTANDING CHEMISTRY

UNDERSTANDING CHEMISTRY

Robert J. Ouellette
THE OHIO STATE UNIVERSITY

Harper & Row, Publishers
New York, Hagerstown, San Francisco, London

Sponsoring Editor: John A. Woods
Special Projects Editor: Carol J. Dempster
Project Editor: Lois Lombardo
Designer: Rita Naughton
Production Supervisor: Will C. Jomarrón
Compositor: William Clowes & Sons Limited
Printer/Binder: Halliday Lithograph Corporation
Art Studio: Vantage Art Inc.

UNDERSTANDING CHEMISTRY

Library of Congress Cataloging in Publication Data
Ouellette, Robert J Date-
Understanding chemistry.

Includes index.
1. Chemistry. I. Title.
QD31.2.086 540 75-30718
ISBN 0-06-044968-3

CONTENTS

PREFACE

The primary purpose of this text is to provide an approach to learning chemistry for students who have not previously had a chemistry course and whose science background is very limited. The secondary objective is to present the subject matter in sufficient detail and at a level on which the student may study and learn without the need of a lecturer to provide additional or clarifying material.

In length the text is designed to be covered in one semester. If it is used in a one-quarter course, some chapters would have to be cut or even deleted. Chapter 14, The Gas Laws, and Chapter 20, Nuclear Chemistry, are the only two chapters which could be deleted without loss of subject matter vital to the completion of other chapters. Any cutback within chapters would have to be selectively chosen to fit local course requirements.

Individual study by students who have had no previous chemistry background is facilitated by several features of the text:

(1) Statements of learning objectives at the start of each chapter.
(2) A chapter introduction that frequently employs analogies to non-chemical examples and places the material in perspective with previous chapters.
(3) Carefully defined terms and important concepts appearing in boldface.
(4) A chapter summary of terms.
(5) An emphasis on the vocabulary of chemistry that is accomplished by the introduction of a list of terms in the learning objectives, the use of boldface within chapters for these terms, a restatement of the terms in a summary, questions dealing with the terms at the end of each chapter, and a glossary at the end of the text.

(6) Detailed outlines and descriptions of problem solving starting from the rudiments of mathematics.
(7) Solved examples within the chapter that employ the techniques outlined in the chapter.
(8) A listing of problems at the end of the chapter that are labeled according to type and content and that follow the order of appearance of the subject matter within the chapter.

I am grateful to Dr. Carol Dempster of Harper & Row, Publishers for her suggestions and advice for rewriting the manuscript after the initial draft. Her scientific background and editorial skills have significantly contributed to the final presentation.

My wife, Margaret, made the work load easier by reading and correcting the first draft of the manuscript. Her ability to find and fill in missing words and incomplete thoughts and even to correct and reorder paragraphs are a tribute to her patience and understanding. In addition, she read all galley and page proofs.

I thank Carol Rose for her typing of the manuscript at various stages. Her talents include both high typing speed and an ability to decipher my handwriting.

Finally, in spite of the assistance received, I must take responsibility for the complete text. Although I trust that the errors are few, no text in its first edition escapes unscathed from criticism. Comments from teachers and especially students would be most welcomed. My mailing address is

The Ohio State University
Department of Chemistry
4042 Evans Laboratory
141 West 18th Avenue
Columbus, Ohio 43210

R.J.O.

PREFACE TO THE STUDENT

In this book you will encounter numerous theories, laws, symbols, and terms that are unfamiliar. Indeed the very language used may be quite different from any you have experienced in other college courses. **There is no easy way to learn chemistry. You must start memorizing facts as they are encountered and developing an understanding of each facet of chemistry as the story unfolds.** There is much interdependent material, and an absence of study for one week can have a serious effect on your overall performance in the course. You will not be able to read several chapters the night before an exam and do justice to the course.

Merely reviewing chapter material, even on a regular basis, is not enough for success in chemistry. **It is necessary to practice using the terms, symbols, and equations. Problems must be done regularly to develop proficiency in methods of approach.** Only by working out many, many problems can a full understanding of chemistry be acquired.

In this book there are numerous solved problems. **Attempt to do each one as you proceed through the book.** If you find it necessary to depend on the author's solution, return to the previous subject matter and learn why you were unable to solve the problem. Then try another problem of the same type to see if you have improved your ability.

Assigned material should be read several times. A first reading may give an overall impression and indicate what areas will need more careful study. A second reading should be dedicated to acquiring as complete an understanding as possible. The student who studies ahead of class will be in the best position to learn from the session. You can listen in a more relaxed manner and actively think about what is being said. Furthermore you will be prepared to listen closely to that subject matter which is the most troublesome to you. If all the material presented in class is new, you will be able to only transcribe words, which on the whole have little meaning. Attempting to learn from notes that were taken without any understanding is a futile process. **Prepare before every class and chemistry will not be the difficult subject that it is reputed to be.**

R.J.O.

UNDERSTANDING CHEMISTRY

INTRODUCTION TO CHEMISTRY

LEARNING OBJECTIVES

When you finish this chapter you should be able to:

(a) illustrate the impact of chemistry on the quality of our lives.
(b) distinguish the specialty areas of chemistry.
(c) demonstrate the scientific approach toward a specific problem.
(d) relate scientific observations to scientific law.
(e) differentiate between fact, hypothesis, and theory.
(f) match the following important terms with their corresponding definitions.

chemistry
hypothesis
law
matter
model
organic
scientific method
theory

Much of the material in this chapter will be familiar to you. You already have an understanding of some of the elementary principles of science as a result of your secondary school studies. Your awareness of chemistry has, no doubt, been expanded by both the press and the communications media.

Most of the things we use in daily life are not present in that form in nature. Practically all the products and goods of a modern society are the result of transformations of naturally occurring chemicals into useful materials by processes of chemistry. For example, oil is vitally important to an industrialized society because it is a major source of the chemicals used to form products as diverse as medicine, fertilizer, and plastics. The conversion of oil into these required products is accomplished by chemistry.

Each of the substances produced by chemists affects our lives in many ways. Most are beneficial but some are eventually shown to be harmful. Vinyl chloride, a major industrial chemical used in the production of vinyl plastics, has recently been shown to be a cancer-producing agent. Its use is being carefully monitored and evaluated.

In this chapter a brief summary, review, and overview of chemistry as it affects our lives are given. This introduction provides a common starting point for all students regardless of the extent of their previous chemistry background. The application of the scientific method to the understanding of natural phenomena is discussed.

1.1 WHAT IS CHEMISTRY?

The word chemistry in the author's younger years evoked an image of highly trained individuals in white lab coats who, by their efforts, were making the world a better place in which to live. The chemist was more knowledgeable and could make decisions that the average citizen could not and did not need to comprehend. Today students often have diametrically opposed viewpoints from those of the previous generation. Chemistry has altered their lives in many ways and it often seems not for the overall good. Chemists, and indeed all scientists, are not viewed today with the same awe and respect they were formerly. Furthermore, the citizen wants to know what chemistry and its related technology are going to do next. No doubt there is concern for the environment, the quality of our life, the air we breathe, the water we drink, the food we eat, and the limited energy sources of earth. But mere concern for our earth and its people, plants, and animals does not necessarily produce solutions. Indeed, many concerned individuals do not have sufficient knowledge of the sciences to decide rationally how the environment can best be served. To be informed citizens and voters, people will have to learn the fundamentals of science. The starting point in chemistry is to ask the question, "What is chemistry?"

CHEMISTRY
Matter
composition
structure
reactions

Chemistry is a science that deals with the composition, structure, and reactions of matter. Matter occupies space and has mass. Matter can be perceived by one or more senses and constitutes any physical body of the universe. The study of chemistry involves many complexities because of (1) the vast quantities of types of matter that are known, (2) the details of the structures for each type of matter that must be learned and understood, and (3) the virtually limitless transformations that various types of matter may undergo when they interact with each other.

The list of questions that chemistry will be employed to answer is one that is continually expanding and indeed will probably never grow shorter. Why does man age? Why do certain individuals have genetic abnormalities? How can clean efficient sources of energy be developed? What makes a strawberry red? What is responsible for the delightful aroma of a fine wine?

The chemist is interested principally in the composition and structure of matter. Only if sufficient knowledge is gained about the structure of the various types of matter can the practical consequences of chemistry be successfully and fully developed. The structure of matter is submicroscopic in scale and cannot be seen by the most powerful optical microscopes. However, the chemist has developed a host of detailed mental pictures or models of matter. These models are developed by inference from observations on large quantities of matter. A chemist studies the structure of matter in much the same way that a detective seeks to unravel a mystery—by accumulating circumstantial evidence (Figure 1.1). We realize that none of us will ever see the detailed structure of matter and hence will never know positively whether we are right or wrong. Nevertheless, we seek to derive such an overwhelming body of circumstantial evidence that the solution of the mystery is almost without doubt. Even when we are confident that the mystery is solved, the case is never closed. Each new bit of evidence is used to check over the solution to

Figure 1.1 The chemistry detective.

see if it is still valid. The process will never end as long as chemists are curious about the structure of matter in greater detail.

1.2 CHEMISTRY—BENEFICIAL OR MALEFICENT

This generation more than any that has preceded it is very aware of the tremendous impact that science in general and chemistry in particular have on the quality of our lives. For years the chemist could point with pride to the many contributions chemistry had made to the advancing technology and improved quality of life. Today chemists are more defensive. They hasten to point out that in reality the problems of technology are due to a complex mixture of social, economic, and individual factors, and are not purely chemical in nature.

Chemical technology in the plastics industry has revolutionized our way of living. We no longer have to worry about chipping or breaking dishes; china plates have been replaced with virtually indestructible plastic plates. Plastic containers of all sizes and shapes allow for convenient storage of foods. A sandwich wrapped in a plastic bag is guaranteed to maintain its freshness. The variety of plastics produced has allowed metal to be replaced in certain gears of machines. One of the major components of automobiles is plastic in its varied forms. In 1972, the American automobile industry used over 1.25 billion pounds of plastics.

Of course there is another side of the coin in the story of plastics. Although plastic packaging materials are inexpensive and convenient, their wide use has added immeasurably to the litter problem. Plastics have been designed for commercial reasons to have high resistance to sun, air, and water. Organisms cannot feed on and break down plastics to form substances that our environment can use naturally. Currently, plastics research is directed toward products that have acceptable shelf lives but are biodegradable in a reasonable time period. The chemicals can then be returned to nature for eventual reuse either by natural or by industrial processes.

Is the chemist responsible for roadside litter?

Chemicals for agriculture have made the United States the most efficient producer of food in the world. Without fertilizers and pesticides, we would have to employ many more people to produce food for this country and might not be able to export food to less fortunate countries. For several decades the United States with its well-developed agricultural chemical plants, has shown the way to feed the millions of inhabitants of this planet.

Again, as in the case of plastics, agricultural chemicals have some serious detrimental effects on the environment. Chemical fertilizers are washed into streams and lakes and contribute to the growth of algae at the expense of other aquatic life. Ultimately the balance of the water life is altered so severely that normal life ceases for all practical purposes. Perhaps the most notorious agricultural chemical is DDT, which is a Dr. Jekyll turned Mr. Hyde. In 1939, DDT was viewed as the chemical savior of mankind. It was and still is vital to the control of the malaria mosquito. In addition, DDT kills many agricultural pests. By destroying the natural pests of various crops, it has increased crop yields in all parts of the world. However, there are probably few who are unaware of the effect that the accumulation of this persistent and long-lived pesticide in the environment has on many animals. When one animal feeds on another it acquires that prey's lifetime DDT intake. Since each predator feeds on many prey, the very stable DDT successively accumulates to a higher percentage in each predator in the animal food chain. In many animals high levels of DDT interfere with their natural processes. For example, the shells of newly laid eggs of certain birds do not have sufficient structural strength to enable the offspring to survive. One solution for the problems caused by the use of DDT is to develop new agricultural chemicals that are sufficiently long-lasting in the environment to allow them to be useful but not so long-lived that accumulation occurs in either plants or animals.

The boll weevil causes 200 million dollars damage to cotton crops in a year.

Medicinal drugs have allowed the medical profession to alleviate suffering and extend life for individuals who only a few decades ago would have either existed in pain or died. There is no doubt that chemistry and medicine will be wedded for all future time. The picture is not without its unfortunate features. Side effects of drugs administered even by qualified physicians do cause some suffering and occasional death. The dependence of some members of society on medicinal drugs has become so great it approaches a national scandal. In addition, the younger generation has become so accustomed to a society in which legal drugs are widely used that for some it is easy to slip into addiction to illicit drugs.

Numerous other examples of both the good and bad effects of chemical technology on our lives could be cited. However, the reader is probably aware of many of them. Should chemistry be viewed as good or bad on balance? The author believes that a single judgment of this type is impossible and indeed inappropriate. Chemical knowledge, like all knowledge, is neither good nor bad; only the uses of knowledge can be judged. Technological problems have been created by people and they must be solved by people applying technology sensibly and with a strong social consciousness. The chemist cannot ignore a development in the laboratory that may tomorrow give us a wonder drug but also create problems in the decades ahead. Humanity would never progress if we were inactive for fear of the unknown future. Similarly, a scientist cannot hide a fact that might be presently viewed as detrimental to society in some of its potential uses if the same information properly applied in the future might be useful to mankind. We have to have faith in our ability to control technology for the good of the planet.

1.3 THE SCOPE OF CHEMISTRY

The field of chemistry is so broad that a wide variety of specialists are included under the umbrella of the term chemist. At one time lines could be drawn between various sciences. However, as illustrated in Figure 1.2, chemistry merges in one direction into physics and in another into biology. There are both chemical physicists and physical chemists who are interested in the same aspects of a natural phenomenon. Similarly, there are molecular biologists and biochemists studying the very same problems with but small differences in emphasis. In between these two ends of the chemistry spectrum there are many subdivisions or specialties.

At one time chemistry was regarded as sharply divided into organic and inorganic chemistry. The term *organic* meant pertaining to plant or animal

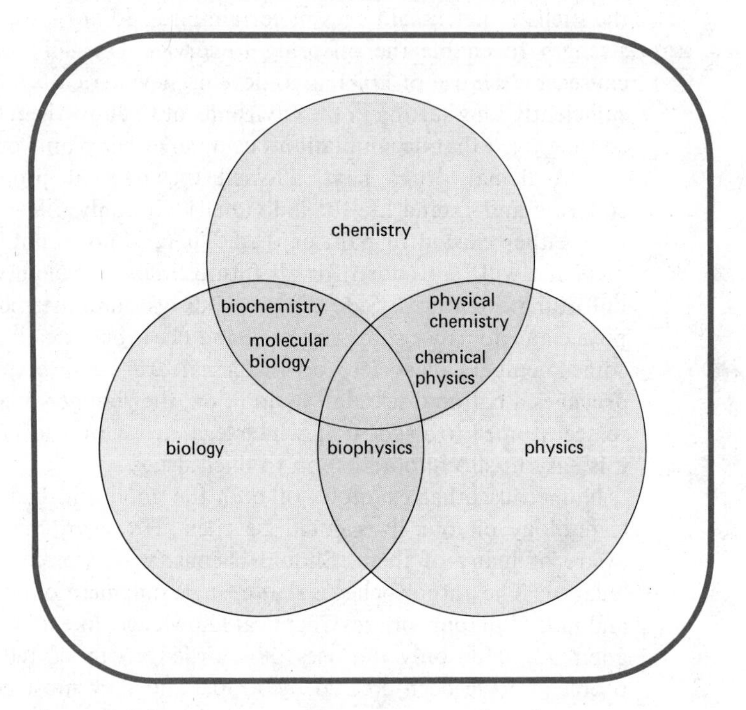

Figure 1.2 Common areas of interests among the sciences.

organisms and the matter of which they are composed. The matter contained in living sources was thought to contain a mysterious "vital force" which distinguished it from inanimate matter. All these organic substances contain carbon, and **organic chemistry today involves any carbon-containing substance regardless of whether it has ever been associated with life or not. Inorganic chemistry is the study of all substances other than those classified as organic.**

Other subdivisions of chemistry include analytical chemistry, biochemistry, electrochemistry, geochemistry, physical chemistry, and radiochemistry. Within each of these and many other areas of chemistry there are innumerable specialties. In the future, as the problems become more complex and new situations arise, there will undoubtedly be more areas defined. Two such areas might be astrochemistry and ecochemistry.

1.4 THE SCIENTIFIC METHOD

Observations and laws

All natural sciences are dependent on observations for their growth. **The method of making observations concerning a problem of interest, of cataloging observations, and of ultimately using these observations in the solution of the problem, is called the scientific method.**

In the development of science, qualitative observations were made very early. Human beings probably have been making general observations about their surroundings since their appearance on earth. Conveyance of these observations required communication, and written records gradually accumulated. Quantitative observation, measurement, is the major stepping-stone from speculation, exaggeration, and omission to the orderliness required for an understanding of one's surroundings. Once such measurements have been made, recorded, and delineated with respect to the conditions under which they were obtained, it is then possible for other individuals to reproduce the observations. The ability to reproduce an observation is a requirement in the development of science. Reproducibility ensures that an observation is a fact and not the creation of the observer's prejudices or desires.

Sufficient numbers of quantitative observations of the regularities in the behavior of the universe lead to the so-called laws of nature. **A law is simply an explicit statement of fact that is already inherent in the information obtained by observations.** Nothing new in understanding nature is obtained by stating a law; the law merely summarizes what has been observed (Figure 1.3).

A law is an explicit statement of fact about information obtained by observation.

Each of us could state laws that affect our lives. For example, we could say that our feet will descend and touch the floor when we arise from bed in the morning. We have made the observation many times and have complete confidence that this will occur every morning (Figure 1.4). Thus our law states a belief that there is an orderliness to nature and that under the same set of circumstances we can observe the same phenomenon again and again. Nature is not capricious. No scientist questions his faith that natural phenomena can be repeated time and time again. Occasionally it will appear that nature has failed us as identical circumstances lead to different outcomes. On closer examination, however, we find that the purported identical circumstances are in reality different. The apparent caprice in nature is usually only a reflection of our inexperience or ignorance of all factors controlling a natural phenomenon.

There is always the chance that an observation will be made that is radically different from that predicted by a law. Indeed, the law stands as a vulnerable target because one inconsistent prediction may lead to its revision or complete rejection. If such an event should occur, a search would commence for a new law to provide a more general statement that could account for all previously observed facts and the new observation. An example of such a

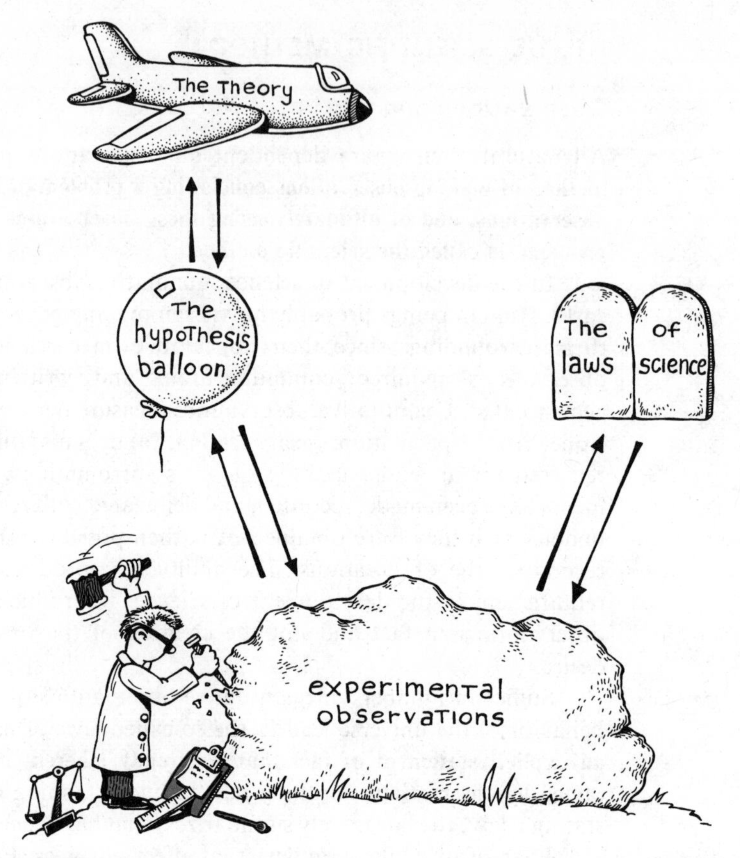

Figure 1.3 The scientific method.

change is in the law of conservation of mass: **Matter does not increase or decrease in mass during chemical reactions.** The observation had been tested and retested over nearly 200 years. However, with the advent of the atomic age, it was observed that matter is destroyed in a nuclear reaction and energy is produced. Further investigation showed that whenever the same quantity of matter is destroyed in a nuclear reaction, a definite amount of energy is produced. Thus the law of conservation of mass is replaced by the law of conservation of mass and energy: **Mass and energy cannot be destroyed but only interconverted between each other.**

Models

Science goes beyond the observation of facts and the formulation of these facts into laws. There is a natural human drive to learn about the intimate details of how and why matter behaves as it does. Scientists find it convenient to use models for natural phenomena. The term **model refers to an idea that**

corresponds to what is responsible for the natural phenomena. A model may be a mental picture of a phenomenon; but to communicate with others, verbal or written descriptions of the model are necessary. For precise communication and in order to remove ambiguities of language, geometrical or mathematical models commonly are used.

Picture models which are relatively easy to understand will be used often in this text. The student should realize that there are limitations in such models that keep them from describing all the more complex aspects of nature. Furthermore, the model must not be regarded as more important than the experimental facts. The experimental facts reflect nature; the model is a human creation. Finally, it should be noted that a scientific model is not a small-scale working version of a larger entity, as model planes are miniature representations of actual planes. Scientific models are visualizations of things that we usually cannot see. The models are imagined and are not replicas of the real thing.

Can a model be used in fields of study such as economics?

Models are conceived or postulated by an inductive process and are used to make predictions. There are no simple ways to arrive at models; they simply evolve as facts are gathered and rechecked by scientists. If a model enables the scientist to make accurate predictions, it is said to be validated. If it does not allow accurate predictions, alteration or complete replacement of the model may be required.

Scientists employ a general pattern for testing their models which is called the scientific method. All available experimental measurements and observations must be used to create a model. New experiments are then devised and carried out which will test the predictions of the model. Finally, the model may be modified or replaced as necessitated by the new observations (Figure 1.3).

The terms hypothesis and theory are often used in science. **A hypothesis refers to a tentative model, whereas a theory describes a model that has been tested and validated many times.** The dividing line between the two is arbitrary and cannot be precisely defined.

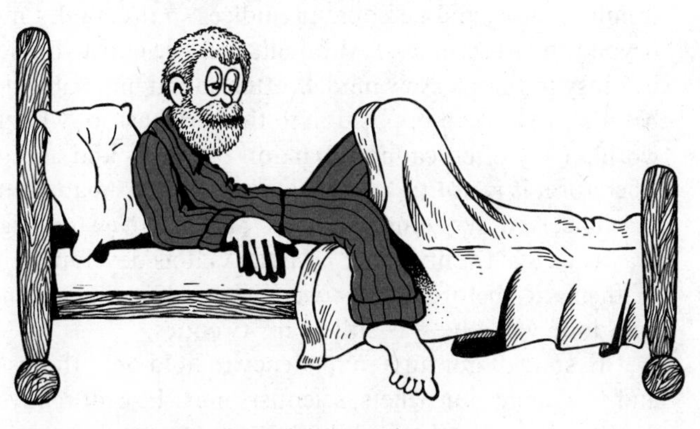

Figure 1.4 The first law of the morning.

As an example of the postulation of a model, consider a box containing an object of unknown shape. If the box is tilted front to back, the observer may detect the sensation of a rolling object. A hypothesis of the shape of the object might be that it is a sphere (Figure 1.5). However the object might in reality be a cylinder. As a test of the sphere hypothesis, the observer might tilt the box side to side. The object which is cylindrical is perceived to slide and not to roll. The sphere model is faulty because a spherical object would still roll side to side. Thus the observer might then hypothesize that a solid cylindrical object is responsible for the observation. However, the actual object could be a solid cylinder, a hollow cylinder, a cylinder with rounded

Figure 1.5 The development of a model.

ends, or a rolling-pin-shaped cylinder. Thus the model which is hypothesized may not bear a direct resemblance to the real object except in the functions already perceived. Only with the development of many other experimental tests could one formulate a theory about the details of the shape of the object inside the box.

How does a hypothesis differ from a theory?

A scientist must be ready to remove himself or herself from previous training, ideas, and personal prejudices so that faulty models are not retained beyond their usefulness. Although every scientist should realize this difficulty it is easy to forget. New models often sound untenable to established scientists because they are in opposition to the very basis of what they believe about the world. They often cannot abandon old ideas and accept radically new ones. Therefore, it is not too surprising to find that many important creative ideas are conceived by young scientists. Sir Isaac Newton suggested the concept of gravity at age twenty-four; Evariste Galois developed the theory of groups in mathematics before he was twenty-one; Albert Einstein's most original work was done while he was still in his twenties.

In spite of our human tendency to hold onto things that are familiar to us and to defend our beliefs, scientists must be ever ready to allow their models to be examined and tested by the entire scientific community. All data can be reviewed and new data accumulated to test the model. Over the long run, a theory will be either accepted or rejected on the basis of its merit. The collective judgment of the scientific community prevails until challenged by new evidence. The new evidence, either in support of or in challenge to a model, is never long in coming in modern science. Any important model proposed in the scientific literature will immediately be scrutinized in detail, and there are always individuals ready to challenge a model by devising critical tests of its validity.

SUMMARY

Our world consists of many different substances. More and more of the materials we deal with are being produced by chemical companies for specialized applications. Each of these chemicals must be evaluated for its long-term effect on our environment.

The natural sciences are based on ***experimental observations****, natural* ***laws****, and the application of these laws to the solution of problems. Science involves, in addition, a search for an understanding of what is responsible for natural phenomena. Science uses* ***models*** *to account for natural phenomena. A tentative model is a* ***hypothesis*** *while an established model is a* ***theory****.*

QUESTIONS AND PROBLEMS

Chemicals and Our World

1. Examine food labels such as on a bottle of soda, a jar of jam, or a loaf of bread. List the chemical additives contained in these products. Insofar as possible, determine the purpose of each substance.
2. Examine the contents of commercial products contained in the kitchen cabinet and the medicine cabinet, found among garden supplies, and used with the workbench. Do the chemical names listed on the labels have any meaning to you? Put the list aside and check it again after completing your course in chemistry.
3. What are the pros and cons of long-lasting pesticides?
4. Examine your surroundings and determine the ways in which plastics are used.

Science and the Scientific Method

5. What is a scientific law? List some of the scientific laws that you may have encountered before this course in chemistry.
6. Do you think that chemistry can be an isolated science separate from economic, political, and social concerns?
7. Describe how chemistry differs from biology and physics.
8. Examine your college catalog and determine what major fields require chemistry in their program.

9. Although chemistry is important in the solution of pollution problems, the fields of biology, physics, and even the social sciences are also involved. Outline some of the ways in which scientists in these areas can contribute to the solution of pollution problems.

10. Have you ever encountered examples of scientific models? Consider what the terms atom, molecule, electron, and DNA mean to you.

11. Comment on the statement: In science a wrong theory can be valuable and better than no theory at all.

2 MEASUREMENT

LEARNING OBJECTIVES

When you finish this chapter you should be able to:

(a) determine the number of significant figures in a given number.

(b) express and round off the results of mathematical operations to the proper number of significant figures.

(c) organize and use factor units in problem solving.

(d) list the metric units and their common prefixes for mass, length, and volume.

(e) interconvert a given measurement in one prefixed metric unit into another prefixed unit.

(f) convert degrees Fahrenheit to degrees Celsius and vice versa.

(g) convert degrees Celsius to degrees Kelvin and vice versa.

(h) distinguish between the forms of energy.

(i) calculate the density of a substance from the mass and volume.

(j) match the following terms with their corresponding definition.

density
energy
measurement
metric system
significant figures
specific gravity
thermometer

In everything we do we must communicate. The form this communication takes depends on the nature of our activity. Certainly a musician describes a musical composition differently than an architect describes a building. The description in either case can be qualitative or quantitative. Each description will employ a certain vocabulary or set of measurements appropriate to the field.

Chemistry, too, has its own vocabulary and types of measurements which must be understood for effective communication of chemical principles. The laws which were established during the development of modern chemistry are based on well-designed experiments carried out under carefully measured conditions. For chemistry then, as well as for other sciences, measurements are of particular importance. In order to study matter and its changes, chemists must measure quantities of both matter and energy. Specifically, they keep track of volume, mass, length, temperature, and energy measurements. They also find it necessary to know the reliability of their measurements.

Nearly all scientists today employ the metric system to measure matter. Length in this system is measured in meters, volume in liters, and mass in grams. Temperature is measured on either the Celsius or Kelvin scale, while energy is given in calories. In this chapter we will learn about the size of these measurements and the mathematics involved in relating them to one another.

2.1 SIGNIFICANT FIGURES

Prior to examining the system of measurement used in chemistry, we must first consider the significance of any numerical quantity which we observe, measure, or are given. Are all of the digits in a number necessarily reliable or well established? **The number of digits in a number which give us reliable information is called the number of significant figures.** Some numbers we encounter are exact quantities and we do not concern ourselves with the number of significant figures. Thus the number of items in a dozen is exactly 12. However, when a ruler is used in a measurement of a distance, there is an uncertainty in the exactness of the measurement, and the quantity must be stated so as to indicate its reliability. This reliability is given by the number of significant figures.

What is the mathematical significance of the number 9000 when discussing the number of soldiers killed on the beaches of Normandy on D day? Were exactly 9000 soldiers killed or is the number 9000 reliable to the nearest 10 soldiers, nearest 100 soldiers, or nearest 1000 soldiers? In short, does the number 9000 contain three, two or one significant figures?

The number of significant figures in a quantity is equal to the number of digits which give reliable information.

When a child raises three fingers in response to a question about her age, you know that she has passed her third birthday but not her fourth birthday;

she has stated her age to one significant figure. As the child grows older she may start making distinctions such as, "I'm almost seven," or, "I'm six and a half." Such statements more accurately express age than does the use of a single figure and are a start toward using two significant figures. When the child turns 16 and can drive a car, the age of the child is now known to two significant figures, and only then will the state issue a driving license. In later years this woman may prefer to have it known that she is "in her thirties." We conclude that she is older than 30 but not yet 40. Thus the age is known only to the nearest 10 years or one significant figure in the tens place.

In all sciences it is necessary to state quantities numerically and simultaneously give the precision of the number. If a sample is said to have a mass of 9 grams (g), the scientist means that to the nearest gram the sample contains 9 g of matter. The mass may be between 8.5 and 9.5 g. If it were less than 8.5 g or more than 9.5 g, the mass would be reported as 8 g or 10 g, respectively. The mass is precise to one significant figure. If the sample were placed on a more accurate balance, the mass might be found to be 8.8 g. This quantity has two significant figures and we know that the actual mass is less than 8.85 g but more than 8.75 g. In each of the two cases cited, the last significant figure is only approximate. The degree of precision of each number is given by the number of significant figures. **When we write down a number, it is necessary to make sure that the number does indeed contain the same number of significant figures as the reliability of the measurement.** No measurement should ever be expressed or written as more reliable than the actual measurement.

How reliable do you think the placement of a football by an official is?

Zeros are sometimes significant figures and are other times merely used to place the number on the measurement scale being used. Students often have trouble determining which zeros in a quantity are significant. Suppose two different samples are weighed on a balance which can give the mass reliably to 0.001 gram. If the mass of one sample is found to be 5.000 grams, the sample mass is between 4.9995 grams and 5.0005 grams and the number of significant figures is four. Thus the three zeros following the decimal point are significant. If the other sample has a mass of 0.015 gram, the first zero after the decimal point is not a significant figure; this zero only gives the magnitude of the number. The sample actually has a mass between 0.0145 gram and 0.0155 gram and the number 0.015 contains only two significant figures.

Numbers containing zeros, such as 10,000, are difficult to interpret and their significance may depend on the source of the quantity. The number might contain five significant figures and so represent a quantity between 9,999.5 and 10,000.5. If the quantity refers to the number of people in a crowd, the accuracy of the number may be between 9500 and 10,500. For this reason it is convenient to use notations involving powers of ten. For the quantity 10,000 known to the nearest unit, the number is expressed as 1.0000×10^4. For the number of people in a crowd known only to the nearest thousand, we use 1.0×10^4.

How many significant figures are there in the national debt of the United States?

Unless the power of ten notation is used, we normally assume that zeros at the end of numbers indicate only the relative magnitude of the number and are not significant figures. Thus 10,000 is viewed as containing only one significant figure. If the quantity must represent five significant figures, then a bar may be placed above all significant zeros, as in $1\overline{0,000}$. If the number is significant only to the nearest hundred, the number is written $1\overline{0,0}00$. The rules of significant figures are summarized in Table 2.1.

TABLE 2.1 rules of significant figures

1 **All digits 1 through 9 inclusive are significant.** Thus 14.9 contains 3 significant figures while 125.62 contains 5 significant figures.

2 **Zero is significant if it appears between two nonzero digits.** Thus 306, 30.6, 3.06, and 0.306 all contain three significant figures.

3 **A terminal zero to the right of a decimal in a number greater than one is significant if it is expressing a reliable measurement.** Thus 279.0, 27.90, and 2.790 all contain four significant figures.

4 **A terminal zero to the right of a decimal point in a number less than one is significant if it expresses reliable information.** Thus 0.2790 contains four significant figures if indeed the value has been shown to be zero in the ten-thousandth place.

5 **A zero which is used only to fix the decimal in a number less than one is not significant.** Thus 0.456, 0.0456, 0.00456, and 0.000456 all contain only three significant figures.

6 **Terminal zeros in an integer are not significant unless accompanied by a bar above them.** Thus 450 contains only two significant figures while $45\bar{0}$ contains three significant figures.

Problem 2.1

What are the number of significant figures in each of the following numbers? (Give the rules used for your answer.)

1. 5041	**2.** 5401	**3.** 5410
4. $541\bar{0}$	**5.** 0.5401	**6.** 0.5410
7. 0.05401	**8.** 0.05410	**9.** 54.10

Solution

1. 4 (1, 2)	**2.** 4 (1, 2)	**3.** 3 (1, 6)
4. 4 (1, 6)	**5.** 4 (1, 2)	**6.** 4 (1, 4)
7. 4 (1, 2, 5)	**8.** 4 (1, 4, 5)	**9.** 4 (1, 3)

2.2 MATHEMATICAL OPERATIONS AND SIGNIFICANT FIGURES

Whenever two or more quantities representing measurements and expressed to their proper number of significant figures are added, subtracted, multiplied, or

divided, the resulting answer cannot be expressed to any more reliability than the least significant quantity used. The rules governing the reliability of numbers derived from mathematical operations are given in the next three subsections.

Addition and subtraction

When adding or subtracting numbers, the answer must not contain any significant figures beyond the place common to all of the numbers. Thus the sum of the numbers 12.2 and 13.31 is 25.5 and not 25.51. Only the tenths place is common to both numbers. The hundredths place is not given in 12.2 and is not known. The same rule applies to subtraction. The difference of the numbers 13.31 and 12.2 is 1.1 and not 1.11. In each case cited, the number of significant figures reflects the reliability of the resultant number based on the reliability of the quantities used in obtaining it.

12.2
13.31
25.5 Why?

Problem 2.2 What is the sum of the numbers given? Express the answer to the proper number of significant figures.

$25.1 + 15 + 14.15$

Solution The place common to all numbers is the units place. A line can be placed at this point to avoid using the numbers to the right in the answer.

25	.1
15	
14	.15
54	.25

ANSWER

54

Problem 2.3 What is the sum of the numbers given, expressed to the proper number of significant figures?

$0.5432 + 0.11 + 1.311$

Solution

0.54	32
0.11	
1.31	1
1.96	42

ANSWER

1.96

Problem 2.4 What is the difference of 16.29 and 3.168?

Solution

$$\begin{array}{r|l} 16.29 & \\ -\ 3.16 & 8 \\ \hline 13.12 & 2 \end{array}$$

ANSWER

13.12

Multiplication and division

$$\begin{array}{r} 201 \\ \times 3 \\ \hline 600 \end{array}$$ Why?

$$3\,\overline{)603}\ = 200$$ Why?

When multiplying or dividing numbers, the answer must not contain more significant figures than the least number of significant figures used in the operations. Thus the product of 201 × 3 is 603, but the answer can be expressed to only one significant figure since 3 has only one significant figure. The proper answer is 600 in which only the 6 is significant. Similarly, the quotient 603/3 is not 201 but rather 200.

Problem 2.5 What is the product of these quantities?

$$304 \times 11$$

Solution

$$304 \times 11 = 3344$$

However, the answer can be expressed to only two significant figures, or 3300.

ANSWER

3300

Problem 2.6 What is the quotient of these quantities?

$$\frac{1760.1}{25}$$

Solution Division yields the number 70.4. However, since the divisor contains only two significant figures, the quotient must be expressed as $7\bar{0}$.

Rounding off

In the preceding problems of this section, the nonsignificant figures were discarded. All of the discards were mathematically valid because the examples were carefully chosen so that the nonsignificant figures were less than 5. However, if the nonsignificant figure is 5 or greater than 5, the figure cannot be discarded and must be used in rounding off to the required number of significant figures. Hence if 25.21 must be expressed to three significant figures, the answer is 25.2 as the 1 in the hundredths place is less than 5. However, if the number 25.2 must be expressed to one significant figure, the answer must be rounded off to 30 as the nonsignificant number 5.2 is greater than 5.

25.21	four significant figures
25.2	three significant figures
25	two significant figures
30	one significant figure

If the number to be rounded off is 5, the 5 is dropped and the last significant digit is left the same if it is even and the last significant digit is increased by one if it is odd.

Problem 2.7 What is the sum of the following numbers?

$$10.7 + 17.43 + 3.56$$

Solution

10.7	
17.4	3
3.5	6
31.6	9

Rounding off to the tenths place, the answer is 31.7.

Problem 2.8 What is the product of the following numbers?

$$5.61 \times 21$$

Solution The product is 117.81. Since 21 contains only two significant figures, only two are allowed in the answer. The seven cannot be dropped but must be used to round off to the correct answer 120. Remember that the zero in 120 is not significant.

2.3 THE FACTOR UNIT SOLUTION OF PROBLEM SOLVING

Prior to introducing the metric system and its application to chemistry, a description of a very useful approach to problem solving will be presented. It is often necessary to convert one measured quantity into another equivalent quantity having different units. This is accomplished by a conversion factor, which is a multiplier having two or more units associated with it. **The approach of using conversion factors in the solution of problems is called the factor unit method.** The approach is called the factor unit method because the factor is numerically equivalent to unity, and when a quantity is multiplied by the factor, it is changed into an alternate but equivalent number having different units. The factor unit method is very simple and is widely employed to solve problems by developing the proper mathematical relationships using the units of measurement as a guideline. The factor unit method coupled with a systematic approach to problem solving will be necessary for a successful experience with chemistry.

Always label quantities with their proper units.

The outline of the suggested problem solving technique is as follows:

1 **Examine the data given and note the units associated with all numbers.**
2 **Determine what is asked for and what units are desired.**
3 **Write down the vital data given, along with the units, to the left of an equality sign.**
4 **Write down a symbol for the desired unknown, along with the units, to the right of the equality sign.**
5 **Develop the conversion factors and their units which, when multiplied by the known data, will give the desired unknown.**
6 **Check your work to see that the units are equal on both sides of the equation.**
7 **Carry out the arithmetic and check the answer for mathematical reasonableness.**
8 **Check that the proper number of significant digits are used.**

In order to illustrate the factor unit method, let us pretend that you have taken your young sister and four of her friends to an amusement park for the day. At the end of a successful and happy day, you are passing the last vending machine, and they all decide to ask for 15-cent candy bars. A rapid examination reveals that you have 75 cents in change. Now, can you purchase this last treat for all five children, or do you tell them that they will spoil their supper if they eat more now? While you probably could ascertain whether the candy bars are within your financial limitations without the factor unit method, the use of the factor unit method with familiar quantities is the best way to start off this presentation.

Multiplication of a quantity by 1 or its equivalent changes the quantity into its equivalent.

You know that the number of candy bars within your purchasing power is the unknown quantity. Furthermore, you have as your given quantity 75 cents. Therefore, you write:

$$75 \text{ cents} \times \text{factor} = ?\ \text{candy bars}$$

In addition, we know from the clamor of the children that a single candy bar costs only 15 cents.

$$15 \text{ cents} = 1 \text{ candy bar}$$

Dividing both sides of this statement by 15 cents gives a factor, F_1, which is equal to one.

$\frac{1 \text{ candy bar}}{15¢}$

$$\frac{15 \text{ cents}}{15 \text{ cents}} = 1 = \frac{1 \text{ candy bar}}{15 \text{ cents}} = F_1$$

This term of 1 candy bar divided by 15 cents is just a mathematical way of saying that 1 candy bar costs 15 cents.

Alternatively, dividing both sides of the original statement by 1 candy bar gives factor F_2, which is also equal to one.

$\frac{15¢}{1 \text{ candy bar}}$

$$F_2 = \frac{15 \text{ cents}}{1 \text{ candy bar}} = \frac{1 \text{ candy bar}}{1 \text{ candy bar}} = 1$$

The term is a mathematical way of saying 15 cents buys 1 candy bar.

Both F_1 and F_2, then, express the same information but their focus is different; F_1 focuses on the candy bars and F_2 focuses on the money. Remember that both F_1 and F_2 are equal to 1. Multiplication of a number by 1 leaves the number mathematically unchanged. Its form may be different but it is still equivalent to the original number. Thus multiplication of F_1 by 75 cents will give a number equivalent to 75 cents.

Since you know how much money you have and you are interested in candy bars, F_1 has the proper focus for you. Indeed, multiplication of 75 cents by F_1 gives an answer in terms of candy bars.

$$75 \text{ cents} \times F_1 = 75 \text{ cents} \times \frac{1 \text{ candy bar}}{15 \text{ cents}} = 5 \text{ candy bars}$$

Multiplication of 75 cents by F_2 gives an answer in units which are nonsensical for you.

$$75 \text{ cents} \times F_2 = 75 \text{ cents} \times \frac{15 \text{ cents}}{1 \text{ candy bar}} = \frac{1125 \text{ cents}^2}{1 \text{ candy bar}}$$

Thus only the use of F_1 gives a reasonable answer in terms of the desired units. Furthermore, we know that 5 is a numerically reasonable number of candy bars which can be purchased for 75 cents.

If you were the vending machine owner, F_2 would have the proper focus for you. Suppose you know that you have 100 candy bars in the machine. How much money would this net you?

$$100 \text{ candy bars} = ? \text{ cents}$$

$$100 \text{ candy bars} \times \frac{15 \text{ cents}}{1 \text{ candy bar}} = 1500 \text{ cents}$$

Since both factors express the same information, it is sometimes confusing as to which one to use. Make sure that the units that you are interested in finding out about are on top. If you do use the inappropriate factor, however, your answer will be in strange enough units to alert you that something is amiss.

Problem 2.9 How many dozen eggs are present in a basket containing 66 eggs?

Solution

Step 1: given is 66 eggs
Step 2: sought quantity is ? dozen
Steps 3 and 4:

$$66 \text{ eggs} \times \text{factor} = ?\text{ dozen}$$

Step 5: 1 dozen = 12 eggs

$$\frac{1 \text{ dozen}}{1 \text{ dozen}} = 1 = \frac{12 \text{ eggs}}{1 \text{ dozen}} = F_1$$

$$\frac{1 \text{ dozen}}{12 \text{ eggs}} = F_2 = \frac{12 \text{ eggs}}{12 \text{ eggs}} = 1$$

Steps 6 and 7: Only F_2 will give the proper units and quantity:

$$66 \text{ eggs} \times \frac{1 \text{ dozen}}{12 \text{ eggs}} = 5.5 \text{ dozen}$$

Step 8: There are two significant figures in both 66 and 5.5. Note that the factor can have as many significant figures as desired, since it is an exact value.

Problem 2.10 The Indianapolis 500 is run on a track 2.5 miles long. How many inches long is the track?

Solution

Step 1: given is 2.5 miles
Step 2: sought quantity is ? inches
Steps 3 and 4:

$$2.5 \text{ miles} \times \text{factor} = ?\text{ inches}$$

Step 5: In order to convert miles into inches, the factor would have to be inches/mile. We do not know that factor. We do know factors relating miles and feet, and also feet and inches.

1 mile = 5280 feet   1 foot = 12 inches.

Considering the first equality, the factors are:

$$F_1 = \frac{1 \text{ mile}}{5280 \text{ feet}} \qquad F_2 = \frac{5280 \text{ feet}}{1 \text{ mile}}$$

Steps 6 and 7: If we use F_2, the units of miles will cancel and the distance will be given in feet and not inches as desired.

$$2.5 \text{ miles} \times \frac{5280 \text{ feet}}{1 \text{ mile}} \times \text{factor} = ?\text{ inches}$$

Now we consider the two factors relating feet and inches.

$$F_3 = \frac{1 \text{ foot}}{12 \text{ inches}} \qquad F_4 = \frac{12 \text{ inches}}{1 \text{ foot}}$$

Only F_4 will allow us to cancel the units of feet and obtain an answer in inches.

$$2.5 \text{ }\cancel{\text{miles}} \times \frac{5280 \text{ }\cancel{\text{feet}}}{1 \text{ }\cancel{\text{mile}}} \times \frac{12 \text{ inches}}{1 \text{ }\cancel{\text{foot}}} = 158{,}400 \text{ inches}$$

The answer is reasonable as we know that there are many feet in a mile and many inches in a foot. Therefore, a product of the two quantities will give a large number.

Step 8: Since the quantity 2.5 contains only two significant figures, the answer can have only two significant figures, and it is best expressed by rounding off and using exponential notation.

ANSWER

1.6×10^5 inches

2.4 WHAT IS A MEASUREMENT?

Every day we measure quantities such as time, distance, volume, weight, and the number of items in a collection. Mundane measurements as well as the most exacting scientific experiments depend on the same concept. **All measurements are made by use of or comparison to a standard measuring device.** Spark gaps in the spark plugs of a car involve a prescribed distance that is set on the basis of a unit of length in wide use. The pound of hamburger in a supermarket display case is measured by a scale that is set to a standard of weight that we have come to accept socially. The prescription of the quantity of a drug to be administered to a patient by a nurse involves accepted standard units. The extremely precise measurements involved in the construction, launching, and control of space satellites are also dependent on standards that we have all agreed to use.

As will be pointed out in subsequent discussions of the specific units of measurements currently in use, the choice of standards is arbitrary. Our civilization, as did all civilizations in the past, developed the dimensions that were convenient. The only significant difference between present-day measurements and those of past civilizations is that today's are more precise. Primitive societies may have used word equivalents of one, two, few, or many. The concept of one thousand as opposed to one million was probably beyond comprehension to early hunters and was perhaps not even important. It mattered only that there were many birds in the sky or few deer in the forest and that they were more than the number of fingers on a person's hand. Distance may have been expressed in terms of the number of steps or days required to traverse it. People initially made measurements in terms of human dimensions. The actual quantity involved then depended on the stature of the people or the part of the anatomy chosen as a measuring device.

Examples of measurements still in use today that initially were based on human anatomy are the inch, the foot, the fathom, and the cubit. The inch was based on the breadth of the thumb; the foot on the length of a foot, and the cubit on the length of the forearm. Because these quantities vary from individual to individual, they were defined in terms of the dimensions of the king. Of course, this meant that with the death of the king and the ascension to the throne of a new king, one could have hailed, "The king is dead, long live the king, and here come new measurements!" The fathom, which is still in nautical usage, was the distance between the fingertips of a person's outstretched arms. The fathom is now defined more precisely as 6 feet. A cubit was used in biblical terms. Noah's ark, as described in the Book of Genesis, was 300 cubits long, 50 cubits wide, and 30 cubits high. The cubit was probably about 18 inches and is so defined as part of the English system of measure.

It became mandatory with increased trade and commerce between countries and with an advancing technology that more sophisticated and universal measurement systems be developed. Two such systems of measure in use today are the English system and the metric system.

2.5 ENGLISH SYSTEM OF MEASUREMENT

By the eighteenth century, the English-speaking countries of the world had developed the system of measurement with which we are familiar today. The units of distance include the inch, foot, yard, and mile. Other less familiar distance units include the rod, which is $16\frac{1}{2}$ feet, and the furlong, which is 660 feet. To convert from one unit to another, it is necessary to use a variety of conversion factors. Although we consider it to be easy to convert 24 inches to 2 feet because of our early schooling, it is somewhat more tedious to determine the number of yards in 2 miles. Furthermore, the subdivisions of an inch into fractions of $\frac{1}{2}$, $\frac{1}{4}$, $\frac{1}{8}$, $\frac{1}{16}$, and $\frac{1}{32}$ introduce more complexities than most people care to admit.

How many $\frac{1}{32}$ inches are there in $\frac{1}{2}$ inch?

The variety of measurements is probably no more evident than in units of volume, where bushel, peck, gallon, quart, pint, cup, tablespoon, and teaspoon are used. Although U.S. citizens may defend our English system of volume measurement, one only need ask how many teaspoons there are in a gallon to illustrate the awkwardness of the wide variety of factors interrelating volume measurements. Even weight measurements are unwieldly; the ton, pound, ounce (both troy and avoirdupois), grain, and carat are used.

2.6 METRIC SYSTEM

A committee was established by the order of the French National Assembly in 1790 to develop a system of measurement to be used nationally. Prior to that time, different provinces and even isolated communities often used different measurements. After nine years the metric system was developed and adopted by the French government as their official system. The logical aspects of this system for expressing both large and small quantities gained in acceptance until it is now used virtually worldwide. Both common everyday measurements and the precise measurements required in the sciences are easily expressed in the metric system. Only in Great Britain, Canada, and the United States is the English system still used. Great Britain and Canada are now in the process of converting to the metric system. The United States is taking slow steps in that direction and may be the last bastion of the English system.

Prepare for a metric world now!

The standard units of the metric system are the meter (m) for length, the gram (g) for mass, and the liter for volume. Subdivisions and multiples of the standard units of the metric system employ only factors of ten. Prefixes are used to indicate the size of the unit relative to the standard unit. The use of this system is illustrated in Table 2.2. You *must* learn these units and their equivalents in order to use the factor unit method of solving problems.

TABLE 2.2

metric units of length, volume, and mass

Prefix	Relation to basic unit	Length	Volume	Mass
kilo	$\times 1000$ (10^3)	kilometer (km)	kiloliter (kl)[a]	kilogram (kg)
(none)	$\times 1$ (10^0)	meter (m)	liter (l)	gram (g)
deci	$\times 0.1$ (10^{-1})	decimeter (dm)	deciliter (dl)[a]	decigram (dg)[a]
centi	$\times 0.01$ (10^{-2})	centimeter (cm)	centiliter (cl)[a]	centigram (cg)
milli	$\times 0.001$ (10^{-3})	millimeter (mm)	milliliter (ml)[b]	milligram (mg)
micro	$\times 0.000001$ (10^{-6})	micrometer (μm)	microliter (μl)	microgram (μg)

[a] These units are not widely used.
[b] The milliliter occupies a volume of one cubic centimeter (cc). Both ml and cc (or cm^3) may be used interchangeably.

In order to compare measurements taken in different units, it is necessary to convert all measurements to the same unit. This conversion involves the use of the factor unit method and will be presented in the next section. However, prior to a discussion of this approach, you should take note that the various units of length, volume, or mass within the metric system differ from one another only by factors of ten. We are familiar with this kind of relationship in our currency system (Table 2.3). For example, we know that

TABLE 2.3

the metric dollar

Metric prefix	Relation to basic unit	Metric dollar	Common usage
kilo	$\times 1000$	kilodollar	grand, kilobuck
(none)	$\times 1$	dollar	dollar, buck
deci	$\times 0.1$	decidollar	dime
centi	$\times 0.01$	centidollar	cent
milli	$\times 0.001$	millidollar	mill

\$1.50 is the same thing as 150 cents or that \$0.25 is 25 cents. A cent is one hundredth of a dollar, just as a centimeter is one hundredth of a meter.

Problem 2.11 How many cents are there in \$2.27?

Solution Although you may automatically move the decimal two places to the right to obtain the correct answer 227 cents, the factor unit method may be used for practice in preparation for problems involving less familiar units.

Steps 1 through 4:

$$\$2.27 \times \text{factor} = ?\ \text{cents}$$

Step 5: The factors possible are:

$$F_1 = \frac{\$1.00}{100\ \text{cents}} \qquad F_2 = \frac{100\ \text{cents}}{\$1.00}$$

Steps 6 and 7:

$$\cancel{\$}2.27 \times \frac{100\ \text{cents}}{\cancel{\$}1.00} = 227\ \text{cents}$$

Step 8: There are three significant figures in both \$2.27 and 227 cents.

2.7 THINKING METRIC

In Section 2.3, you learned how to use the factor unit method in solving problems involving units with which you are familiar. If you have not "cheated" and avoided the method, you now are prepared to think and do metric problems.

Problem 2.12 Convert 6.7 m into millimeters.

Solution

$$6.7\ \text{m} \times \text{factor} = ?\ \text{mm}$$

The factors interrelating meter and millimeter are derived from the fact that:

$$1\ \text{m} = 1000\ \text{mm}$$

$$F_1 = \frac{1\ \text{m}}{1000\ \text{mm}} \qquad F_2 = \frac{1000\ \text{mm}}{1\ \text{m}}$$

The solution to the conversion requires the use of F_2 in order to cancel the units and convert meters to millimeters.

$$6.7\ \cancel{\text{m}} \times \frac{1000\ \text{mm}}{1\ \cancel{\text{m}}} = 6700\ \text{mm}$$

The answer has the correct units and is reasonable as there are many millimeters in a meter. Lastly, the answer is expressed with only two significant figures as the measured quantity has two significant figures.

Problem 2.13 Convert 97 cg into kilograms.

Solution

$$97 \text{ cg} \times \text{factor} = ?\text{ kg}$$

We do not know a direct relationship between centigram and kilogram. Therefore, we proceed stepwise by first converting centigrams into grams and then grams into kilograms. From the relation 100 cg = 1 g, we may write two factors:

$$F_1 = \frac{100 \text{ cg}}{1 \text{ g}} \qquad F_2 = \frac{1 \text{ g}}{100 \text{ cg}}$$

Using F_2, the centigrams are canceled and the units are now in grams.

$$97 \text{ cg} \times \frac{1 \text{ g}}{100 \text{ cg}} \times \text{factor} = ?\text{ kg}$$

Next knowing that 1 kg = 1000 g, we may write:

$$F_3 = \frac{1 \text{ kg}}{1000 \text{ g}} \qquad F_4 = \frac{1000 \text{ g}}{1 \text{ kg}}$$

Only F_3 provides the correct cancellation of units.

$$97 \text{ cg} \times \frac{1 \text{ g}}{100 \text{ cg}} \times \frac{1 \text{ kg}}{1000 \text{ g}} = 0.00097 \text{ kg}$$

The number of significant figures is two. Furthermore, we know that the answer is numerically correct since 97 cg is a small fraction of the larger unit, the kilogram.

Have you really used the factor unit method?

After working with the factor unit method for a while, you should be able to "shift" decimal points just as you can with dollars and cents. However, make sure that you fully understand the method and do not rely on artificial devices or rules which may fail on you in a crisis situation like an examination.

2.8 METRIC AND ENGLISH EQUIVALENTS

In Section 2.6, you were introduced to the metric terms meter, liter, and gram without any examples of how large these quantities are or how they compare to the inch, quart, and pound of the English system. In Table 2.4, the

TABLE 2.4 **metric–English equivalents**

Dimension	Metric unit	English unit
Length	1.00 m	39.4 in.
	2.54 cm	1.00 in.
	1 km	0.6 mile
Volume	1.00 liter	1.06 qt
	0.94 liter	1.00 qt
Mass	454 g	1.00 lb
	1.0 kg	2.2 lb

metric-English equivalents are given. You should learn these equivalents. Practice until you are able to do conversions from metric to English systems and vice versa easily. Representations of the equivalency between metric and English systems are given in Figures 2.1 and 2.2.

How many liters are there in a gallon?

Problem 2.14 A recipe in a cookbook calls for 2 lb of hamburger. When the United States converts to the metric system, how many grams of hamburger will you have to purchase to make the recipe?

Solution We know that 454 g = 1 lb. Therefore, the two factors are 454 g/1 lb and 1 lb/454 g. Only the former factor will result in proper cancellation of units and result in a number of the correct magnitude.

$$2\ \cancel{\text{lb}} \times \frac{454\ \text{g}}{1\ \cancel{\text{lb}}} = 908\ \text{g}$$

The correct answer to one significant figure is 900 g.

Problem 2.15 Road signs in kilometers are being posted on some interstate highways. You see a sign which indicates that Columbus is 61 kilometers away. What is the distance in miles?

Solution A kilometer is a shorter distance than a mile, and we know the equality 1 km = 0.6 mile. The factors are 1 km/0.6 mile and 0.6 mile/1 km. Only the latter factor gives the proper cancellation of units and answer of correct magnitude.

$$61\ \cancel{\text{km}} \times \frac{0.6\ \text{mile}}{1\ \cancel{\text{km}}} = 36.6\ \text{miles}$$

The correct answer rounded off to two significant figures is 37 miles.

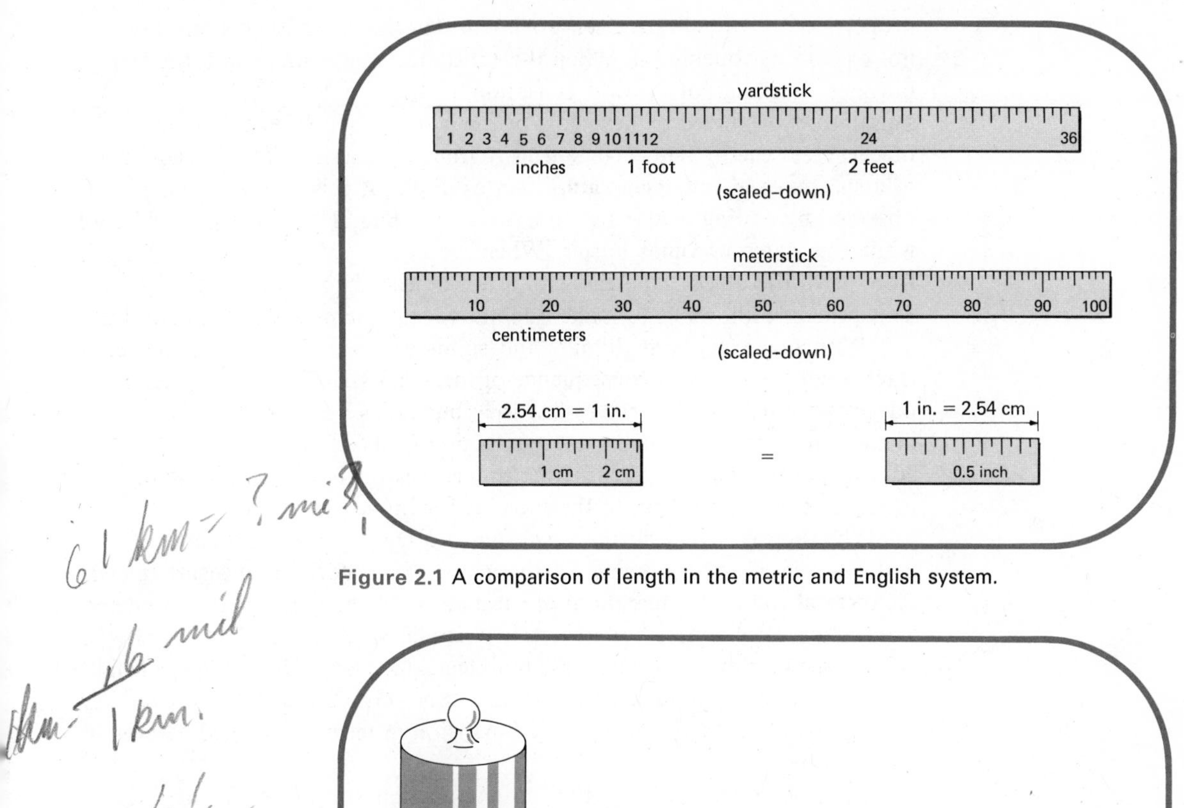

Figure 2.1 A comparison of length in the metric and English system.

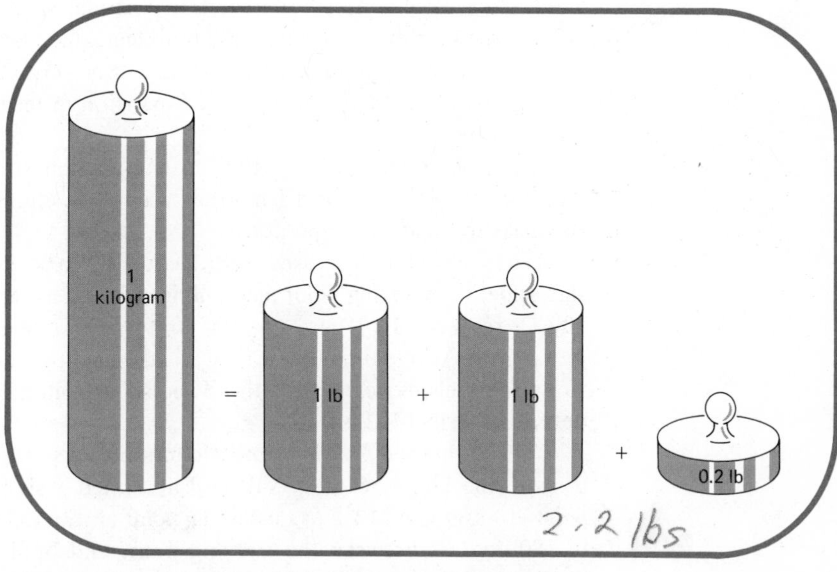

Figure 2.2 A comparison of the kilogram and the pound.

2.9 TEMPERATURE

Many properties of matter depend on temperature. The volumes of solids, liquids, and especially gases are affected by temperature. For example, a gas expands greatly in volume when heated. For this reason it is necessary to specify the temperature of a substance before discussing any of its physical properties. In a qualitative sense we can determine whether a substance is hot or cold by touching it. When the substance is in contact with the skin, a burning, freezing, or neutral sensation is transmitted to the brain. This sensation is produced by heat flow or the lack of it between the skin and the object. **Heat energy flows spontaneously from a substance of high temperature to a substance of low temperature.** An example of this phenomenon can be observed by adding cold water to a hot frying pan. The frying pan will cool while the water becomes hotter. When we touch a cold object, heat flows from us to the object, whereas with a hot object, the reverse process occurs. This principle allows us to make qualitative statements about temperature.

With few exceptions, liquids and solids expand when heated and contract when cooled. As a consequence of thermal expansion, it is necessary to put expansion joints in bridges lest they buckle as a result of the difference in the amount by which the various components of the bridge expand or contract. Advantage is taken of the differences in the expansion of various metals in the construction of thermostats for home heating units.

The thermal expansion of a substance often is used to define the temperature scale of a thermometer. **A thermometer is a device to measure (meter) the thermal energy (temperature) of substances.** In the mercury thermometer, which consists of a glass bulb sealed to a capillary tube, a small change in the volume of the mercury in the glass bulb leads to a large change in the height of the mercury column in the capillary tube. Gradations on the capillary tube indicate the degree of mercury expansion in terms of defined degrees of temperature.

Because water is one of the most important substances in the maintenance of life, it is not surprising that water was chosen to provide scales for the measurement of temperature.

0°C = 32°F
100°C = 212°F

The Celsius scale is used most frequently in scientific work. **On the Celsius scale, the freezing point and boiling point of water are defined as 0°C and 100°C, respectively.** There are 100 uniform degree intervals between these two points. At one time this scale was called centigrade, but it is now referred to as the Celsius scale after the Swedish astronomer, Ander Celsius, who developed it in 1742.

The Fahrenheit scale is most commonly used in nonscientific applications in the United States. Gabriel Fahrenheit assigned **32°F as the freezing point of water and 212°F as the boiling point of water.** This arrangement results in 180 degrees between the freezing point and boiling point of water. The Celsius and Fahrenheit scales are compared in Figure 2.3.

The temperature of an object can be expressed on both the Fahrenheit and the Celsius scales because the position of the mercury column is independent of the scales placed on the thermometer. A given temperature on one scale can be converted to the other scale simply by remembering the basis of their origin. There are 180 Fahrenheit units for every 100 Celsius units because these are the differences between the two reference points. Therefore, the Celsius degree is $\frac{180}{100} = \frac{9}{5}$ as large as the Fahrenheit degree.

$$\frac{9 \text{ divisions °F}}{5 \text{ divisions °C}}$$

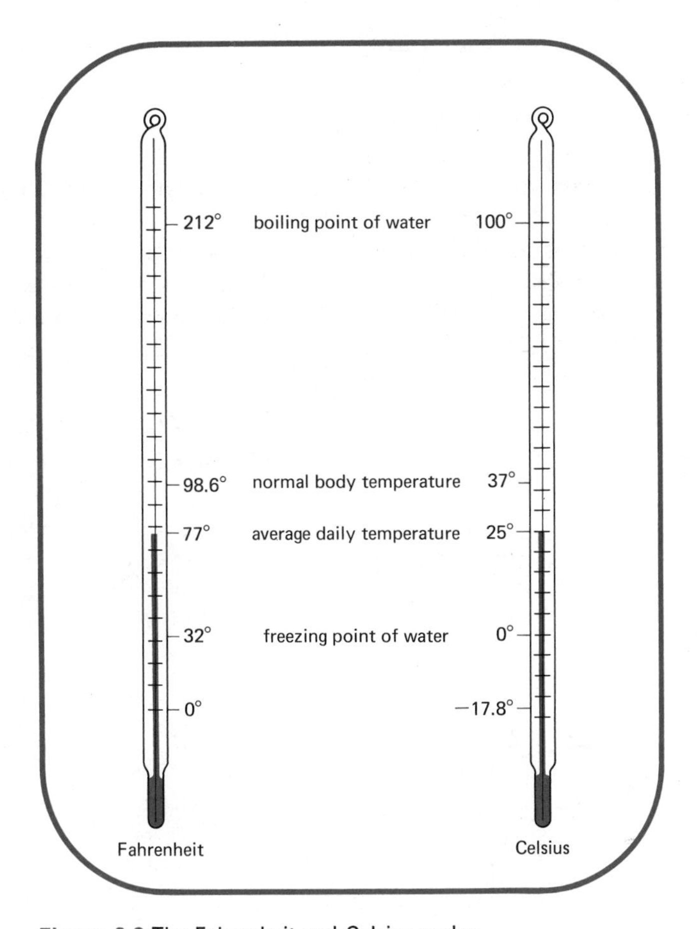

Figure 2.3 The Fahrenheit and Celsius scales.

This factor, properly applied with the knowledge of the temperature equivalents of the two reference points, allows conversion from one scale to the other.

In order to convert Celsius units to Fahrenheit units the procedure is:

$$\text{°C above or below freezing point of water} \times \frac{9 \text{ units °F}}{5 \text{ units °C}} = \text{°F above or below freezing point of water}$$

Since the freezing point of water is 32°F, this quantity must be added to the number of Fahrenheit units obtained in order to get the temperature in Fahrenheit degrees.

$$(°C \times \tfrac{9}{5}) + 32 = °F$$

By rearranging the equation, an alternate form yielding Celsius degrees can be obtained.

1.8
100 | 180
100
800

$$°C \times \tfrac{9}{5} = °F - 32$$

or

$$°C = \tfrac{5}{9}(°F - 32)$$

Note that in the last equation, 32 must be subtracted from °F in order to obtain the number of Fahrenheit units above or below the freezing point of water. After this conversion the number of Fahrenheit units is multiplied by $\frac{5}{9}$ to obtain the temperature as Celsius degrees.

There is another important temperature system called the Kelvin scale. On this scale, 0°C is 273°K. **The lowest temperature attainable is −273°C, or 0°K.** The size temperature units are identical on both the Kelvin and Celsius scale. Thus we may write.:

$$°K = °C + 273 \quad \text{or} \quad °C = °K - 273$$

The Kelvin and Celsius scales are compared in Figure 2.4.

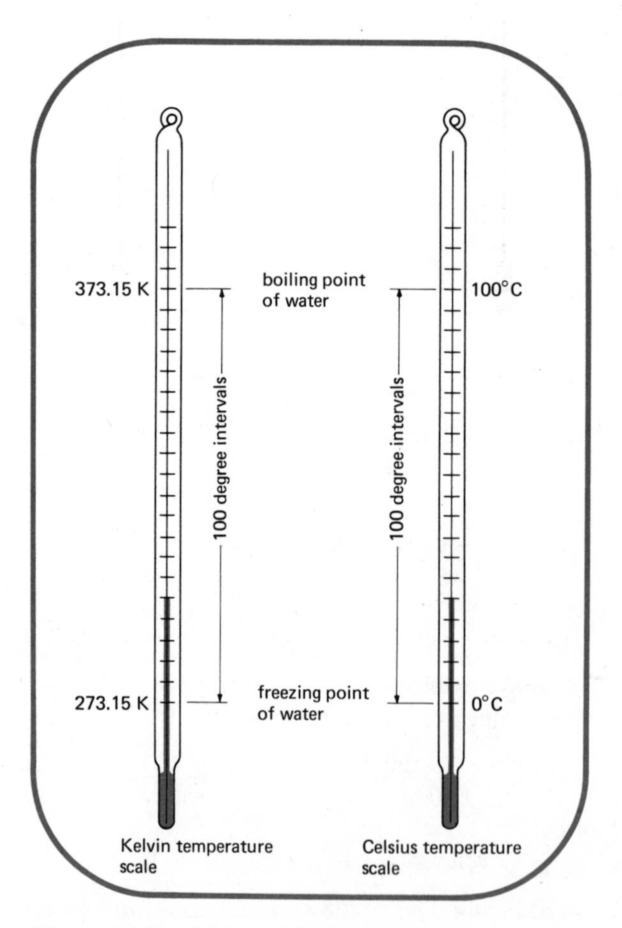

Figure 2.4 The Kelvin and Celsius scales.

Problem 2.16 The normal body temperature is 98.6°F. What temperature is considered normal by a European doctor?

Solution You may substitute directly into either formula, but the following is the most efficient.

$$°C = \frac{5}{9}(°F - 32.0)$$
$$°C = \frac{5}{9}(98.6 - 32.0)$$
$$°C = \frac{5}{9}(66.6)$$
$$°C = 37.0$$

You may note in future studies that many hospitals, biology labs, and biochemistry labs carry out experiments at 37°C, which is the normal body temperature.

Problem 2.17 Many early chemical experiments in European laboratories were reported at a room temperature of 20°C. How does this compare with modern U.S. laboratories?

Solution Substituting into the following equation, we have:

$$°F = \frac{9}{5}°C + 32$$
$$°F = (\frac{9}{5} \times 20) + 32$$
$$°F = 36 + 32$$
$$°F = 68$$

This value is lower than the average room temperature of 72°F in U.S. laboratories.

2.10 ENERGY

Energy is the ability to do work. Various types of energy are encountered in chemistry: potential energy, kinetic energy, heat energy, electrical energy, and chemical energy.

Potential energy refers to the energy that a substance possesses because of its position relative to a second possible position. A wound watch spring possesses potential energy because it can work as it unwinds. Similarly, water behind a dam possesses potential energy. This energy can be released if the water is made to turn a turbine while going through the dam. The energy released by both systems described is transformed into kinetic energy of some object (the gears of a watch or the blades of a turbine) that is moving. **Thus the energy possessed by a moving object is called kinetic energy.**

What type of energy is present in a football in flight?

Chemical energy is the potential energy stored in substances that can be released during a chemical reaction. As the subject of chemistry is developed in this text, it is shown that the chemical energy stored in various substances determines the reactions that they will undergo.

The energy required for or generated in chemical reactions may be in the form of either heat or electrical energy. **A reaction that occurs with the evolution of heat is exothermic, whereas a reaction in which heat energy is required from the surroundings is endothermic.**

The Calorie

The most common unit of heat energy is the **calorie** (cal). **It is the amount of energy required to raise the temperature of 1 g of water 1°C from 14.5°C to 15.5°C.**

The kilocalorie (kcal), equivalent to 1000 cal, is the unit used in dietary tables, although it is referred to as a Calorie (Cal). Thus the dietetic Calorie is the amount of energy required to raise 1 kg of water by 1°C.

Each person has caloric requirements that depend, among other things, on body size, metabolic rate, and physical activity. A rough estimate of a minimum caloric intake to maintain a sedentary patient in a hospital bed is 1.0 Cal/hr for each kilogram of body weight, or 1.0 Cal kg^{-1} hr^{-1}.

Problem 2.18 How many Calories are required for a 150-lb patient in a hospital in one day?

Solution First, the mass equivalent of 150 lb must be determined.

$$150\ \not{lb} \times \frac{1.0\ kg}{2.2\ \not{lb}} = 68\ kg$$

Next, the number of Calories required per hour is calculated.

$$68\ \not{kg} \times \frac{1.\bar{0}\ Cal}{\not{kg}\ hr} = 68\ \frac{Cal}{hr}$$

Finally, the number of Calories required per day is calculated to two significant figures.

$$68\ \frac{Cal}{\not{hr}} \times 24\ \frac{\not{hr}}{day} = 1600\ \frac{Cal}{day}$$

Specific heat

Whenever heat energy is added to matter, the temperature increases. However, each type of matter requires a different number of calories to produce a given change in temperature per given mass. This quantity is called **the**

specific heat and is the number of calories required to raise 1.00 g of a substance 1.00°C. In Table 2.5, the specific heats of several common substances in cal g^{-1} $°C^{-1}$ are given.

TABLE 2.5

specific heats of common substances

Substance	Specific heat cal/(g °C)[a]
water	1.00
alcohol	0.581
sugar	0.299
aluminum	0.212
salt	0.204
iron	0.108
silver	0.0557
gold	0.0312
lead	0.0293
uranium	0.0280

[a] The units are read as calories per gram per degree Celsius.

2.11 HOW MUCH MATTER?

Weight *vs.* mass

Weight refers to the force exerted on a quantity of matter by the gravitational attraction of the earth. Because weight is actually a force on a mass, weight and mass are not identical. The mass of a specific object is a constant which is independent of the gravitational field in which it is located. Whereas the weight of a substance may vary depending on its geographical location on earth, its mass is unchanged. Matter at the equator weighs less than in the Northern Hemisphere because the earth is not a perfect sphere and bulges at the equator. The object at the equator is farther from the center of the earth and the gravitational force of attraction is less. A more dramatic example of the effect of gravity on weight has been shown by astronauts on the moon. A moon rock weighs only one-sixth that of its weight on earth because the lunar gravity is one-sixth that of the earth's. On the return trip to earth, the moon rock is almost weightless. However, the mass of the rock is the same regardless of its location. Although the terms *mass* and *weight* often are used interchangeably, mass is the real measure of the quantity of matter.

What is the mass of an 18 g object on the moon compared to its mass on earth?

Problem 2.19 A linebacker weighs 220 lb. What is his mass in kilograms?

Solution

$$220\ \cancel{\text{lb}} \times \frac{1\ \text{kg}}{2.2\ \cancel{\text{lb}}} = 1\bar{0}0\ \text{kg}$$

Density

In describing matter, both extrinsic and intrinsic properties are used. **Extrinsic properties are qualities that are not characteristic of the type of matter but depend only on the particular sample.** The extrinsic properties of a given sample of water are its volume and mass. These properties do not define the substance water from the chemist's viewpoint. **Intrinsic properties are qualities that are characteristic of any sample of a particular substance regardless of the size, shape, or mass of the sample.** An example of an intrinsic property of water is its density. **Density is defined as mass per unit volume.** At 4°C, water has a density of 1 g/ml regardless of the size of the water sample. The larger the volume, the larger will be the mass of the sample, and the ratio remains a constant.

Density is an intrinsic property.

Because a variety of units can be used in measuring the volume and mass of a substance, its density can be expressed in various units. For example, the density of water is 1.00 g/ml, 62.4 lb/ft^3, or 8.35 lb/gal. Usually the density of liquids and solids is expressed in grams per milliliter (g/ml) because the resultant values avoid small fractions and extremely large numbers. For the same reason the densities of gases are expressed in grams per liter. A list of the densities of some common substances at 25°C is given in Table 2.6.

TABLE 2.6 densities of common substances at 25°C and atmospheric pressure

Liquids (g/ml)		Solids (g/cm^3)		Gases (g/liter)	
alcohol	0.79	iron	7.86	carbon dioxide	1.80
bromine	3.12	gold	19.3	carbon monoxide	1.25
ether	0.74	lead	11.3	hydrogen	0.03
olive oil	0.92	rock salt	2.2	helium	0.16
turpentine	0.87	sugar	1.59	methane	0.66
water	1.00	uranium	19.0	oxygen	1.43
mercury	13.53	wood, balsa	0.12	nitrogen	1.25

Problem 2.20

A $15\bar{0}$-cm^3 sample of iron has a mass of 1179 g. What is the density of the iron?

Solution

Density is defined as mass per unit volume. Thus the only mathematical operation which will give the proper ratio of units is:

$$\text{density} = \frac{\text{mass}}{\text{volume}} = \frac{1179\ \text{g}}{15\bar{0}\ \text{cm}^3} = 7.86\ \text{g/cm}^3$$

Problem 2.21 A urine sample has a density of 1.030 g/ml. What is the mass of a 5$\bar{0}$-ml sample of urine?

Solution

$$1.030\ \frac{\text{g}}{\cancel{\text{ml}}} \times 5\bar{0}\ \cancel{\text{ml}} = 51.5\ \text{g}$$

Rounding off to two significant figures because of the volume measurement, the answer is 52 g.

Problem 2.22 The density of sulfuric acid is 1.83 g/ml at 20°C. A chemist needs 91.5 g of sulfuric acid. What volume must be measured to obtain the requisite mass?

Solution

$$91.5\ \cancel{\text{g}} \times \frac{1.00\ \text{ml}}{1.83\ \cancel{\text{g}}} = 50.0\ \text{ml}$$

Note that in order to solve the problem, the reciprocal of the density must be used to provide for cancellation of units.

As indicated in the discussion of temperature, the physical properties of substances are dependent on temperature. Density is one such property. Most substances expand upon heating and therefore the density will decrease as the temperature increases. For this reason the density of mercury at 25°C is given as $d^{25} = 13.53$ g/ml. The 25 used as a superscript on the symbol d representing density refers to the temperature at which the density was determined.

Specific gravity

Because water has a density of 1.00 g/ml at 4°C and is widely distributed on earth, it has been chosen as a reference material to compare the relative densities of other materials. **The mass of a given volume of a substance divided by the mass of the same volume of water is the specific gravity of the substance.** Because the specific gravity is a ratio of two numbers having identical units, the specific gravity is without units. A 10.0-ml sample of alcohol has a mass of 7.9 g. Because the same volume of water has a mass of 10.0 g, the specific gravity of alcohol is 7.9/10.0 = 0.79. Of course, since the density of water is 1.00 g/ml, the specific gravity of a substance is numerically nearly equal to its density in grams/milliliter.

The specific gravity of liquids is easy to measure with a device called a hydrometer. Filling station attendants use a hydrometer to determine the specific gravity of the liquid in automobile batteries and antifreeze in radiators. The higher the specific gravity of the liquid, the higher the hydrometer tube will float. Hydrometers are also used to check the specific gravities of liquids in industrial plants. A common test used in hospitals and clinics is the determination of the specific gravity of urine. The normal range is 1.003 to 1.030, with an average of 1.018. Examples of hydrometers are illustrated in Figure 2.5.

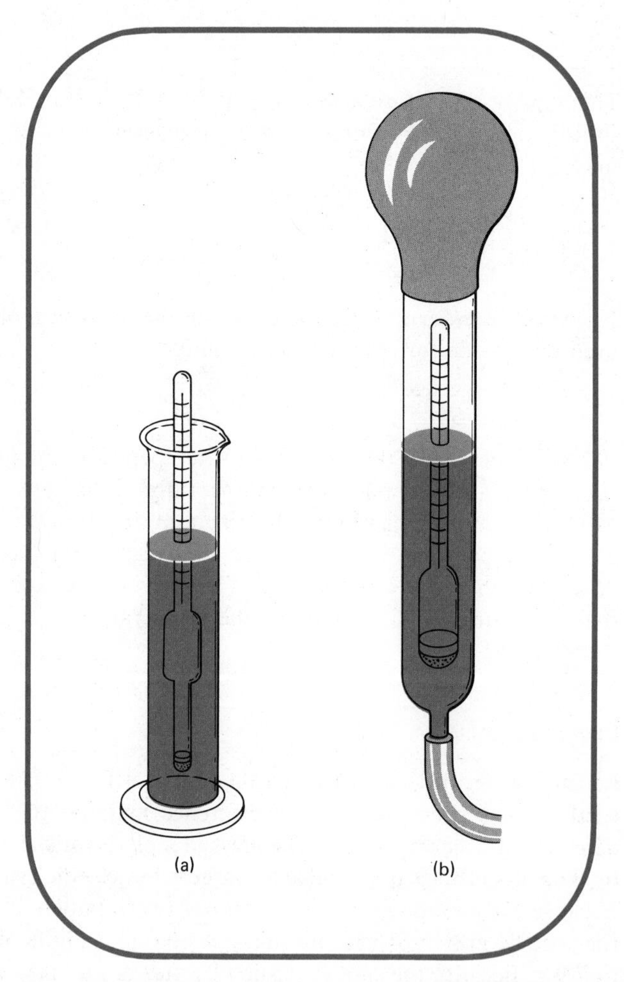

Figure 2.5 Hydrometer tubes used in the determination of (a) specific gravity of urine and (b) specific gravity of antifreeze. The hydrometer sinks until it displaces an amount of fluid equal to its own weight. The volume of the hydrometer has been calibrated to read the density that corresponds to the ratio of its weight to the displaced volume. This gives a direct reading of the fluid density.

Problem 2.23 The specific gravity of the liquid chloroform is 1.49 at 20°C. What is the mass of 50.0 ml of chloroform?

Solution Since the specific gravity is 1.49, the density is 1.49 g/ml.

$$1.49\ \frac{\text{g}}{\cancel{\text{ml}}} \times 50.0\ \cancel{\text{ml}} = 74.5\ \text{g}$$

SUMMARY

Numerical quantities must be expressed to the proper number of ***significant figures****. Any mathematical operation can only yield an answer which is as accurate as the numerical quantities used in the calculation.*

The ***factor unit method*** *when used with a systematic approach to problem solving produces a solution which can easily be shown to be correct. Units may be treated the same as numbers in mathematical operations.*

All ***measurements*** *are made by use of or comparison to a standard measuring device. The* ***metric system*** *is used in chemistry. The meter, gram, and liter are the standard measures for length, mass, and volume, respectively. Subdivision and multiple units of the standard unit employ only factors of ten.*

The ***Celsius*** *and* ***Kelvin*** *scales are used in scientific work to express temperature. The* ***calorie*** *is used to express quantities of energy.* ***Chemical energy*** *is potential energy stored in substances which can be released during a chemical reaction.*

Density *is the mass per unit volume of a substance.* ***Specific gravity*** *is a ratio of the density of a substance divided by the density of water.*

QUESTIONS AND PROBLEMS

Definitions

1. Define the following terms.
(a) calorie **(b)** Calorie
(c) thermometer **(d)** potential energy
(e) kinetic energy **(f)** density
(g) specific gravity **(h)** specific heat
(i) energy

2. Distinguish between the following similar terms.
(a) Celsius and Fahrenheit temperature
(b) Celsius and Kelvin temperature
(c) density and specific gravity
(d) mass and weight
(e) calorie and Calorie

3. What is a significant figure? Outline in your own words the rules of significant figures.

4. What is meant by rounding off? Outline in your own words the rules for rounding off.

5. What is a measurement?

Significant figures

6. How many significant figures are present in each of the following numbers?
(a) 4210 **(b)** 0.7
(c) 1.070 **(d)** 39,751
(e) 0.00861 **(f)** $560{,}706{,}\bar{0}00$
(g) 0.00010 **(h)** 3.09

7. Round off each of the following to four significant figures.
(a) 0.051916 **(b)** 51.054
(c) 139.155 **(d)** 6.1449
(e) 0.71351 **(f)** 0.0057823
(g) 1.4850 **(h)** 189.4

8. Carry out the following mathematical operations and express the answer to the proper number of significant figures.
(a) $97.45 - 10.3$ **(b)** $16.49 + 0.512 + 1.3$
(c) 6.023×2.0 **(d)** 0.01×59.5
(e) $0.5555 \div 11$ **(f)** $1000 \div 25$
(g) $1.234 + 1.01 + 0.015$

Metric interconversions

9. Carry out each of the following.
(a) 59 mm into cm **(b)** 153 cm into km
(c) 325 ml into liters **(d)** 5 liters into ml
(e) 211 mg into kg **(f)** 7 kg into cg

10. An adult of average activity inhales $1\bar{0}{,}000$ liters of air a day. What is the equivalent volume in milliliters?

11. A pound has a mass of exactly 0.4535924 kg expressed to seven significant figures. What is the equivalent mass in milligrams?

Metric and English interconversions

12. How many milliliters are there in a gallon?

13. A French male chauvinist enjoys objects of dimensions 91, 66, 91 centimeters. What are the object's measurements in inches?

14. A speed sign on the approach to Mexico City reads 40 km/hr. Should you drive your American-made car at 25 mph or 65 mph?

15. If 20 kg is the baggage allowance on an international air flight, how many pounds are you allowed to carry?

16. A snail travels at 1 cm/min. What is its velocity in furlongs per fortnight? A furlong is 660 feet and a fortnight is 14 days.

17. A good U.S. sprinter runs the hundred-yard dash in 9.1 seconds. Assuming that the sprinter accelerates instantaneously and runs at a constant speed during the race, how long would it take the sprinter to run 100 meters?

Miscellaneous interconversions

18. A Tennessee walking horse is generally 15 to 16 hands high. A hand is 4 inches. How high is a Tennessee walking horse in feet?

19. A heavyweight boxer from Great Britain weighs more than 13 stones. A stone is 14 pounds. What is the minimum weight in pounds of a heavyweight boxer?

20. What were the dimensions of Noah's ark in feet (see page 24)? Compare this length to the length of the flagship *Santa María* used by Columbus in his voyage to the new world. Although the flagship ran aground off Cap Haitien, Haiti, and was destroyed, the length was thought to be approximately 90 feet.

Temperature

21. Convert each of the following temperatures into Celsius degrees.
(a) 203°F **(b)** 14°F
(c) −58°F **(d)** $-4\bar{0}$°F
(e) 122°F **(f)** 257°F

22. Convert each of the following temperatures into Fahrenheit degrees.
(a) $5\bar{0}$°C **(b)** 125°C
(c) 5°C **(d)** $-2\bar{0}$°C
(e) $-1\overline{00}$°C **(f)** $2\overline{00}$°C

23. Convert each of the following temperatures into Kelvin degrees.
(a) 25°C **(b)** 237°C
(c) 0°C **(d)** −100°C
(e) −237°C **(f)** $1\overline{00}$°C

24. Silver melts at 961°C. What is its melting point in degrees Fahrenheit?

25. What is the equivalent of 0°K on the Fahrenheit scale?

26. Thermometers containing alcohol and a red pigment are used to measure very low temperatures. Alcohol freezes at −117°C. What is the coldest temperature in degrees Fahrenheit that such a thermometer could record?

Energy

27. How many calories are required to heat each of the following systems to the indicated temperature? Assume that the definition of a calorie applies to any one-degree temperature interval.
(a) 1.0 g of water from 25°C to 75°C
(b) 25 g of water from 25°C to 26°C
(c) 25 g of water from 25°C to 75°C

28. Approximately 0.05 cal is required to heat 1 g of tin from 0°C to 1°C. How many calories are required to heat 1 kg of tin from 0°C to 100°C?

29. One ounce (approximately $1\frac{1}{4}$ cups) of a cereal yields 112 Cal on oxidation. How many kilograms of water could be heated from 20°C to 30°C by "burning" the cereal?

30. The specific heat of silver is 0.0557. How many calories are required to heat 1.00 kg of silver from 10°C to 30°C?

Weight and mass

31. Calculate your mass in kilograms.

32. Mexico City is 7500 ft above sea level. Would you weigh more or less in Mexico City as compared to New York City? Explain why.

33. What advantages would a man with heart disease have in living on the moon, assuming that all necessary life-support systems could be transported there for him?

34. The specific gravity of a urine sample is 1.02. What is the mass of 100 ml of this material?

35. A cube of gold that is $2.\overline{00}$ cm on a side has a mass of 154.4 g. What is the density of gold?

36. The density of hydrogen gas at 25°C and normal atmospheric pressure is 0.09 g/liter. The Hindenburg dirigible occupied a volume of 1.9×10^8 liters. How many kilograms of hydrogen did the Hindenburg contain?

37. Calculate the volume of each of the following samples.
(a) 15.72 g of iron; $d^{25} = 7.86$ g/cm^3
(b) 135.3 g of mercury; $d^{25} = 13.53$ g/cm^3
(c) 42.5 g of uranium; $d^{25} = 19.0$ g/cm^3
(d) 158 g of alcohol; $d^{25} = 0.79$ g/ml
(e) 2.86 g of oxygen; $d^{25} = 1.43$ g/liter

38. Calculate the mass of each of the following samples.
(a) $2\overline{0}$ ml of bromine; $d^{25} = 3.12$ g/ml
(b) $15\overline{0}$ cm^3 of balsa; $d^{25} = 0.12$ g/cm^3
(c) 5 liters of carbon dioxide; $d^{25} = 1.80$ g/liter
(d) $1\overline{00}$ ml of olive oil; $d^{25} = 0.92$ g/ml

39. The mass of premature babies is customarily determined in grams. If a "premie" has a mass of $15\overline{00}$ g, what is its weight in pounds?

Miscellaneous problems

40. A 160-lb adult has about 5 liters of blood. Each cubic millimeter of blood contains 5 million red blood cells. How many red blood cells does the 160-lb adult have?

41. Approximately 3200 red blood cells placed side by side would measure 1 in. What is the diameter of the red blood cell in millimeters?

42. Alpha Centauri is 4 light years distance from earth. Light travels at 1.86×10^5 miles per second. What is the number of miles separating Alpha Centauri and earth?

43. The speed of light is 186,000 miles per second. What is the corresponding value in centimeters per second?

3 MATTER

LEARNING OBJECTIVES

When you finish this chapter you should be able to:

(a) distinguish between pure substances and mixtures.
(b) differentiate homogeneous mixtures from heterogeneous mixtures.
(c) relate elements and compounds.
(d) compare the three states of matter.
(e) distinguish between physical and chemical changes.
(f) differentiate physical properties from chemical properties.
(g) express a chemical change in terms of a chemical equation.
(h) recognize the experimental consequences of the law of conservation of energy and the law of conservation of mass.
(i) match the following terms to their proper definitions.

chemical change
chemical property
chemical reaction
compound
element
heterogeneous
homogeneous
mixture
phase
physical change
physical property
product
reactant
state

Before you start to read this chapter, daydream a bit and slowly look at the things around you. Scanning the room, you can probably identify a number of different materials such as plastic, wood, paper, glass, metal, cloth, and leather. A glance out of the window may bring into view a cloudy sky, grass, water, brick, stone, asphalt roads, and another person. If you were to extend the range of your observations to the rest of our planet, the number of different objects "in view" would be infinitely large. Every object viewed consists of a mixture of substances.

People have tried for thousands of years to understand the variety of substances present in our world. A first step in this understanding is to classify and define materials according to their components. In this chapter we will classify matter as either a mixture or a pure substance, and we will look at the differences between these classes. Then we will explore the different kinds of mixtures and the different kinds of pure substances. For example, some pure substances are elements, the fundamental building blocks of all matter. Other pure substances are compounds because they are combinations of elements.

The organization of matter presented in this chapter will involve numerous terms which will not be discussed with precision or in detail. As you proceed with your study of chemistry and the subject matter expands, these terms will become better understood.

3.1 CLASSIFICATION OF MATTER

Most matter that we encounter every day is a complex mixture. Air contains five gaseous components in reasonably high percentage concentration and several additional ones in low percentage concentration. Gasoline is a mixture of at least 20 liquids, and the composition varies according to locality and season. Practically all food consists of an incredible number of components. The human body probably contains hundreds of thousands of substances.

There are relatively few substances in common use that are considered pure. The copper used in electrical wiring is very pure. However, most other metals are mixtures of several substances. Although steel, for example, consists mainly of iron, it may also contain manganese, chromium, and carbon. Some medicines consist of a single substance although mixtures of compounded drugs are more common. Sugar is extremely pure, but table salt contains small quantities of additives among which is sodium iodide, important in the prevention of goiter. On the limited list of pure substances encountered in everyday life are distilled water, baking soda, and moth balls.

Classification of matter was a difficult problem for early scientists (Figure 3.1). Only occasionally did they find matter in pure form free of other substances. The great complexity of mixtures of substances occurring in nature created a roadblock in the development of the science of chemistry.

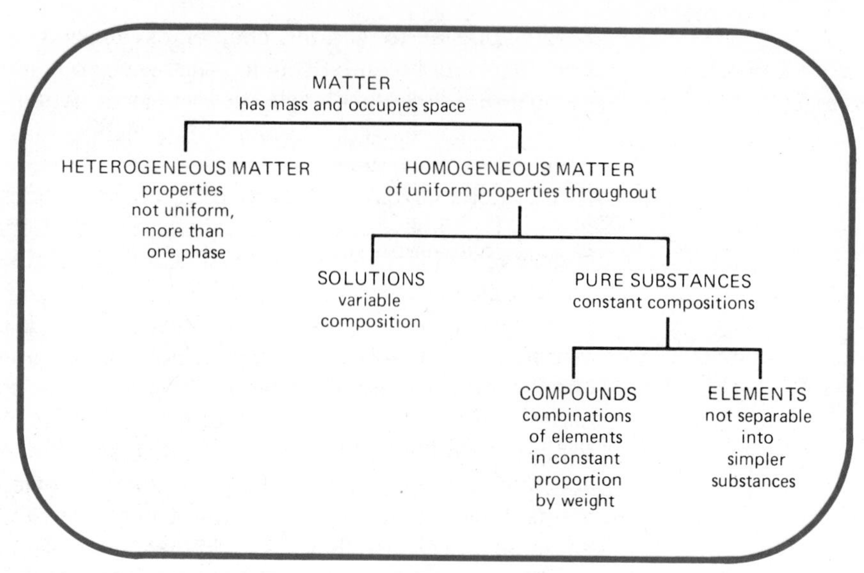

Figure 3.1 Classification of matter.

It was necessary to isolate pure substances from mixtures before these substances could be identified and their properties determined.

Mixtures

A mixture which is uniform in structure and composition is homogeneous.

Two types of mixtures exist, heterogeneous and homogeneous. **A heterogeneous mixture contains parts that are unlike or without interrelation. A homogeneous mixture is uniform in structure or composition throughout.** Heterogeneous mixtures are easily identified because they can be seen to contain two or more substances. A pepperoni and cheese pizza is a heterogeneous mixture of a number of identifiable components. Even the pepperoni is a heterogeneous mixture. The physical differences between the two or more substances contained in a heterogeneous mixture allow for separation of the mixture into its components.

Each of the different parts of a heterogeneous mixture constitutes a phase. **A phase is a part of a system throughout which the properties are uniform.** For example, an oil and vinegar salad dressing consists of two phases regardless of whether the oil and vinegar are allowed to settle in the bottle or are shaken vigorously to form thousands of mixed droplets.

Can a heterogeneous mixture consist of one phase?

Heterogeneous mixtures are nonuniform and consist of two or more phases. They can be of variable composition. Obviously an oil and vinegar mixture can contain any proportion of oil and vinegar. A marbled steak can contain varying amounts of fat and protein. The properties of the heterogeneous mixture therefore can vary from portion to portion. For example, different bites of the pepperoni and cheese pizza taste differently depending on whether they include a pepperoni sample. As will be seen, mixtures in general vary in composition but pure substances do not.

Homogeneous mixtures are single phase systems. Examples of homogeneous mixtures are air, a saltwater solution, and brass, which is a mixture of copper and zinc. The distinction between heterogeneous and homogeneous mixtures is one of uniformity of properties. A homogeneous mixture and any fraction of it have the same properties; a heterogeneous mixture is nonuniform, and its properties vary between phases.

Variability of composition is common to homogeneous mixtures as well as to heterogeneous mixtures. For example, a cup of coffee, which is a homogeneous mixture, may contain varying proportions of coffee, sugar, and cream dissolved in water. The different cups of coffee around a table are all homogeneous mixtures but their composition varies from one cup to another depending on individual preferences in taste.

A single phase or a homogeneous system may be either a pure substance or a homogeneous mixture. **A pure substance cannot be separated into components by physical means. A physical separation is any process by which the components of a mixture are separated without altering the identity of any of them.**

Pure substances

Pure substances can be divided into two classes, *elements* and *compounds*. Elements form the smaller class. The idea that there are certain basic substances, elements, from which other materials are composed, has been around since the time of the early Greek civilization. Aristotle thought that all matter consisted of four elements: earth, air, fire, and water. However, elements were not formally defined until Robert Boyle did so in 1661. His definition represents the evolution of the idea that the universe is composed of relatively few simple substances. Boyle's classic definition can be stated concisely: An element is a substance that cannot be constructed from or decomposed into simpler substances. This definition stood the test of time until relatively recently when it was recognized that elements are composed of even more elementary substances (these are discussed later). As a result of the discovery that elements can be decomposed into simpler substances, the definition had to be modified as follows: **Elements are substances that cannot be constructed from or decomposed into simpler substances by normal chemical processes.**

At one time water was thought to be an element. Why is this incorrect?

A compound is a combination of elements which cannot be broken down by physical means. Compounds can be broken down into their constituent elements only by chemical reactions. In addition, compounds are of definite composition and are thus distinguished from homogeneous mixtures, which may be of variable composition. An example of a compound is water, which consists of 88.81 percent oxygen and 11.19 percent hydrogen. The elements hydrogen and oxygen may be mixed in all proportions to form many homogeneous gaseous mixtures. However, none of the homogeneous gaseous mixtures has the properties of water, including the one that happens to be 88.91 percent oxygen and 11.19 percent hydrogen. Water contains hydrogen and oxygen combined with each other in a special way; it is distinguished from a mixture by the fixed proportions of its constituent elements.

A general definition for all compounds is as follows: **A homogeneous substance composed of two or more elements in a fixed proportion by mass is called a compound.**

3.2 STATES OF MATTER

Matter exists in three states: gaseous, liquid, and solid. In our common experience almost every substance is thought of in terms of a single state. Oxygen is a gas, oil is a liquid, and iron is a solid. However, most matter can exist in all three states if the proper conditions of temperature and pressure are attained.

GAS
LIQUID
SOLID

Water is one of the few substances that most people have observed in all three physical states. Although it is most often perceived as a liquid, at normal atmospheric pressure water exists as a solid below 0°C and as a gas above 100°C. Whereas the terms ice and steam are used to describe the solid and gaseous states of water, it should be emphasized that to the chemist the substance is still water. Water in any of the three states still consists of the elements hydrogen and oxygen combined in a certain way (Figure 3.2).

Iron is usually thought of as a solid. At 1535°C, iron is a liquid. At this temperature or higher, liquid iron can be poured from the giant caldrons of the steel industry. If iron is heated to 3000°C at ordinary pressure, it becomes a gas. Nevertheless, most people still regard iron as a solid because this is its physical state under normal conditions.

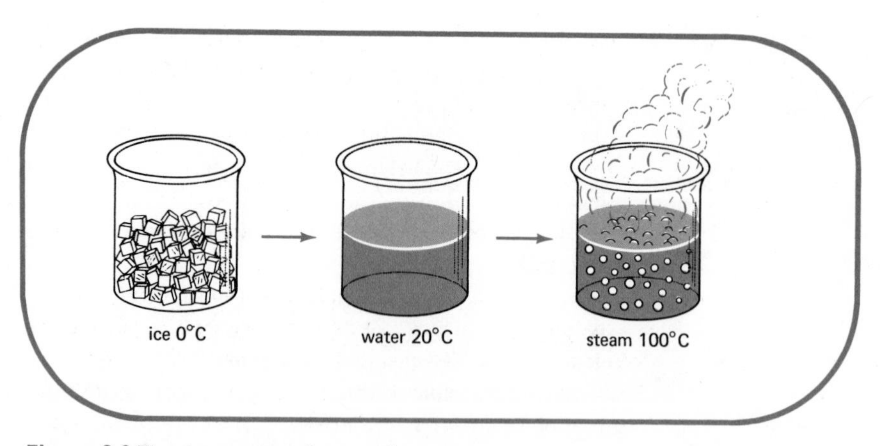

Figure 3.2 The three states of matter for water.

As a contrast to the properties of iron, consider oxygen. Oxygen is commonly thought of as a gas responsible for the maintenance of life. However, at −183°C, oxygen exists as a liquid at ordinary pressure. If oxygen is cooled to below −218.4°C, it becomes a solid. In neither state could oxygen support life as we know it.

Our viewpoint on the physical states of various types of matter is conditioned by our earth experience. Carbon dioxide on Mars can exist as a solid and may be responsible for the Martian polar caps. As a contrast, consider the average temperature of 177°C of the planet Mercury. If water exists on Mercury, it must be present only as a gas. From the foregoing it should be concluded that temperature has a marked effect on the physical state of a substance. Therefore, in the following discussion of the physical states, an average or normal room temperature is assumed.

Solids

Solids are characterized by their definite volume and shape, which are independent of the size and the shape of the container in which they are placed. The shape of a solid is fixed, and it does not flow to fit the shape of a container as does a liquid. Solids, for all practical purposes, are incompressible and resist deformation of their structure. For example, the volume of a given piece of coal beneath the enormous pressure of the earth is the same as it is on the surface of the earth.

Densities in the range of 1 g/cm^3 to 20 g/cm^3 usually characterize solids. Metals such as lead (density of 11.3 g/cm^3) and gold (density of 19.3 g/cm^3) are among the densest solids.

Most solids when heated will expand slightly. For a temperature increase of 1°C, the expansion is usually less than 0.01 percent. Such a small expansion, of course, allows for the use of solids such as wood, metal, and even plastic in construction.

Diffusion of solids is extremely slow and requires eons to proceed a few centimeters. This fact allows geologists to determine the ages of rock deposits in the earth and to define periods of the development of earth from the boundaries between rock layers.

Liquids

How does a solid differ from a liquid?

Liquids possess a constant volume at a specified temperature and pressure but do not have a characteristic shape. They will fill a container from the bottom up.

The densities of liquids are generally less than those of solids and range from 0.5 g/ml to 1.5 g/ml, with few exceptions. One notable exception is mercury, a liquid metal with a density of 13.5 g/ml. Of course, if the other normally solid metals are melted, their densities also would be large.

Liquids are only slightly compressible but usually more so than are the corresponding solids. Their negligible compressibility allows them to be used to transmit pressure, as in the case of hydraulic fluids. If hydraulic fluids were compressible, part of the pressure exerted by a foot on a brake pedal would be used to compress the liquid rather than stop the car.

Most liquids expand slightly when heated. The average increase in volume is 0.1 percent per degree Celsius. This increase is about ten times greater than for solids.

Gases

Unlike liquids and solids, gases have no characteristic shape or volume and can be contained in any size or shape vessel. The volume of solids and liquids have clearly visible boundaries. A gas, on the other hand, fills completely any container in which it is placed and its boundaries are the walls of the container. If a colored gas such as bromine is placed in a transparent container, the gas can be seen to be distributed evenly throughout the container. If the same quantity is transferred to a larger or differently shaped container, the gas distributes itself uniformly throughout the new container. The intensity of the color of the gas will diminish, of course, in the larger container.

At normal atmospheric pressure and temperature, the densities of gases are in the range of 0.0002 g/ml to 0.004 g/ml. Therefore, it is more convenient to state the densities of gases in terms of grams per liter (0.2 g/liter to 4 g/liter).

Gases are highly compressible. Under high pressure the volume of a gas can be decreased by a factor of nearly 1000. This is why industry can place large volumes of gases in cylinders. The gas in tanks used by a welder would occupy a volume several thousand times larger than that of the tank if the gas were at normal atmospheric pressure.

Gases undergo considerable expansion when heated under constant pressure. The density of gases thus decreases when heated. It is for this reason that hot air rises.

Submicroscopic features of matter

Although no proof will be given at this point, the features of the submicroscopic nature of matter are well established. **All states of matter consist of small individual particles called atoms or aggregates of atoms bonded together to form molecules.** In the solid state these particles are closely packed together in a regular array to form geometric shapes which are often manifested in the shape of the crystal of the solid (Figure 3.3). These particles are strongly attracted to each other.

In the liquid state the particles move relative to one another but still are held together by strong attractive forces. The mobility of the particles with a lesser degree of internal cohesion is responsible for the fluidity of the liquid. On the average, the particles in the liquid state are farther apart than in the solid state, as illustrated in Figure 3.3.

The particles in the gaseous state move essentially independently of one another. The energy of the gaseous particles is high enough to overcome the attractive forces that hold them together in the liquid and solid states. Compared to the liquid and solid state, the gaseous state contains particles that are far apart from one another, as illustrated in Figure 3.3. The actual volume occupied by the gaseous particles is about one-thousandth that of the volume in which the gas is contained.

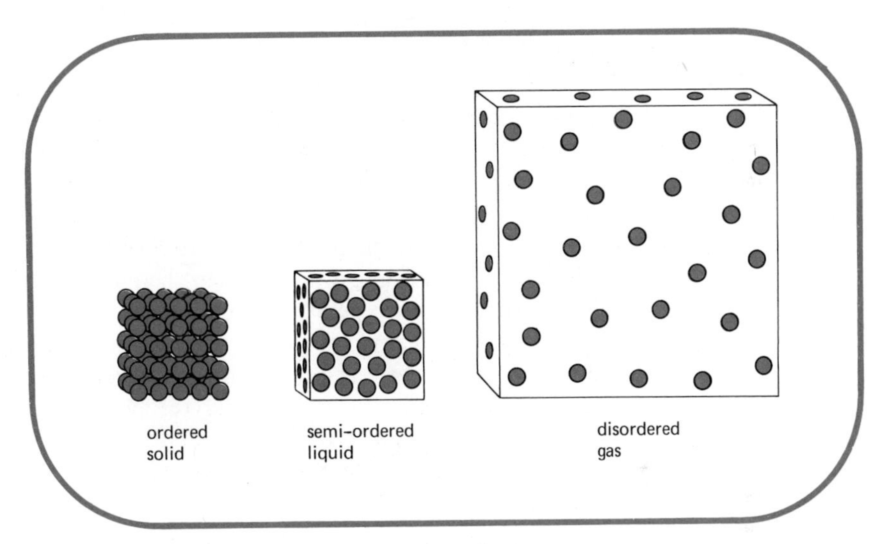

Figure 3.3 The degree of order of the states of matter.

3.3 PROPERTIES OF MATTER

Every substance can be identified by a set of properties which are characteristic of that substance. Several substances may have a few or even many properties in common, but no two substances are alike in all respects. These properties are conveniently divided into physical and chemical properties.

Physical properties

Physical properties are characteristics that are determined without altering the chemical composition of matter. Pure water is a colorless, odorless, and tasteless liquid which freezes at 0°C and boils at 100°C. Both the qualitative adjectives and the quantitative numbers describe physical properties of water. Similarly, chlorine can be described as a yellow green gas, with a suffocating odor and a sharp sour taste, that boils at −34.5°C and freezes at −101.6°C.

Thus physical properties are akin to the description 5 feet 6 inches tall, weight 160 pounds, blue eyes, and brown hair, which describe this author. Although any or all of these physical features of the author may be duplicated in a person reading this book, additional physical characteristics would certainly distinguish between the author and the reader. For example, both the amount and style of hair might be different so that the author and the reader could be distinguished. If the reader is slightly balding and has short hair, then another physical characteristic such as foot size, waistline, or any of innumerable dimensions could be considered. In addition, skin tint and even body odor are physical characteristics.

Chemical properties

Can you name one physical and one chemical property of water?

Chemical properties of a substance describe the changes in chemical composition that take place during chemical reactions with other substances. Water will react violently with sodium metal to produce hydrogen gas and the substance known as sodium hydroxide. Chlorine will react with sodium to form sodium chloride, known as table salt. There are many chemical properties which will be given throughout this text. There are fewer good analogies for a human equivalent of chemical properties than there are for physical properties. In the company of other people, an individual could be shy and retiring, or gregarious and boastful, but these features would still best be described as physical because the person remains physically unchanged. A person is chemically changed by the chemical reactions which occur in the metabolism of food and its conversion into body fat and protein. Human skin is chemically changed when it is burned.

Physical changes

Physical changes of matter alter its appearance without a change in its chemical composition. The most common examples are changes in state, as in the case of water when it changes from solid to liquid. The individual particles of water are identical in steam, liquid water, and ice. Similarly, the metal filament in a light bulb changes appearance when the electricity is turned on. The glowing filament is still chemically the same; when the light is turned off, the filament is as it originally appeared (Figure 3.4).

Chemical changes

When a small current is passed into a flashbulb, which consists of fine wires of magnesium metal in a gaseous atmosphere of oxygen, a flash results. The appearance of the contents of the bulb changes and the chemical identity has been altered. A delicate powdery substance known as magnesium oxide has been formed. This substance has very different physical and chemical properties from the original materials. A chemical change has occurred (Figure 3.4).

Figure 3.4 Examples of physical and chemical changes.

Mercuric oxide is an orange red powder which, when heated, results in the formation of colorless oxygen gas and silvery liquid metal mercury. Mercury and oxygen have different physical and chemical properties from mercuric oxide. A chemical change has occurred. Therefore, a **chemical change results in the formation of new or different substances.**

3.4 CHEMICAL EQUATIONS

In order to record the chemical transformations that matter undergoes, chemists use chemical equations. At this point we shall use word equations to describe chemical changes. As we progress the equations will be altered in appearance and made more quantitative. In the case of the flashbulb, the word equation is:

magnesium plus oxygen produces magnesium oxide plus heat and light

For the mercuric oxide experiment described above, a similar word equation is:

mercuric oxide plus heat yields mercury plus oxygen

A more compact equation for this latter chemical change is:

$$\text{mercuric oxide} \xrightarrow{\Delta} \text{mercury} + \text{oxygen}$$

The arrow is equivalent to "produces" or "yields." The Greek letter delta above the arrow indicates heat is required for the reaction to occur. For most of this text, the delta will be omitted since our prime concern will be largely with matter and its changes.

When energy is discussed, some additional symbols will be presented to make the discussion more quantitative. Thus, although electrical energy is used in the flashbulb, and heat and light are produced, the chemical equation is simply represented as:

magnesium + oxygen → magnesium oxide

REACTANT → PRODUCT

In an equation, the substances to the left of the arrow are the starting materials or reactants; the substances that result from the chemical reaction are called products and are on the right side of the arrow. In the flashbulb experiment, both magnesium and oxygen are reactants whereas magnesium oxide is a product. There may be one or several reactants or one or more products in a chemical reaction. In the mercuric oxide experiment there is only one reactant but there are two products. Note that oxygen is a reactant in one reaction and a product in another reaction. Any substance may be a reactant or a product.

3.5 LAW OF CONSERVATION OF MASS

It has been shown repeatedly during the last two centuries that the products of a chemical reaction have the same total mass as the sum of the masses of the starting materials. These observations have led to the general acceptance of the law of conservation of mass: **There is no experimentally detectable gain or loss of mass during an ordinary chemical reaction.** The general concept involved in this law was stated first in 1756 by the Russian scientist M. V. Lomonosov. However, it was the French chemist Antoine Lavoisier who gradually persuaded the scientific community to accept the concept of conservation of mass in 1774 by his careful research on the mass relationships between oxygen and metals in compounds such as tin and mercury oxide. Usually Lavoisier is given exclusive credit for the law of conservation of mass, a fact that probably reflects the high regard shown for his unique contribution in developing an understanding of combustion.

The flashbulb provides an excellent example of the law of conservation of mass since both the reactants and products are sealed in the bulb and cannot escape. If the mass of the bulb is determined both before and after the flash, it can be demonstrated that the chemical reaction occurs without change in mass (Figure 3.5). In the experiment depicted, the two flashbulbs had the same mass prior to flashing one of them. Lavoisier, of course, did not have flashbulbs and had to find ways to contain the reactants and products quantitatively. An example of Lavoisier's results would be an observation that 100.0 g of mercuric oxide when heated yields 92.6 g of mercury and 7.4 g of oxygen. The sum of the mass of the products is equal to the mass of the reactant.

There is no experimentally detectable gain or loss of mass during an ordinary chemical reaction.

Problem 3.1 Methane, a compound of carbon and hydrogen, will burn in oxygen to yield water and carbon dioxide. In order to completely burn 100 g of methane, 400 g of oxygen is required. The water produced in this reaction has a mass of 225 g. How much carbon dioxide is produced?

Solution

The total mass of the reactants is 100 g + 400 g = 500 g. Since the reaction consumes the entire amount of methane, then the sum of the masses of the products must also equal 500 g. Since 225 g of water is produced, the mass of carbon dioxide must be 500 g − 225 g = 275 g of carbon dioxide.

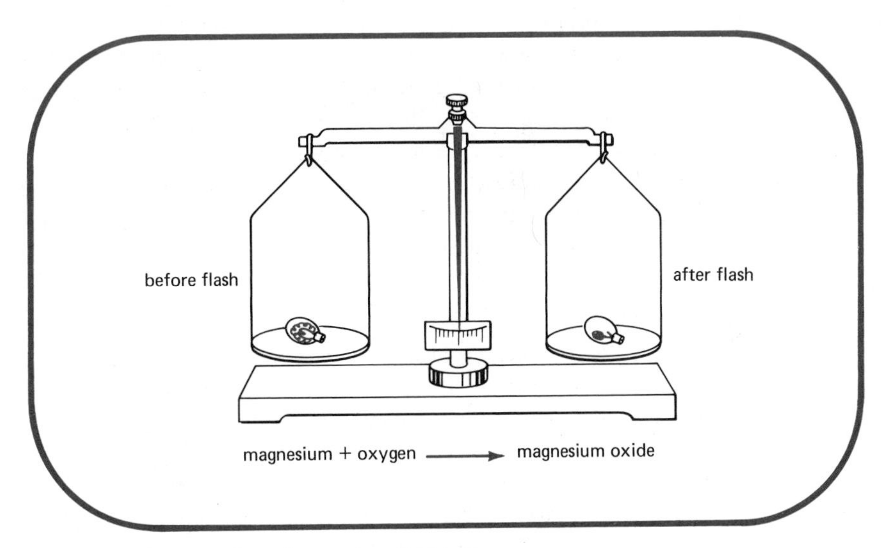

Figure 3.5 Conservation of mass in a flashbulb.

3.6 LAW OF CONSERVATION OF ENERGY

As indicated in Chapter 2, there are several forms in which energy can appear. In addition to the electrical and heat energy with which we are familiar, one very important but less well recognized form of energy is chemical energy. Energy can be stored in matter during a chemical reaction. If the reverse of that transformation can be made to occur, the original reactants are regenerated and the energy originally stored is released.

Radiant energy of the sun is transformed into chemical energy in the conversion of carbon dioxide and water into sugar in the process of photosynthesis. When sugar is metabolized by animals to yield carbon dioxide and water, the chemical energy is released.

In the electrolysis of water, an electric current causes the water to chemically decompose into hydrogen gas and oxygen gas. The electrical energy is absorbed in the reaction and the products hydrogen and oxygen possess more chemical energy than the reactant water. If hydrogen and oxygen are burned to produce water, the energy is released as heat. Thus electrical energy can be transformed into chemical energy, which in turn can be transformed into heat energy. The heat energy liberated by a given mass of hydrogen and oxygen is exactly equal to the electrical energy originally used to produce it from water. Therefore, energy can be transformed from one form to another and is not lost. **Energy can be neither created nor destroyed but its form can be changed. This is the law of conservation of energy.**

In the last half-century, the laws of conservation of mass and energy have had to be qualified. There is an experimentally detectable loss of mass in nuclear reactions. The Einstein equation, which states that mass m and energy E are interconvertible according to the formula $E = mc^2$, where c is the velocity of light, is found to account for the loss of mass. The conversion of mass to energy in nuclear reactions accounts for the tremendous quantities of energy released. One microgram (0.000001 gram) of matter is equivalent to 2.5×10^9 calories of energy.

Because energy and mass are interconvertible, the laws of conservation of mass and energy can be combined. **The total mass and energy in a chemical reaction is conserved.**

SUMMARY

Most matter exists as ***homogeneous mixtures*** *or* ***heterogeneous mixtures****. Homogeneous mixtures have uniform properties and consist of a single* ***phase*** *while heterogeneous mixtures have nonuniform properties.*

Pure substances are either ***elements*** *or* ***compounds****. Elements are substances that cannot be constructed from or decomposed into simpler substances by normal chemical processes; compounds can be broken down into their constituent elements by chemical reactions.*

There are three ***states*** *of matter: gas, liquid, and solid. A change in the state of a substance does not alter its chemical identity. The properties of each state can be described in terms of small individual particles called* ***atoms*** *and* ***molecules****.*

All matter can be characterized by unique sets of ***chemical properties*** *and* ***physical properties****. Chemical properties are characteristics of the changes that a substance undergoes in chemical reactions. Physical properties are determined without altering the chemical composition of the substance.*

Chemical reactions are described in terms of ***chemical equations*** *in which the starting materials are identified as* ***reactants*** *and the materials produced are known as* ***products****. Each chemical reaction occurs so that the* ***law of conservation of mass*** *and the* ***law of conservation of energy*** *are obeyed.*

QUESTIONS AND PROBLEMS

Definitions

1. Define each of the following terms.

(a) matter **(b)** homogeneous
(c) heterogeneous **(d)** phase
(e) mixture **(f)** state

(g) chemical property **(h)** physical property
(i) element **(j)** compound
(k) reaction

2. Tell the difference between each of the following terms.
(a) reactant and product
(b) compound and element
(c) physical and chemical change
(d) heterogeneous and homogeneous mixture
(e) compound and homogeneous mixture

Physical and chemical changes

3. Indicate whether the process described is an example of a physical or chemical change.
(a) chopping of wood **(b)** burning of wood
(c) boiling of water **(d)** melting of iron
(e) rusting of iron **(f)** burning of gasoline
(g) melting of wax **(h)** burning of sulfur
(i) grinding beef into hamburger **(j)** digesting hamburger
(k) putting sugar into tea **(l)** burning toast
(m) cooking an egg **(n)** lighting a match

4. Distinguish between a physical and a chemical change.

Physical and chemical properties

5. What physical properties distinguish a gaseous mixture containing 88.81 percent oxygen and 11.19 percent hydrogen from the compound water?

6. Why are ice, water, and steam not classified as different substances?

7. What are some of the physical and chemical properties of this page?

States of matter

8. How would you describe the term state of matter to someone who has not had a course in chemistry?

9. Classify each of the following substances as a solid, liquid, or gas at room temperature.
(a) salt **(b)** oxygen
(c) water **(d)** sugar
(e) mercury **(f)** sulfur
(g) iron **(h)** antifreeze
(i) cooking gas **(j)** brass

Elements, compounds, and mixtures

10. Make a list of materials which are used every day by you. Identify them as mixtures, elements, or compounds.

11. How might each of the following mixtures be separated?
   (a) iron filings and sand
   (b) alcohol and water
   (c) oil and water
   (d) salt and water

12. Classify each of the following according to whether it is a mixture, an element, or a compound.
   (a) graphite —carbon —
   (b) diamond
   (c) vodka
   (d) dry ice
   (e) salt
   (f) LOX (liquid oxygen)
   (g) sugar
   (h) sugared tea
   (i) peanut butter sandwich
   (j) beer
   (k) water
   (l) molten iron

Chemical equations

13. What does a chemical equation describe?

14. Look up the definition of photosynthesis and write a word equation for the reaction.

15. What are the reactants and products in the following reaction?

$$\text{cooking gas} + \text{oxygen} \rightarrow \text{water} + \text{carbon dioxide}$$

Laws of chemistry

16. Describe in your own words the following laws.
   (a) conservation of mass
   (b) conservation of energy
   (c) conservation of mass and energy

17. One pound of sulfur will burn in air to produce two pounds of sulfur dioxide. How many pounds of oxygen are consumed in the process?

18. A 50-g sample of lead reacts with oxygen to give 57.8 g of a substance called litharge. How much oxygen reacted with the lead sample?

19. If 100 kcal of energy is released when a sample of hydrogen and oxygen burn, how much energy will be required to electrolyze the water that is formed to regenerate the hydrogen and oxygen?

20. How much energy could be obtained from the destruction of one gram of matter in a nuclear reaction?

use Einstein's equation $E = mc^2$

4 ELEMENTS AND COMPOUNDS

LEARNING OBJECTIVES

When you finish this chapter you should be able to:

(a) recognize the chemical symbols for the common elements.

(b) inventory the elements present in the atmosphere, the earth's crust, and the human body.

(c) illustrate the chronology of the discovery of the elements.

(d) distinguish between metals and nonmetals by their chemical and physical properties.

(e) demonstrate the laws of definite proportions and multiple proportions.

(f) match the following terms to their proper definitions.

chemical symbol
compound
element
metal
nonmetal
semimetal

There are several million pure substances in our world. If each one were unique and bore no resemblance or relationship to any other substance, the development of an understanding of pure matter would be exceedingly difficult and could be handled only by the most intelligent individuals. Fortunately, the class of pure substances consists of 106 simple materials called elements. The remaining millions are compounds which contain these elements in combination with each other. Their combinations are governed by two laws known as the law of definite proportions and the law of multiple proportions.

That 106 elements can be responsible for the millions of known compounds should not be surprising. Our English language is based on 26 simple items known as letters. These letters are combined in a variety of ways to give thousands of words. Word formation is governed by certain rules which are part of our language.

In chemistry the elements are the chemical alphabet and the compounds are the words or combinations of elements. In this chapter you will first learn about the basic building blocks of matter, the elements. Then you will learn the rules governing the combination of elements to form compounds.

4.1 ELEMENTS

Compounds consist of combinations of elements in definite proportions.

Several million pure substances have been isolated or prepared by chemists. Of these, however, only 106 cannot be broken down by chemical means to component simpler substances. **These 106 simple substances are called elements. All of the other pure substances can be shown to consist of combinations of elements in definite proportions and are known as compounds.** In the development of chemistry, a number of substances were at one time thought to be elements. However, as new physical and chemical techniques were developed, it was shown that these substances were either mixtures or compounds. Quicklime was believed to be an element until Sir Humphrey Davy showed that the passage of an electric current through melted lime produces calcium and oxygen. Therefore, quicklime is actually a compound of calcium and oxygen. This discovery occurred in 1807.

Names of elements

Like many words in any language, the chemical names for the elements find their roots in other languages. Some elements have names that describe their properties in a language other than English. The metal bismuth originally was named from the German words *weisse masse* which means *white mass*. Through a period of contractions and changes of pronunciation, the name changed from *wismat* to *bismat* and finally to *bismuth*.

Some elements have been named after their place of discovery. Germanium, discovered in 1886 by Winkler, and francium, discovered in 1939 by Perey, indicated nationalistic allegiances. Among the more curious examples of naming elements by geography are the elemental names yttrium, ytterbium, erbium, and terbium. These are all based on the town of Ytterby in Sweden. Even in recent years the practice has not ceased. The elements berkelium, californium, and americium are among the elements discovered by scientists at the University of California at Berkeley.

Another practice in naming elements involves using the name of an individual for honorary purposes. Whereas gadolinium, named after the Finnish chemist Gadolin, and samarium, named after a Russian mine official, may be somewhat obscure, there are nevertheless important scientific figures who have been commemorated in elemental names. Curium, einsteinium, fermium, hahnium, and lawrencium all honor contributors to our knowledge of radioactive elements and the nuclear age. Rutherfordium and mendelevium honor scientists who provided much of the foundation for our modern concept of the atom.

Symbols of the elements

Chemists have long used symbols to represent the elements. These symbols are a form of chemical shorthand. The early alchemists used figures as symbols to maintain the secrecy of the early science. Some of these symbols are listed in Figure 4.1. Currently used symbols are more prosaic and more logical. They are similar to abbreviations such as Vt. for the state of Vermont, Dr. for doctor, and Ms. for women.

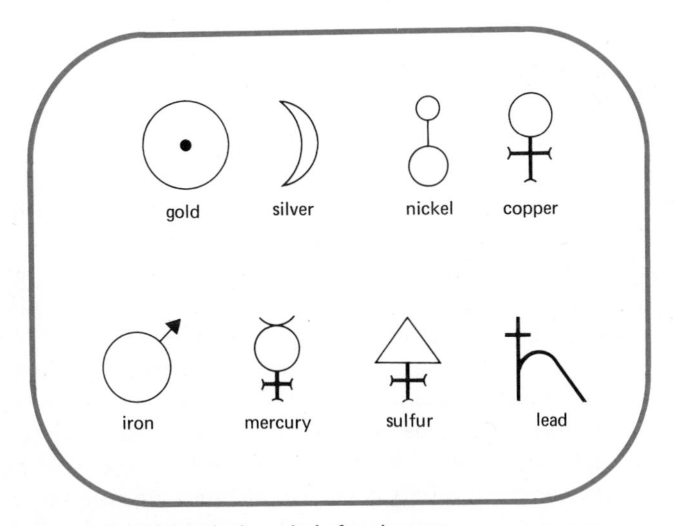

Figure 4.1 Alchemical symbols for elements.

Eleven of the 106 elements have symbols derived from older names and bear no resemblance to the currently used name. These elements and their symbols are listed in Table 4.1. The first letter of the currently accepted name is used as a symbol for 12 of the elements. Several of these are listed in Table 4.1. You should start to learn the symbols for those elements described in this text as they appear. The remaining elements and their symbols are listed on the inside cover of this text. Note that the symbol is capitalized if only one letter is used.

TABLE 4.1

symbols of some elements

Present name	Symbol	Former name
antimony	Sb	stibium
boron	B	
carbon	C	
copper	Cu	cuprum
fluorine	F	
gold	Au	aurum
hydrogen	H	
iodine	I	
iron	Fe	ferrum
lead	Pb	plumbum
mercury	Hg	hydrargyrum
oxygen	O	
potassium	K	kalium
silver	Ag	argentum
sulfur	S	
sodium	Na	natrium
tin	Sn	stannum
tungsten	W	wolfram
uranium	U	
vanadium	V	

The majority of the elements have symbols consisting of two letters, only the first of which is capitalized. There are 44 elements whose symbols consist of the first two letters of the name. There are 20 symbols consisting of the first and third letter of the name, whereas, 19 consist of the first letter and some other letter contained in the name. The necessity for using other than the first or first two letters contained in the name of the element is the result of limitations on the number of letters in the alphabet and similarities in the names of the elements. Consider the example of carbon, calcium, cadmium, and californium, whose symbols are C, Ca, Cd, and Cf, respectively. It is evident that a variety of choices of letters are necessary to provide symbols for these four elements whose first two letters are identical.

C	carbon
Ca	calcium
Cd	cadmium
Cf	californium

4.2 ELEMENTAL ABUNDANCE

Which elements are the most common and important to man? In answering such a question one must define the area to be considered. If the entire universe is considered, then 90% of matter is hydrogen. It is the fusion of hydrogen that provides the energy of the sun and other stars. Helium is the second most abundant element, amounting to 9%, and the remaining 104 elements make up only 1% of the universe, with oxygen, neon, carbon, and nitrogen next in order of decreasing abundance.

If only the surface of the earth is considered, the abundance of the elements is considerably different. The percent composition by weight listed in Table 4.2 is for a 10-mile thick shell of the earth, the atmosphere, and the ocean. Only ten elements make up approximately 99% of this small but, for us, very important part of the universe. If the entire earth is considered, then the five most abundant elements are iron, oxygen, silicon, magnesium, and nickel. Clearly the massive core of the earth has a far different composition than the segment we inhabit.

iron
oxygen
silicon
magnesium
nickel

Another way of deciding which elements are most important to humans is to consider the composition of the human body. The percent composition of the human body is given in Table 4.2. The chemical composition of humans is very different from the composition of the environment. Approximately 25

TABLE 4.2

the abundance of elements in various parts of the environment

Element	Earth crust, sea, and air	Total earth	Atmosphere	Human body
hydrogen	1.9		0.00005	10.0
oxygen	49.20	29.5	20.9	65.0
carbon	0.08		0.03	18.0
nitrogen	0.03		78.0	3.0
calcium	3.39	1.1		2.0
potassium	2.40			0.2
silicon	25.70	15.2		
magnesium	1.93	12.7		0.04
phosphorus	0.11			1.1
sulfur	0.06	1.9		0.2
aluminum	7.50	1.1		
sodium	2.64			0.1
iron	4.71	34.6		
titanium	0.58			
chlorine	0.19			0.1
argon			0.93	
boron				
nickel	0.02	2.4		
neon			0.0018	

elements are present in the human body but only 10 of these occur in quantities greater than 0.1%. Many elements, although present only in trace quantities, are vital to life. Cobalt occurs in vitamin B_{12}, and zinc, manganese, and magnesium are part of a variety of enzymes that are important in body functions.

4.3 THE DISCOVERY OF THE ELEMENTS

Some of the substances we now accept as elements were already known in ancient times. Among these elements are carbon, copper, gold, iron, lead, mercury, silver, sulfur, and tin. Prior to 1700, the alchemists were able to isolate arsenic, antimony, bismuth, phosphorus, and zinc. Because, their experimental techniques were limited, only the more easily obtainable elements were added to the list of elementary substances.

Between 1775 and 1830, the list of elements increased markedly, as illustrated in Figure 4.2. This rapid growth from 15 to 50 elements was due to the development of both chemical theory and experimental techniques. Theory provided clues for searching for elements. New theory predicted the existence of elements with certain properties. Since it is easier to find something once

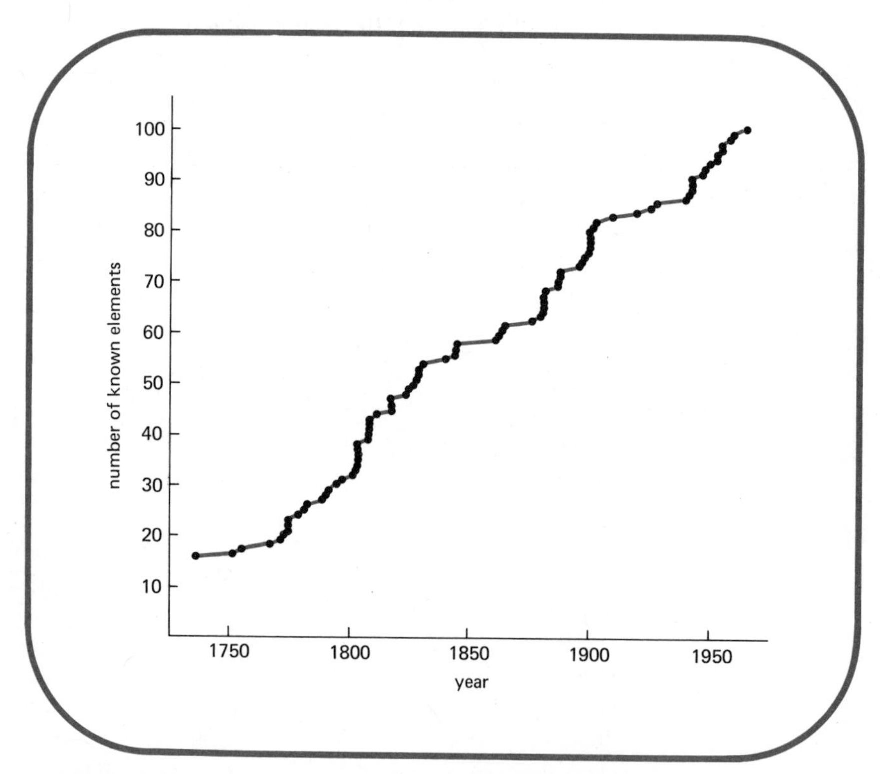

Figure 4.2 Chronology of the discovery of the elements.

you have an idea of what it is like, elements were discovered at a faster pace. In addition, the fact that the search for elements based on theory was so fruitful gave credence to the theory. The advance in experimental techniques is well illustrated by the sharp increase in the number of known elements in 1807–1808. Sir Humphry Davy discovered that a voltaic cell could be used to produce elements that are so reactive they occur only in compounds in nature. He isolated barium, calcium, magnesium, potassium, sodium, and strontium. Other sharp rises in the curve in Figure 4.2 are due to similar advances in experimental technique.

In the early part of this century, very few new elements were discovered. Since 1940, 16 elements have been added to the list. These elements are not naturally occurring on earth but have been obtained in nuclear reactions. They are radioactive and undergo nuclear fission to produce other elements. We still consider these substances to be elements in spite of the fact that they do yield other elements. The energies involved in such transformations are much higher than those of ordinary chemical reactions in which compounds can be decomposed into their constituent elements. The distinction between the decomposition of compounds into elements and the nuclear fission of elements into other elements will become clearer as you proceed in your study of chemistry.

4.4 METALS AND NONMETALS

ELEMENTS
Metals Nonmetals

All of the elements may be classified into two groups, metals and nonmetals. The classification is based on both physical and chemical properties. You are more familiar with metals not only because they are widely used but also because they are more numerous than nonmetals.

First we will consider the distinction between metals and nonmetals based on physical properties. At this time the physical properties are the more familiar to you. A listing of the physical properties of metals and some specific examples are given in Table 4.3.

TABLE 4.3
properties of metals

Property	Example
malleable (can be beaten into sheets)	gold leaf
ductile (can be drawn into wire)	copper wire
high luster	silverware
electric and thermal conductor	copper
solid (except for mercury)	tin, zinc, platinum
high density	lead (d^{20} = 11.3 g/cm^3)
high melting point	iron melts at 1535°C
high boiling point	iron boils at 3000°C
hardness	iron as in steel

Nonmetals do not have these properties. Five of the nonmetals, carbon, phosphorus, sulfur, selenium, and iodine, are solids. They are generally brittle, nonlustrous, with low melting points. They are nonconductors of electricity and poor conductors of heat. Bromine is the only liquid nonmetal at room temperature and normal atmospheric pressure. The remaining nonmetals—nitrogen, oxygen, fluorine, chlorine, hydrogen, helium, neon, argon, krypton, xenon, and radon—are gases.

Metals can be distinguished from nonmetals by their chemical properties. **Metals rarely combine with each other to form compounds. Nonmetals form many compounds with each other.** For instance, carbon and oxygen yield carbon dioxide; carbon and chlorine form carbon tetrachloride; sulfur and oxygen give sulfur dioxide; and hydrogen and oxygen yield water. Metals and nonmetals commonly combine with each other to form compounds as listed in Table 4.4.

TABLE 4.4 **compounds of metals and nonmetals**

Compound	Metal	Nonmetal
litharge	lead	oxygen
alumina	aluminum	oxygen
lime	calcium	oxygen
limestone	calcium	carbon and oxygen
pyrite (fool's gold)	iron	sulfur
galena	lead	sulfur
potash	potassium	oxygen and hydrogen
hypo (photographic)	sodium	sulfur and oxygen
saltpeter	sodium	nitrogen and oxygen
baking soda	sodium	carbon, hydrogen, and oxygen

There are some elements which do not fit completely into either the metal or nonmetal class. **Elements which have some properties of both metals and nonmetals are called semimetals or metalloids.** The semimetals are boron, aluminum, silicon, germanium, arsenic, antimony, and tellurium. The electrical conductivity of this group is less than for metals but they do have other metallic properties. Aluminum, for example, has a metallic luster, is quite malleable, and can be used in construction when mixed with other metals.

4.5 COMPOUNDS

From the relatively small number of known elements, approximately 4 million compounds have been identified. Many millions more are possible. That only 4 million are known is a reflection of the fact that many elements are unreactive and form few if any compounds. In addition, many elements occur in relatively low abundance and have not been examined extensively. Compounds containing more than four different elements are rare. If this were not the case, the number of possible compounds would soar beyond comprehension.

The compounds that contain carbon and hydrogen are far more numerous than all of the other compounds of the other elements. Because they occur naturally in living material, **carbon-containing compounds are called organic compounds.**

4.6 LAW OF DEFINITE PROPORTIONS

After some pioneering research by Lavoisier, additional careful analyses of chemical changes were reported by Joseph Proust and Jeremiah Richter. They observed that not only was matter conserved in chemical changes but that the quantities of each element involved also remained unchanged. In addition, they found that the composition of a specific compound is independent of its source, providing it has been rigorously purified. The composition of pure water, for example, is always 88.81 percent oxygen and 11.19 percent hydrogen, whether it is obtained by the purification of seawater or rainwater. The composition can be determined by breaking the compound down into its constituent elements. Alternatively, the constituent elements can be recombined in the proper amounts by appropriate chemical reactions to produce the desired compound. Proust and Richter's observations can be summarized in the *law of definite proportions:* **When elements combine to form compounds, they do so in definite proportions by mass.**

An illustration of the law of definite proportions is shown by the results of several experiments given in Figure 4.3. Lead and sulfur combine to form

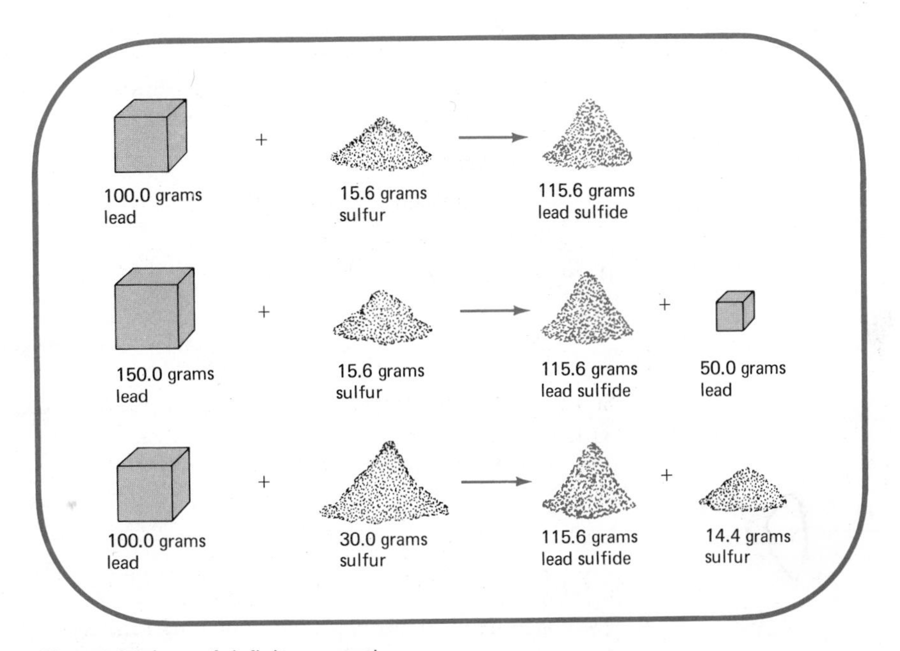

Figure 4.3 Law of definite proportions.

the compound called lead sulfide. If 100.0 g of lead is heated with 15.6. g of sulfur, then 115.6 g of lead sulfide results (the law of conservation of mass). No lead or sulfur remains. If 150.0 g of lead is heated with 15.6 g of sulfur, then 115.6 g of lead sulfide results but 50.0 g of lead is unreacted. Lead and sulfur can only combine in the ratio of 100.0 g of lead per 15.6 g of sulfur to form lead sulfide. As further verification of the law of definite proportions, consider the experiment in which 100.0 g of lead is heated with 30.0 g of sulfur. Again 115.6 g of lead sulfide results, but in this case 14.4 g of sulfur remains unreacted.

What is the law of definite proportions?

Problem 4.1

The reaction of 60.7 g of gaseous chlorine with 39.3 g of metallic sodium results in the formation of 100.0 g of sodium chloride. No uncombined elements remain. What will occur if 60.7 g of chlorine is allowed to react with 46.2 g of sodium?

Solution

According to the law of definite proportions, 60.7 g of chlorine will react with exactly 39.3 g of sodium and no more. Therefore, 46.2 − 39.3 = 6.9 g of sodium will remain.

Problem 4.2

A 1.00 g sample of hydrogen reacts completely with 7.93 g of oxygen to produce water. What will occur when 2.00 g of hydrogen reacts with 17.0 g of oxygen?

Solution

According to the law of definite proportions, elements combine in definite proportions by mass. We know that the ratio:

$$\frac{7.93 \text{ g oxygen}}{1.00 \text{ g hydrogen}}$$

represents the exact quantities required to form water. In order to determine the quantity of oxygen required for 2.00 g of hydrogen, we write:

$$\frac{7.93 \text{ g oxygen}}{1.00 \text{ g hydrogen}} \times 2.00 \text{ g hydrogen} = 15.9 \text{ g oxygen}$$

Since 17.0 g of oxygen is available, there is more than enough oxygen to react with the hydrogen. Therefore, 17.0 − 15.9 = 1.1 g of oxygen will remain. The amount of water formed will be 2.00 + 15.9 = 17.9 g.

4.7 LAW OF MULTIPLE PROPORTIONS

Two or more elements often combine under different experimental conditions to produce different compounds. Experimentally, it has been found that there is a simple relationship between the masses of the elements involved in forming the compounds. For example, the combustion of carbon in the presence of sufficient oxygen produces a gas that is nonpoisonous and noncombustible. If an insufficient amount of oxygen is available, a poisonous combustible gas is produced. Analysis of the noncombustible gas, carbon dioxide, reveals that for every 1.000 g of carbon there is combined with it 2.670 g of oxygen. In the case of the combustible gas, carbon monoxide, every 1.000 g of carbon is combined with 1.335 g of oxygen:

carbon dioxide: 1.000 g C + 2.670 g O
carbon monoxide: 1.000 g C + 1.335 g O

The ratio of the mass of oxygen combined with the same mass of carbon in the two compounds is 2.670/1.335, or 2/1. A statement of the experimental facts described above is known as the *law of multiple proportions:* **When two or more elements combine to form more than one compound, the masses of one element that combine with a fixed mass of another element are in the ratio of small whole numbers.**

The experimental basis of the law of multiple proportions involved much careful work by many chemists. An interesting historical note on the original determination of the composition of carbon dioxide involves the experimental conditions of the French chemist Dumas and the Belgian chemist Stas. In order to obtain the purest form of carbon and thus achieve accurate results, they used diamond as their source of carbon! Other less expensive experimental results are some of those of Berzelius listed in Table 4.5. The metals, lead, iron, and copper, all combine with oxygen to form compounds called oxides. From the data listed in Table 4.5, it can be seen that the masses of one element combined with a fixed mass of another element are in the ratio of small whole numbers within the proper number of significant figures.

What is the law of multiple proportions?

TABLE 4.5

examples of the law of multiple proportions

Element	Oxide	Mass of oxygen per 100 grams of metal	Ratio of masses of oxygen	Smallest whole number ratio
lead	brown oxide	15.6 g		
	yellow oxide	7.8 g	2.00/1.00	2/1
iron	red oxide	44.2 g	1.50/1.00	3/2
	black oxide	29.6 g		
copper	red oxide	12.4 g	0.500/1.00	1/2
	black oxide	25.0 g		

Problem 4.3

Assume that 100.0 g of sulfur combines with 146.4 g of oxygen to yield a compound, A. If 100.0 g of sulfur also will combine with 97.8 g of oxygen to yield a second compound, B, show that the law of multiple proportions accounts for these results.

Solution The 100.0 g of sulfur is common to both experiments described; thus it is possible to directly compare the ratio of the masses of the second element, oxygen. The ratio is 146.4/97.8 = 1.50 or 3/1. The data are consistent with the law of multiple proportions.

SUMMARY

*There are **106** known **elements** which are named in a variety of ways. Some of the names are derived from languages other than English. Each element is represented by a shorthand called a **chemical symbol**.*

*The amounts of the elements present in any part of the universe vary. Hydrogen is the most abundant element in the universe, whereas iron, oxygen, silicon, magnesium, and nickel are most abundant on earth. Most elements are classified as either **metals** or **nonmetals**, with metals forming the larger class. A limited number of **semimetals** have properties intermediate between metals and nonmetals.*

*Compounds consist of elements combined in definite proportions by mass as stated in the **law of definite proportions**. Two or more compounds may be formed from the same elements. However, the **law of multiple proportions** states that the masses of one element that combine with a fixed mass of another element are in the ratio of small whole numbers.*

QUESTIONS AND PROBLEMS

Definitions

1. Define each of the following terms.
(a) element
(b) compound
(c) chemical symbol
(d) metal
(e) nonmetal
(f) semimetal

2. What is the law of definite proportions?

3. What is the law of multiple proportions?

4. How is an element distinguished from a compound?

Element names

5. List the names of elements named for famous scientists.

6. List the names of elements named by United States scientists to honor their country, state, and university.

Element symbols

7. List as many elements as you can which have a single letter as a symbol.

8. Why are symbols used to represent elements?

9. Why is it necessary to use two letters for symbols of some elements?

10. Why are some of the elements represented by symbols which bear no resemblance to their English name?

11. List as many elements as you can which have symbols which do not resemble their name.

12. Match the following names with their symbols.

(a) carbon	**(1)** O
(b) iodine	**(2)** Ca
(c) potassium	**(3)** Ag
(d) sulfur	**(4)** Sn
(e) calcium	**(5)** I
(f) gold	**(6)** Fe
(g) tin	**(7)** K
(h) iron	**(8)** C
(i) oxygen	**(9)** S
(j) silver	**(10)** Au

Element abundance

13. What is the most abundant element by mass in the universe?

14. What are the five most abundant elements in the entire earth?

15. What are the two most abundant elements in the earth's crust?

16. What are the two most abundant elements in the atmosphere?

17. What are the three most abundant elements in the human body?

Metals and nonmetals

18. How many elements are known?

19. How many elements are metals?

20. How many elements are nonmetals?

21. How many elements are semimetals?

22. Name one liquid metal and one liquid nonmetal.

23. Name two solid nonmetals.

24. Name four gaseous nonmetals.

Compounds

25. Give three examples of compounds which contain oxygen combined with a nonmetal.

26. Give two examples of compounds containing a metal combined with oxygen.

27. Give two examples of compounds containing a metal combined with sulfur.

28. How can millions of compounds be formed from only 106 elements?

Law of definite proportions

29. A 50.0-g sample of lead reacts completely with oxygen to give 57.8 g of lead oxide.
(a) How much oxygen reacted with the lead?
(b) How much oxygen would be needed to react with 25.0 g of lead?
(c) What will happen if 50.0 g of lead has 10.0 g of oxygen available for reaction?
(d) What will happen if 60.0 g of lead has 7.8 g of oxygen available for reaction?

Law of multiple proportions

30. Hydrogen and oxygen combine to give water but, under different experimental conditions, to give hydrogen peroxide. In water, 1.00 g of hydrogen is combined with 7.93 g of oxygen, while in hydrogen peroxide, 1.00 g of hydrogen is combined with 15.9 g of oxygen. Show that these data are consistent with the law of multiple proportions.

31. Two different compounds of nitrogen and hydrogen have been found to have the following mass compositions: one compound contains 41.62 g of nitrogen combined with 1.00 g of hydrogen; the second compound contains 4.64 g of nitrogen combined with 1.00 g of hydrogen. Show from these data that the two compounds are in agreement with the law of multiple proportions.

5

ATOMS, MOLECULES, AND IONS

LEARNING OBJECTIVES

When you finish this chapter you should be able to:

(a) list the postulates of Dalton's atomic theory.
(b) identify the three subatomic particles by name, mass and charge.
(c) describe the atomic arrangement of electrons, protons, and neutrons.
(d) express the number of subatomic particles contained in an atom by an elemental symbol.
(e) represent molecules by their molecular formula.
(f) distinguish between the ions formed by losing or gaining electrons.
(g) calculate the molecular weights of molecules.
(h) match the following terms to their proper definitions.

anion
atom
atomic weight
atomic number
cation
diatomic
electron
ionic compound
isotope
molecule
molecular weight
neutron
polyatomic
proton

We saw in the last chapter that every pure substance is identifiable by its own unique set of properties. But what is it that is responsible for those properties? How small a piece of that substance will still have those properties? Rock sugar can be ground up into the small pieces of powdered sugar and it still tastes the same. Then, too, when sugar is put into water, the crystals disappear and the water acquires the taste of sugar. This indicates that the sugar has been broken into pieces that are far too small to see, but they still have the properties of sugar. In the past, scientists believed that matter was continuous. That is, they thought that a substance could be broken down into smaller and smaller pieces indefinitely. Any piece of a substance, no matter how small, would still have the identity of that substance. We now know that this is not true. There comes a point when the substance can no longer be divided without changing its characteristics. This point is reached when we encounter one of the three types of submicroscopic particles—atoms, molecules, and ions. An understanding of these particles is necessary to our conception of matter and its behavior. We will discuss these particles in this chapter.

In this chapter you will also learn that atoms, molecules, and ions consist of even more fundamental subatomic particles called electrons, protons, and neutrons. These subatomic particles will be discussed in greater detail in Chapters 6 and 7.

5.1 DALTON AND THE ATOMIC THEORY

How is the concept of the indivisible and indestructible atom related to the law of definite proportions?

About 460 B.C., Democritus, a Greek philosopher, suggested that matter consists of a large number of indivisible unchangeable particles. **These particles were called** *atomas*, **meaning** *indivisible* **or** *noncutable*. Today the *atomas* are called atoms. **The theory that matter is composed of atoms is called the atomic theory.** The atoms were perceived by Democritus to be identical except for size and shape. In explaining why changes occur in what are now termed chemical reactions, he suggested that the atoms change their positions relative to each other. Thus, although there is a constant change in the matter of the world, the identity and number of atoms remains unchanged.

It was not until about 1650 A.D., or about 2000 years after Democritus, that an Italian physicist, Gassendi, resurrected the atomic concept and received favorable support from Sir Isaac Newton. A Russian chemist, Mikhail Lomonosov, in the middle of the eighteenth century, was so convinced of the correctness of the particle or atomic theory of matter that he speculated on both the nature and motion of atoms. Indeed, his concept of the kinetic energy of moving atoms was at least a century ahead of his time.

Although the atomic nature of matter was accepted by at least a substantial number of scientists by the beginning of the nineteenth century, this acceptance was based on a loosely defined intuitive sense of its correctness. It remained for an English schoolteacher, John Dalton, to suggest a set of astonishingly simple generalizations to account for the then known laws of chemistry on a quantitative basis. His ideas are collectively called the atomic theory. Although Dalton is customarily given a major role in the development of the theory, it should be pointed out that his ideas were by no means original. His historical role is one of being at the right place at the right time. The rest of the scientific community was ready to test the statements that he made. Certainly the concepts of Democritus could not be tested with the experimental tools of that age.

Some of Dalton's ideas were later proved to be incorrect, but the essentials of the theory are still taught today with only minor modifications. It is interesting to note that although Dalton had a sufficiently remarkable insight into the nature of particulate matter to suggest an atomic theory, he later proved quite resistant to and even obstructive in the development and alteration of his theory by others who were able to verify experimentally that some of his ideas were incorrect.

The atomic theory of Dalton may be summarized as follows:

1 **Elements consist of minute, indivisible, indestructible particles called atoms.**
2 **All atoms of a specific element are identical to each other, that is, they have the same mass and size.**
3 **Atoms of different elements have different masses and sizes.**
4 **In a chemical reaction, atoms of two or more elements combine to form compounds.**
5 **When atoms combine to form compounds, they do so in simple numerical ratios, such as one-to-one, one-to-two, two-to-three, and so on.**
6 **Atoms of two or more elements may combine in different ratios to form more than one compound.**

The first two postulates of Dalton have been modified in light of more modern experimental evidence. We now know, contrary to the first postulate, that the atom is divisible and it is not indestructible. The nuclear age is a consequence of the knowledge gained from splitting the atom. The second postulate is incorrect since many elements exist in two or more forms of slightly differing masses. Such substances are called isotopes of the same element. A discussion of isotopes will be presented in Section 5.5.

5.2 ELECTRONS, PROTONS, AND NEUTRONS

Although atoms of some elements have been "seen" by using an electron microscope, no one has actually seen the subatomic part of the atom. However, the existence of these particles is an accepted article of faith by scientists of all countries. Although the present picture of the atom is immensely

ATOM
Proton
Neutron
Electron

complicated, **the atom essentially can be viewed as being composed of the three subatomic particles, the proton, the neutron, and the electron.** The number of subatomic particles determines the type of atom and hence the identity of the element.

1 amu
1.6603×10^{-24} g

The proton is represented by the symbol *p*; it has a mass of 1.6725×10^{-24} g and a positive charge of 1.60×10^{-19} coulombs. Since these quantities are cumbersome to write and are repeatedly encountered throughout chemistry, we choose to use relative terms to discuss atoms. **A mass of 1.6603×10^{-24} g is called 1 atomic mass unit (amu). Therefore, the proton has a mass of 1.0073 amu or very close to 1 amu.**

$$1.6725 \times 10^{-24}\ \cancel{g} \times \frac{1.0000\ \text{amu}}{1.6603 \times 10^{-24}\ \cancel{g}} = 1.0073\ \text{amu}$$

The charge of $+1.60 \times 10^{-19}$ coulombs is usually replaced by a relative charge of $+1$ without any units. **The proton charge is represented by $+1$.**

The neutron is represented by *n*; it has a mass of 1.6748×10^{-24} g and is electrically neutral. The relative mass of the neutron is 1.0087 amu, or essentially 1 amu.

The electron is represented by e^-; it has a mass of 9.109×16^{-28} g and a negative charge of -1.60×10^{-19} coulombs. On the amu scale, the e^- has a mass of only 0.0005486 amu, which is considered as 0 for most purposes.

$$9.109 \times 10^{-28}\ \cancel{g} \times \frac{1.0000\ \text{amu}}{1.6603 \times 10^{-24}\ \cancel{g}} = 0.0005486\ \text{amu}$$

The charge of the electron is equal in magnitude to the proton but is of opposite sign. Therefore, the **electron charge is represented as -1.**

In Table 5.1, all of the data for subatomic particles are summarized.

TABLE 5.1
properties of subatomic particles

Name	Mass (g)	Relative mass (amu)	Charge (coulombs)	Relative charge
Proton	1.6725×10^{-24}	1.0073	1.60×10^{-19}	1
Neutron	1.6748×10^{-24}	1.0087	0	0
Electron	9.109×10^{-28}	0.0005486	-1.60×10^{-19}	-1

5.3 THE ARRANGEMENT OF THE ELECTRONS, PROTONS, AND NEUTRONS IN THE ATOM

Atoms of different elements differ in the number of subatomic particles which they contain. At this time, a brief description of the arrangement of the subatomic particles in the atom will be given. In the next chapter and in some subsequent parts of this book, atomic structure will be developed in greater detail. Thus, for now, you will learn about the number of residents (subatomic particles), the house (atom) in which they live, and where they spend their time. Details of their lives will be developed in later chapters after you understand who they are.

The fundamental facts abut the subatomic particles within an atom are as follows:

1 **Every proton and neutron is located in the center of the atom, called the nucleus.** Thus the charge of the nucleus is equal to the sum of the charges of the protons. The diameter of the average nucleus is 10^{-13} cm. This small volume contains material of high density, 10^{14} g/ml. Thus a single milliliter of nuclear matter alone would have a mass of approximately 10^8 tons. Incidentally, the so-called black holes postulated by astronomers are thought to consist of nuclear matter alone.

2 **The mass of an atom in amu, or the mass number, is equal to the sum of the number of protons and neutrons contained in the nucleus.** Remember that both the proton and neutron have a mass of essentially 1 amu. The mass of the electrons contribute little to the mass of the atom since they are so light.

How does the atomic number differ from the atomic mass?

3 **The number of protons in the nucleus of an atom is called the atomic number.** Therefore, the number of neutrons contained in a nucleus is equal to the mass number minus the atomic number.

4 **The number of electrons in an electrically neutral atom is equal to the number of protons in the nucleus. If the number of electrons exceeds the number of protons, the atomic particle is negatively charged and is called a negative ion or anion. When the number of electrons is less than the number of protons, the atomic particle is positively charged and is called a positive ion or cation.**

5 **The electrons are located in space about the nucleus.** Electrons possess energy but all the electrons of a single atom do not have identical energies. Each electron is said to exist in an energy level or shell. These shells correspond to areas at considerable distances from the nucleus. These distances are approximately 100,000 times the diameter of the nucleus. Thus the diameter of an atom averages 10^{-8} cm. This structure might be easier to picture if we translate these dimensions to something we are more familiar with. If the atomic nucleus were the size of a basketball, the electron would be an object the size of a period about 20 miles away (Figure 5.1).

5.4 ELEMENTAL SYMBOLS

In Section 5.1, you learned that the smallest unit of an element is an atom of that element. The elemental symbol, then, can refer to an atom, or any number of atoms of that element. Since atoms of one element differ from those of another element by the number of their subatomic particles, it is convenient to have this information about subatomic composition included with the

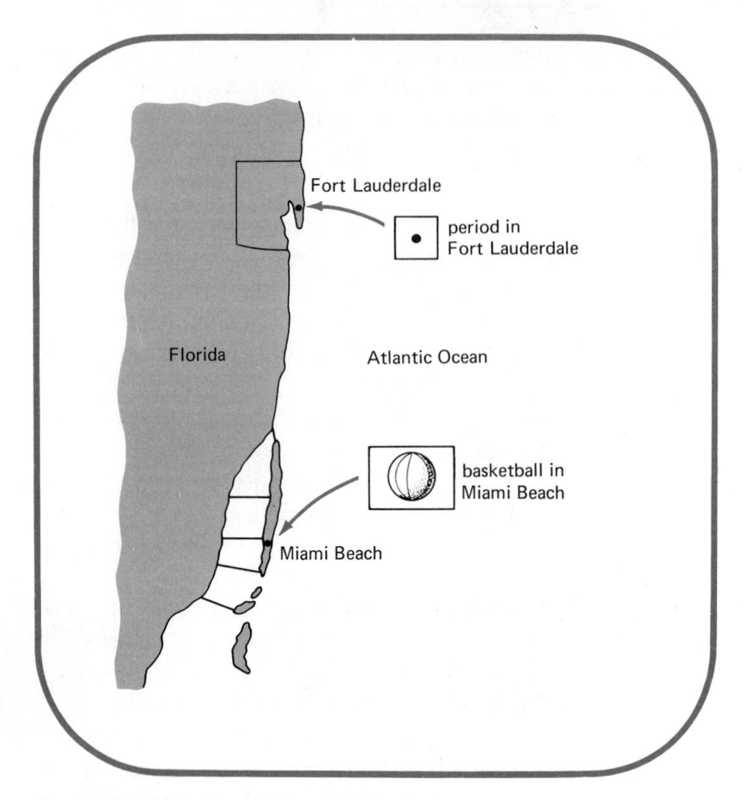

Figure 5.1 An analogy of the relationship between an electron and the nucleus. If the nucleus is the size of a basketball, the electron is a period 20 miles away.

elemental symbol. The symbol for each element is decorated with a couple of numbers to provide this information about the atoms of the element. The following symbolism is used:

$^{A}_{Z}E$ **E is the symbol of the element**
A is the mass number of the element - # of protons + neutrons
Z is the atomic number of the element
↳ # of protons

Recall that the atomic number, Z, is equal to the number of protons in the atom. The mass number, A, is equal to the number of protons and neutrons in the atom. Thus the number of neutrons is equal to $A - Z$.

Now you can write out the constitution of an atom in terms of its subatomic particles by referring to the elemental symbol.

Problem 5.1 What is the significance of $^{1}_{1}H$?

Solution

$$^{1}_{1}H = ^{A}_{Z}E$$

The symbol H stands for the element hydrogen. The superscript 1 for A

means that the sum of the number of proton(s) and neutron(s) in the atom is 1. The subscript 1 for Z means that there is 1 proton in the atom. Therefore, the number of neutrons ($A - Z$) is $1 - 1 = 0$. Since the number of electrons in an atom must be equal to the number of protons, the number of electron(s) is 1.

Problem 5.2 How many protons, neutrons, and electrons are contained in an atom represented by $^{16}_{8}O$?

Solution

$$^{16}_{8}O = ^{A}_{Z}E$$

The elemental symbol represents the element oxygen. Oxygen contains 8 protons as $Z = 8$. Therefore, oxygen also contains 8 electrons as it is electrically neutral. The number of neutrons is given by $A - Z$, or $16 - 8 = 8$.

Problem 5.3 Describe the atom represented by the symbol $^{238}_{92}U$.

Solution

$$^{238}_{92}U = ^{A}_{Z}E$$

The elemental symbol represents uranium. There are 92 protons in uranium as $Z = 92$. Furthermore, uranium also contains 92 electrons. The number of neutrons is given by $A - Z$ or $238 - 92 = 146$.

Problem 5.4 Write the symbol for the atom of gold which contains 79 protons and 118 neutrons. How many electrons are contained in gold?

Solution The symbol for gold is Au. The number of protons is written as a subscript, or $_{79}Au$. Since the mass number is equal to the number of protons and neutrons, or $118 + 79 = 197$, the superscript is 197, or $^{197}_{79}Au$. The number of electrons is equal to the number of protons, or 79 electrons.

A simplified representation of the atom is given by a sphere (the nucleus) which shows the tally of protons and neutrons. The electrons are listed in a partial "orbit" at a distance from the nucleus. Thus $^{16}_{8}O$ may be represented as follows:

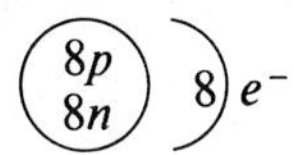

Furthermore, $^{238}_{92}U$ is represented by:

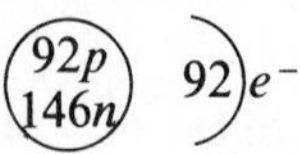

Recall, however, that the actual distance separating a nucleus and its electrons is very large. Furthermore, you are forewarned that not all electrons are the same distance from the nucleus. Details of atomic structure will be provided in Chapter 6.

5.5 ISOTOPES

Isotopes contain the same number of protons.

The second postulate of Dalton was made in the absence of any experimental evidence to the contrary. He thought that each element consisted of atoms of identical mass. However, we now know that atoms of an element can exist with a variety of different masses. These atomic particles, called **isotopes have the same number of protons but differ in the number of neutrons that they contain.** Thus the isotopes of an element have identical atomic numbers but different mass numbers. It is the number of protons in the nucleus (the atomic number) which determines the identity of the element and not the number of neutrons.

An analogy may be drawn between isotopes and identical twins. Identical twins result from the splitting of a single cell formed from the union of one sperm and one egg. As a consequence, identical twins have the same genes or chromosomal makeup. However, the twins need not always be exactly the

Figure 5.2 Identical twins may have different masses. Similarly, isotopes of the same element have different masses.

same weight (Figure 5.2). They are still twins regardless of their weights. In a similar manner, isotopes are the atoms of the same element which differ in their masses.

$^{1}_{1}H$ 1 proton
$^{2}_{1}H$ 1 proton and 1 neutron
$^{3}_{1}H$ 1 proton and 2 neutrons

Hydrogen can exist with 0, 1, or 2 neutrons in its nucleus, although the isotope without any neutrons is the most common in nature. Because the neutron contributes to the mass of the atom, the masses of the three isotopes of hydrogen are different. The isotopes are represented by $^{1}_{1}H$, $^{2}_{1}H$, and $^{3}_{1}H$ and have masses of 1 amu, 2 amu, and 3 amu, respectively.

Problem 5.5 There are three isotopes of oxygen. The atomic number is 8. The isotopes of oxygen contain 8, 9, and 10 neutrons, respectively. How many protons are contained in oxygen? What are the symbols of the isotopes?

Solution The atomic number is equal to the number of protons in the nucleus. Eight protons are present in oxygen. The mass number of an isotope is equal to the sum of the number of protons and neutrons in the nucleus. Thus the mass numbers of the isotopes are $8 + 8 = 16$, $8 + 9 = 17$, and $8 + 10 = 18$. The symbols are $^{16}_{8}O$, $^{17}_{8}O$, and $^{18}_{8}O$.

Problem 5.6 What do the symbols $^{235}_{92}U$ and $^{238}_{92}U$ mean? What relationship exists between the atoms represented by these symbols? How many neutrons are contained in each nucleus?

Solution The subscript 92 indicates that the element represented by U, uranium, contains 92 protons. Because the symbols represent atoms that contain the same number of protons but which differ in mass number, the atoms are isotopes of each other. The number of neutrons in $^{235}_{92}U$ is $235 - 92 = 143$. In $^{238}_{92}U$, the number of neutrons is $238 - 92 = 146$.

5.6 MOLECULES

Elements

The simplest particles having the properties of the gaseous elements helium, neon, argon, krypton, xenon, and radon, are the atoms of these elements. Such elements are said to consist of monatomic particles.

The common gases, nitrogen, oxygen, and hydrogen, consist of particles that contain two atoms; they are said to be diatomic particles. **A particle that consists of two or more atoms that must remain associated in order to possess the properties of the substance is called a molecule.** We see that Dalton's concept of an element is not entirely correct. Some elements such as helium and neon do contain unassociated atoms, but many elements exist as molecules, which in turn are composed of two or more atoms.

Among the elements that are also diatomic are fluorine, chlorine, bromine, and iodine. The molecules of some elements contain many more than two atoms and are said to be polyatomic. One form of phosphorus consists of molecules of four atoms, whereas some sulfur exists in units of eight atoms. **A representation of the molecule of an element involves use of a subscript to the right of the elemental symbol.** The diatomic molecule chlorine is represented as Cl_2, whereas the elemental forms of phosphorus and sulfur are represented as P_4 and S_8, respectively.

Cl_2 subscript indicates that two atoms are present in one chlorine molecule

Compounds

With the exception of ionic compounds (Section 5.7), the molecule is the smallest possible unit of a compound which retains the properties of the compound. **The molecules of compounds contain two or more different types of atoms.** This differs from the molecules of elements which contain only one type of atom. **A representation of a molecule of a compound which indicates the number of each kind of atom contained in the molecule is called the molecular formula.**

Water consists of molecules containing two hydrogen atoms and one oxygen atom. The chemist represents this molecule as H_2O. The subscript 2 to the right of the symbol for hydrogen indicates the number of hydrogen atoms contained in the molecule. No subscript follows the symbol for oxygen which means, by convention, that only one atom of that type is contained in the molecule. The molecule could be represented by OH_2 equally as well. It is merely a matter of convention that chemists have decided to use H_2O instead of OH_2 as the molecular formula for water. Both symbols convey the same information.

The subscripts representing the number of atoms contained in a molecule of a compound are in no way related to the number of atoms present in the free element. Although both hydrogen and oxygen are composed of diatomic molecules, a water molecule contains only one atom of oxygen but two atoms of hydrogen. The two hydrogen atoms present in H_2O are not molecular hydrogen but rather two hydrogen atoms that have chemically combined with oxygen. Molecules of other compounds contain various amounts of hydrogen. Hydrogen combines with fluorine to yield hydrogen fluoride, HF, and with nitrogen to yield ammonia, NH_3.

What does NH_3 mean to you?

The details of why molecules exist will be postponed until Chapter 7. At this point it should be accepted that **there are forces that bind the atoms together in the molecular aggregate. These forces are called chemical bonds.**

Problem 5.7 At a variety of temperatures, the element sulfur may contain 8, 4, or 2 atoms per molecule. What are the symbols for the various molecules?

Solution

Since the goal of this problem is to represent the number of atoms contained in a molecule and not the number of neutrons and protons contained in the nucleus, the use of subscripts and superscripts to the left of the symbol S is unnecessary. A subscript to the right gives the number of atoms per molecule.

ANSWER

S_8 contains 8 atoms per molecule
S_4 contains 4 atoms per molecule
S_2 contains 2 atoms per molecule

Problem 5.8

One of the major components of natural gas is methane. Methane is a compound represented by the molecular formula CH_4. What does this molecular formula mean?

Solution

The molecular formula indicates that the compound contains carbon and hydrogen. Because no number subscript appears to the right of C, there is only one atom of carbon per molecule. There are four atoms of hydrogen per molecule. The smallest unit that has the properties of methane consists of five atoms—one of carbon and four of hydrogen.

Problem 5.9

Methyl mercury, a product of mercury contaminated water, contains 2 atoms of carbon, 6 atoms of hydrogen, and 1 atom of mercury. What is the molecular formula of methyl mercury?

Solution

Without some additional knowledge of chemical conventions, you cannot determine the order in which the elemental symbols should be written. Assuming that the order is carbon followed by hydrogen and then mercury, the molecular formula is C_2H_6Hg.

Problem 5.10

Aspirin, a chemical compound known to the chemist as acetylsalicylic acid, contains 9 atoms of carbon, 8 atoms of hydrogen, and 4 atoms of oxygen. Write the molecular formula.

Solution

Although we do not know the order of the symbols in the molecular formula, let us use C followed by H and finally O. The molecular formula is then given by $C_9H_8O_4$.

5.7 IONS

Cations and anions

As indicated in Section 5.3, atomic particles which contain more or fewer electrons than protons are called ions. **The positively charged ion or cation contains fewer electrons than protons. The negatively charged ion or anion contains more electrons than protons.** These ions are formed from an atom by losing or gaining electrons. In these processes, the atomic identity of the element is unchanged as the number of protons determines the elemental identity. In either type of ion, the charge is a multiple of $\pm 1.60 \times 10^{-19}$ coulombs. Since relative unit charges are used in describing the proton and the electron, the same system is used for ions.

An ion is written with its relative charge as a superscript to the right of the elemental symbol. The superscript bears a positive or negative sign indicating the charge and an integer indicating how many electrons have been removed from or added to the neutral atom.

When an electron is removed from the sodium atom, the sodium ion results. This ion is symbolized as Na^+. Equations representing this process are shown below:

$$(11p)\ 11)e^- \rightarrow (11p)\ 10)e^- + e^-$$

$$Na\begin{pmatrix}11 \text{ protons}\\11 \text{ electrons}\end{pmatrix} \rightarrow Na^+\begin{pmatrix}11 \text{ protons}\\10 \text{ electrons}\end{pmatrix} + 1e^- \text{ (electrons)}$$

$$Na \rightarrow Na^+ + e^-$$

The symbol e^- represents the electron and its relative charge. Since the atomic number of sodium is 11, we know that the atom contains 11 protons and 11 electrons. When an electron is removed from the sodium atom, the resultant sodium ion is positively charged.

Na^+ is a cation obtained by removing one electron from the sodium atom.

When the calcium atom (atomic number of 20) loses 2 electrons, the resultant calcium ion is represented by Ca^{2+}. An equation is given below:

$$(20p)\ 20)e^- \rightarrow (20p)\ 18)e^- + 2\ e^-$$

$$Ca\begin{pmatrix}20 \text{ protons}\\20 \text{ electrons}\end{pmatrix} \rightarrow Ca^{2+}\begin{pmatrix}20 \text{ protons}\\18 \text{ electrons}\end{pmatrix} + 2e^- \text{ (electrons)}$$

$$Ca \rightarrow Ca^{2+} + 2e^-$$

In the case of Ca^{2+}, the ion bears 2 units of positive charge.

As an example of an anion, consider the addition of an electron to the chlorine atom to yield the chloride ion. The atomic number of chlorine is 17.

$$1e^- + (17p)\ 17)e^- \rightarrow (17p)\ 18)e^-$$

Cl⁻ is an anion obtained by adding one electron to a chlorine atom.

$$1e^- + \text{Cl}\begin{pmatrix}17 \text{ protons}\\ 17 \text{ electrons}\end{pmatrix} \rightarrow \text{Cl}^-\begin{pmatrix}17 \text{ protons}\\ 18 \text{ electrons}\end{pmatrix}$$

$$e^- + \text{Cl} \rightarrow \text{Cl}^-$$

In a similar manner, the oxygen atom can accept 2 electrons to yield the oxide ion O^{2-}.

$$2e^- + (8p)\ 8)e^- \rightarrow (8p)\ 10)e^-$$

$$2e^- + \text{O}\begin{pmatrix}8 \text{ protons}\\ 8 \text{ electrons}\end{pmatrix} \rightarrow \text{O}^{2-}\begin{pmatrix}8 \text{ protons}\\ 10 \text{ electrons}\end{pmatrix}$$

$$2e^- + \text{O} \rightarrow \text{O}^{2-}$$

Names of ions

The anions derived from single atoms are named by adding ide to the root of the element's name. The exact root chosen is a matter of convention and must be learned by practice. A list of common anions and cations and their names is given in Table 5.2. The charge of an ion is governed by the electronic structure of the atom which will be discussed in Chapter 6.

TABLE 5.2 common simple ions

Anion	Name	Cation	Name
F^-	fluoride	Li^+	lithium ion
Cl^-	chloride	Na^+	sodium ion
Br^-	bromide	K^+	potassium ion
I^-	iodide	Mg^{2+}	magnesium ion
O^{2-}	oxide	Ca^{2+}	calcium ion
S^{2-}	sulfide	Zn^{2+}	zinc ion

Ions that consist of several atoms held together by chemical bonds similar to those involved in molecules are called polyatomic ions or complex ions. These complex ions differ from molecules in that they bear a charge. The complex ion may have either an excess or deficiency of electrons relative to the atoms making up the complex ion. Negatively charged complex ions are by far the more common. A list of complex ions is given in Table 5.3.

Remember that the superscript on the right indicates the charge on the ion, whereas the subscript represents the number of common atoms occurring in the ion. Thus the nitrate ion, NO_3^-, consists of 1 atom of nitrogen, 3 atoms of oxygen, and 1 electron over and above what the individual neutral atoms would contain.

TABLE 5.3 common complex ions

Formula	Name
NO_3^-	nitrate
NO_2^-	nitrite
SO_4^{2-}	sulfate
SO_3^{2-}	sulfite
HSO_4^-	bisulfate
HSO_3^-	bisulfite
PO_4^{3-}	phosphate
CO_3^{2-}	carbonate
HCO_3^-	bicarbonate
OH^-	hydroxide
CN^-	cyanide
MnO_4^-	permanganate
ClO_4^-	perchlorate
ClO_3^-	chlorate
ClO_2^-	chlorite
ClO^-	hypochlorite
NH_4^+	ammonium
PH_4^+	phosphonium

Compounds containing ions

Because all matter must be electrically neutral overall, the presence of cations in a compound means that anions are also present. In ordinary table salt, sodium chloride, there is an equal number of Na^+ and Cl^- ions in any sample. The chemist's representation is NaCl but there is no actual sodium chloride molecule. In calcium chloride there is one Ca^{2+} for every two Cl^- ions, and the chemist's representation is $CaCl_2$. However, there is no calcium chloride molecule. **Compounds that are composed of ions are called ionic compounds.** Such compounds will be discussed in Chapter 7.

When ions are present in a compound the number of positive charges must balance with the number of negative charges to produce electrically neutral matter. Since the charge on the anion may not always be equal to that on the cation, the numbers of anions will not always equal the number of cations.

$$\text{positive charge} + \text{negative charge} = 0$$

$$\left(\frac{\text{charge}}{\text{cation}}\right)\left(\begin{matrix}\text{relative number}\\ \text{of cations}\end{matrix}\right) + \left(\frac{\text{charge}}{\text{anion}}\right)\left(\begin{matrix}\text{relative number}\\ \text{of anions}\end{matrix}\right) = 0$$

As indicated previously, sodium chloride contains an equal number of sodium ions and chloride ions.

$$\left(\frac{\text{charge}}{\text{sodium cation}}\right)\left(\begin{matrix}\text{relative number}\\ \text{of sodium cations}\end{matrix}\right) + \left(\frac{\text{charge}}{\text{chloride anion}}\right)\left(\begin{matrix}\text{relative number}\\ \text{of chloride anions}\end{matrix}\right) = 0$$

$$(+1)(1) + (-1)(1) = 0$$

In the case of $CaCl_2$, there are two chloride ions for every calcium ion and electrical neutrality results.

$$\left(\frac{\text{charge}}{\text{calcium cation}}\right)\left(\begin{matrix}\text{relative number}\\ \text{of } Ca^{2+}\end{matrix}\right) + \left(\frac{\text{charge}}{\text{chloride anion}}\right)\left(\begin{matrix}\text{relative number}\\ \text{of } Cl^{-}\end{matrix}\right) = 0$$

$$(+2)(1) + (-1)(2) = 0$$

A number of ionic compounds are listed in Table 5.4. A check of the charges on the ions listed in Tables 5.2 and 5.3 will verify that all of the compounds listed in Table 5.4 are electrically neutral. Note that whenever more than one complex ion is required to form a compound, parentheses

TABLE 5.4

names of some ionic compounds

Formula	Name
$LiClO_4$	lithium perchlorate
$LiBr$	lithium bromide
$NaNO_3$	sodium nitrate
$NaHCO_3$	sodium bicarbonate
KCl	potassium chloride
$KMnO_4$	potassium permanganate
K_3PO_4	potassium phosphate
$CaCl_2$	calcium chloride
CaS	calcium sulfide
$CaCO_3$	calcium carbonate
$CaSO_4$	calcium sulfate
$Ca_3(PO_4)_2$	calcium phosphate
MgO	magnesium oxide
MgF_2	magnesium fluoride
$Mg(CN)_2$	magnesium cyanide
$Zn(OH)_2$	zinc hydroxide
$ZnSO_4$	zinc sulfate
NH_4Cl	ammonium chloride
$(NH_4)_2SO_4$	ammonium sulfate
$(NH_4)_3PO_4$	ammonium phosphate

are used to enclose the complex ion. The subscript to the right of the parenthesis indicates the number of such complex ions required to balance the ions of opposite charge. Thus $Ca_3(PO_4)_2$ is an ionic compound that consists of three doubly charged calcium ions for every two triply charged phosphate ions. Additional examples of ionic compounds and their names will be given in Chapters 7 and 9.

Problem 5.11 Sodium oxide is an ionic compound. Write its formulas.

Solution The sodium and oxide ions are Na^+ and O^{2-}, respectively. Therefore, in order to maintain electrical neutrality, there must be two sodium ions for every one oxide ion and the formula must be Na_2O.

$$\left(\frac{\text{charge}}{Na^+ \text{ cation}}\right)\left(\begin{matrix}\text{number of } Na^+ \\ \text{cations}\end{matrix}\right) + \left(\frac{\text{charge}}{O^{2-} \text{ anion}}\right)\left(\begin{matrix}\text{number of } O^{2-} \\ \text{anions}\end{matrix}\right) = 0$$

$$1 \text{ (number of } Na^+ \text{ cations)} = -(-2) \text{ (number of } O^{2-} \text{ anions)}$$

$$\frac{\text{number of } Na^+ \text{ cations}}{\text{number of } O^{2-} \text{ anions}} = \frac{2}{1}$$

5.8 THE MASS OF THE ATOM

As indicated in the discussion of Dalton's atomic theory, it was postulated that each atom has a characteristic mass. We will now consider the question of what the relative and absolute masses are in the case of atoms. The experimental means for determining these quantities are varied and have been developed over many years and will not be described here.

Atomic weight

The atomic weights of the atoms are quantities that give the relative weights in reference to a chosen standard. In other words, the atomic weight of one element indicates how heavy its atom is compared to an atom of another element. Using rounded off values of the atomic weights listed on the inside cover of this book, we can see that the atomic weights of hydrogen and oxygen are 1 and 16, respectively. This tells us that the hydrogen atom is $\frac{1}{16}$ the mass of the oxygen atom. The atomic weight of helium is 4. Therefore, the helium atom is four times the mass of a hydrogen atom but is only $\frac{1}{4}$ that of the oxygen atom.

Unfortunately the term atomic weight has not been changed to atomic mass. The latter term would seem preferable as it eliminates a gravitational dependence. Nevertheless, under the same gravitational effects, the relative weights of two atoms give the relative masses of the same two atoms. We may write:

$$\frac{\text{mass of atom A}}{\text{mass of atom B}} = \frac{\text{atomic weight of A}}{\text{atomic weight of B}}$$

There are no practical units associated with atomic weights, such as grams, pounds, and so on. They are relative terms, but we can designate this by using **awu for atomic weight unit or amu for atomic mass unit.** We need not know what an awu really is in terms of pounds or an amu in terms of grams. If the ratio of two atomic weights is $\frac{1}{16}$, as in the case of hydrogen and oxygen, then we have the information that oxygen is 16 times the mass of hydrogen.

Atomic weights are the average relative weights of a naturally occurring mixture of isotopes.

Since atoms contain an integral number of protons and neutrons, one might expect the atomic weights of elements to be integers and related to each other by whole number ratios. For example, the atomic weights of hydrogen, helium, and oxygen are 1, 4, and 16, respectively, and are related by simple ratios. However, close examination of a table of atomic weights reveals that most of the quantities are not integers and not related to other elements by whole number ratios. This phenomenon is the result of the presence of isotopes in nature. Thus the atomic weight is an "average" value that represents the weighted fraction of the various isotopes which occur naturally. As an example, consider the fact that naturally occurring chlorine contains atoms of $^{35}_{17}Cl$ and $^{35}_{17}Cl$ in the approximate proportion of 4 to 1. The atomic weight of 35.453 is a weighted average of the mass numbers of the isotopes. This average atomic weight can be calculated as follows by using the approximate proportions of the isotopes present in natural chlorine:

weight of four $^{35}_{17}Cl$	= 35 × 4	=	140 awu
weight of one $^{37}_{17}Cl$	= 37	=	37 awu
total weight			177 awu
average weight	= 177/5	=	35.4 awu

At one time the reference of the atomic weight scale was the average weight of the isotopes of oxygen and this average was assigned a value of 16.0000. **In 1961, the scale was changed so that the reference is now the isotope $^{12}_{6}C$ which is assigned a mass of 12.0000.** On this scale, naturally occurring carbon has an atomic weight of 12.0111 as there is some $^{13}_{6}C$ isotope in nature. The average of the isotopes of oxygen gives an atomic weight of 15.9994. However, the new values bear the same relative relationships to each other. Only the absolute values have been slightly altered by choosing a better reference from a scientific point of view.

Could an element other than carbon be used to define atomic weights?

5.9 MASS OF MOLECULES

Molecular weight

In discussing matter and its reactions, we will be considering compounds more frequently than elements because there are many more of them. It is convenient to describe the mass of a molecule in terms of its molecular weight. **The molecular weight of a molecule is the sum of the atomic weights of its component atoms.** Thus, it is a relative term based on the same reference as for atomic weights. The molecule methane, whose molecular formula is CH_4, has a molecular weight of approximately 16.04 awu.

The molecular weight of a molecule is equal to the sum of the atomic weights of the atoms in the molecule.

$$\begin{aligned} \text{molecular weight } CH_4 &= \text{atomic weight C} + 4(\text{atomic weight H}) \\ 16.04 \text{ awu} &= 12.01 \text{ awu} + 4(1.008 \text{ awu}) \end{aligned}$$

Problem 5.12 What is the molecular weight of carbon dioxide, CO_2?

Solution The molecular weight of CO_2 is given by:

$$\begin{aligned} \text{molecular weight } CO_2 &= \text{atomic weight C} + 2(\text{atomic weight O}) \\ 44.01 \text{ awu} &= 12.01 \text{ awu} + 2(16.00 \text{ awu}) \end{aligned}$$

Problem 5.13 Calculate the molecular weight of ethyl alcohol, C_2H_6O.

Solution

$$\begin{aligned} &\text{mass of 2 carbon atoms} &= 2 \times 12.0 &= 24.0 \text{ awu} \\ &\text{mass of 6 hydrogen atoms} &= 6 \times 1.0 &= 6.0 \text{ awu} \\ &\text{mass of 1 oxygen atom} &= 1 \times 16.0 &= 16.0 \text{ awu} \\ \hline &\text{mass of 1 molecule } C_2H_6O & &= 46.0 \text{ awu} \end{aligned}$$

SUMMARY

*All matter is composed of very small particles called **atoms**. Atoms are themselves composed of a specific number of **electrons**, **protons**, and **neutrons**. Protons and neutrons have a mass of **1 atomic mass unit**, whereas electrons have a negligible mass for most purposes. The proton is positively charged and the electron is negatively charged. Protons and neutrons are located in the center of the atom, called the **nucleus**. Electrons are located in space about the nucleus.*

*Every pure substance is composed of its own fundamental particle which is responsible for the properties of that substance. The fundamental particle of some elements is atoms. That of other elements and some compounds is the **molecule**, which is a combination of atoms. The fundamental particles of other compounds are **ions**, which are derived from atoms by either gaining or losing electrons.*

*Every element has atoms with a specific number of protons and an equal number of electrons. If one or more electrons are removed from an atom, a **cation** results. If electrons are gained by an atom, an **anion** results.*

Elemental symbols may have a superscript number and a subscript number placed to the left which represent the ***mass number*** *and* ***atomic number****, respectively.*

Isotopes *of elements contain the same number of protons but differ in the number of neutrons. Most elements consist of a mixture of isotopes.*

Some elements consist of two or more identical atoms bonded together to form ***diatomic or polyatomic molecules****. Molecules of compounds contain two or more different types of atoms. The* ***molecular formula*** *is used to represent the number and kind of atoms present in a molecule. Subscripts to the right of the elemental symbol indicate the number of that atom present in a molecule.*

Ions are charged atomic particles which when associated with each other yield ***ionic compounds****. In these compounds electrical neutrality results from a balance of the proper number of anions and cations.*

Atomic weights *of atoms are quantities that give the relative weights compared to* $^{12}_{6}C$*.* ***Molecular weights*** *are the sum of the atomic weights of each of the constituent atoms times the number of such atoms which are contained in the molecule.*

QUESTIONS AND PROBLEMS

Definitions

1. Define each of the following terms.
- **(a)** atom
- **(b)** molecule
- **(c)** ion
- **(d)** proton
- **(e)** neutron
- **(f)** electron
- **(g)** nucleus
- **(h)** atomic number
- **(i)** mass number
- **(j)** isotope
- **(k)** atomic mass unit
- **(l)** molecular formula
- **(m)** chemical bond
- **(n)** anion
- **(o)** cation
- **(p)** atomic weight
- **(q)** molecular weight

2. Tell the difference between each of the following.
- **(a)** neutron and proton
- **(b)** proton and electron
- **(c)** mass number and atomic number
- **(d)** cation and anion
- **(e)** atomic weight and molecular weight

3. In your own words, write out the postulates of Dalton's atomic theory.

4. Describe the arrangement of the subatomic particles in the atom.

Subatomic particles

5. Prepare a chart giving the relative weight and charge of the three subatomic particles.

6. Why are relative weights of subatomic particles used rather than the absolute masses?

7. Why is the mass of an electron considered zero on a relative mass basis?

8. Where are each of the subatomic particles located in the atom?

9. What are the sizes of the nucleus and the atom?

10. What relationship exists between the number of electrons and protons in an atom?

Elemental symbols

11. What is the atomic number of an element? How is it represented in a symbol?

12. What is the mass number of an element? How is it represented in a symbol?

13. Indicate the number of each subatomic particle present in the following species.
(a) $^{7}_{3}Li$ **(b)** $^{31}_{15}P$
(c) $^{107}_{47}Ag$ **(d)** $^{59}_{27}Co$
(e) $^{69}_{31}Ga$ **(f)** $^{201}_{80}Hg$
(g) $^{199}_{80}Hg$ **(h)** $^{142}_{55}Ce$
(i) $^{22}_{11}Na$ **(j)** $^{17}_{8}O$

14. Write the symbol for each of the following elements which contain the indicated number of protons and neutrons.
(a) fluorine, 9 protons and 10 neutrons
(b) silicon, 14 protons and 16 neutrons
(c) silicon, 14 protons and 14 neutrons
(d) iron, 26 protons and 30 neutrons
(e) uranium, 92 protons and 145 neutrons
(f) potassium, 19 protons and 22 neutrons

Isotopes

15. There is only one form of fluorine in nature. Would $^{19}_{9}F$ be useful in defining an atomic weight scale?

16. The element bromine exists as the isotopes $^{79}_{35}Br$ and $^{81}_{35}Br$ in nature. The atomic weight is 79.9. Approximately how much of each isotope is present in nature?

17. The element magnesium consists of three isotopes of masses 23.99, 24.99, and 25.99. Which of these is the most abundant? The atomic weight of magnesium is 24.3.

18. Boron contains the isotopes $^{10}_{5}B$ and $^{11}_{5}B$ in 18.8% and 81.2% abundance, respectively. What is the average atomic weight of boron?

Molecules

19. Indicate which of the following elements exists as molecules. How many atoms are contained in each molecule?

(a) He (b) H
(c) P (d) N
(e) O (f) S
(g) F (h) Cl

20. What is meant by each of the following formulas?

(a) H_2S (b) HCl
(c) H_2SO_4 (d) C_2H_6O
(e) C_2H_2 (f) $C_6H_{12}O_6$
(g) H_2O_2 (h) N_2

21. A molecule of TNT contains 7 carbon atoms, 5 hydrogen atoms, 3 nitrogen atoms, and 6 oxygen atoms. Write the molecular formula of TNT.

22. A molecule of octane contains 8 carbon atoms and 18 hydrogen atoms. What is the molecular formula of octane?

23. Nicotine has the molecular formula $C_{10}H_{14}N_2$. What does this formula mean to you?

Ions

24. What is an anion? How does it differ from an atom?

25. What is a cation? How does it differ from an atom?

26. Sulfur can gain two electrons to form an ion. What is the symbol of the ion?

27. Scandium (Sc) can lose three electrons to form an ion. What is the symbol of the ion?

28. What is the name of each of the following ions?

(a) O^{2-} (b) Cl^-
(c) S^{2-} (d) Li^+
(e) K^+ (f) Ca^{2+}

29. What is the name of each of the following complex ions?

(a) OH^- (b) NH_4^+
(c) NO_3^- (d) SO_3^{2-}
(e) SO_4^{2-} (f) CN^-
(g) ClO_4^- (h) ClO^-

30. Write the symbols for each of the ions.

(a) hydroxide (b) oxide
(c) sulfide (d) ammonium
(e) chloride (f) sulfate
(g) perchlorate (h) cyanide
(i) phosphate

Ionic compounds

31. Write the correct formula for each of the following compounds.
(a) lithium fluoride **(b)** zinc oxide
(c) sodium cyanide **(d)** magnesium fluoride
(e) zinc cyanide **(f)** sodium nitrate
(g) sodium carbonate **(h)** potassium sulfide

32. Write the correct formula for compounds containing the following ions.
(a) Fe^{3+} and Cl^- **(b)** Na^+ and OH^-
(c) Mg^{2+} and OH^- **(d)** Cd^{2+} and S^{2-}
(e) K^+ and Br^- **(f)** Li^+ and N^{3-}
(g) Ba^{2+} and NO_3^- **(h)** Cs^+ and ClO_4^-

33. Name each of the following ionic compounds.
(a) $Ca(OH)_2$ **(b)** $LiClO_4$
(c) Na_3PO_4 **(d)** K_2SO_4
(e) KNO_3 **(f)** NH_4NO_2
(g) $MgCl_2$ **(h)** LiCN

Molecular weight

34. Using the atomic weights of the elements, calculate the molecular weight of each of the following substances.
(a) H_2 **(b)** O_2
(c) N_2 **(d)** F_2
(e) NH_3 **(f)** N_2H_4
(g) NO **(h)** NO_2
(i) N_2O **(j)** N_2O_3
(k) H_2O **(l)** H_2O_2
(m) HF **(n)** NF_3
(o) NH_4F **(p)** NH_4NO_3

6 ATOMIC THEORY

LEARNING OBJECTIVES

When you finish this chapter you should be able to:

(a) describe the energy levels and subenergy levels within the atom.
(b) determine the number of electrons in an energy level or sublevel.
(c) write the electronic description of elements given their atomic numbers.
(d) represent the valence shell electrons of atoms by electron dot symbols.
(e) compare the number and type of orbitals in a subenergy level.
(f) match the following terms with their corresponding definitions.

electron dot symbol
energy level
orbital
shell
subenergy level
valence electrons

In the laboratory chemists study the properties of substances and the way they interact with one another. Ultimately they would like to be able to predict accurately how two substances will interact when mixed. In the last chapter we found that each pure substance is composed of its own submicroscopic particles. We saw that there were three types of these particles—atoms, molecules, and ions. Since molecules are composed of atoms, and ions are derived from atoms by either gaining or losing electrons, atoms are the submicroscopic units of which all matter is composed. In this chapter we will look more closely at the structure of atoms to see how it affects their behavior.

The atom was pictured as a nucleus containing neutrons and protons with electrons located in the surrounding space a considerable distance away. In this chapter we will learn more about the arrangement of electrons and their energy content. It is these features which determine how atoms interact with one another. In effect, it is impossible to fully understand the chemistry of matter and predict the reactions which will occur between elements without knowing the electronic makeup of their atoms. The material in this chapter is not easy and may take some study. However, it is certainly as important to your understanding of chemistry as anything in this text.

6.1 PRINCIPAL ENERGY LEVELS

The negative electrons in the space about the nucleus are attracted to the positively charged protons in the nucleus. It might be expected that the small electrons would be pulled into the more massive nucleus. However, this does not occur because the electrons have enough energy to travel at high rates of speed and resist this nuclear tug. We can draw an analogy between the electron and nucleus interaction and that which exists when a model airplane travels in a circular path around the operator who uses a set of guide wires to control it. As long as the operator holds onto the wires, the plane will continue to circle. The operator feels a pull against the hand, which is the result of the centrifugal force of the plane. If the guide wires were released, the plane would fly away in a straight line. However, the operator pulls against the guide wires to maintain the orbit of the plane.

You can swing a bucket of water in a circle and avoid spilling water. Why?

In the atom we may think of an electron as a particle whirling in an orbit. The outward centrifugal force is exactly balanced by the inward electrical pull between the oppositely charged electron and the nucleus. An electron in an atom does not have enough energy to completely escape this attractive force and so it finds an orbit where the force it exerts is exactly balanced by the attractive force. As we will see later, the various electrons in an atom do not all have the same energy and thus their location within the atom differs. In general the higher the energy of the electron, the further away from the nucleus will be the orbit.

A convenient model for the atom is our planetary system. The sun may be visualized as the nucleus; the planets in orbits are regarded as the electrons.

The electrons in an atom exist in certain principal energy levels or shells. By this we mean that the electrons possess an energy corresponding to the energy of that level. An analogy might be drawn between energy levels of an atom and the levels of responsibility of the employees of a company. A salesperson works for a company and may make an average salary of \$9000 per year to start with. A salesperson who has been with the company for a while may have reached a new level of responsibility (a larger territory to cover, for example) and a higher salary. This salesperson may make an average annual salary of \$12,000 but is still involved in sales. There are still higher levels of responsibility and salary that salespeople can attain in the company. Suppose the third and fourth levels involved even larger territories and an average annual salary of \$15,000 and \$18,000, respectively. Thus each salesperson in the company is clearly identifiable both in terms of salary and job responsibility. Similarly, an electron in a given energy level has an energy corresponding to that level, while an electron in a different energy level has an energy corresponding to that level. We shall see later how energy and the reactivity or responsibility of the electrons affect the atom.

1 *K*
2 *L*
3 *M*
4 *N*

The principal energy levels in an atom are designated by integers 1, 2, 3, 4, and so on, or by the letters, *K*, *L*, *M*, *N*, and so on. Thus the integer 1 and the letter *K* are alternative ways of identifying an energy level. Similarly, the integer 4 and the letter *N* are equivalent labeling methods (Table 6.1).

TABLE 6.1

electrons and energy levels

Energy level number	Energy level letter	Maximum number of electrons
1	*K*	2
2	*L*	8
3	*M*	18
4	*N*	32
5	*O*	50
6	*P*	72
7	*Q*	98

The energy of the electrons in the energy levels increases in the order $1 < 2 < 3 < 4$, and so on. The more energy an electron has, the more it can resist the nuclear attraction. Therefore, the distance between the electrons and the nucleus increases in the same order. Thus the nearer the electron is to the nucleus, the lower is its energy.

$2n^2$

There are limitations on the number of electrons which can possess a given energy corresponding to a specific energy level. **The limitation on the maximum number of electrons is given by $2n^2$, where n is the number of the**

energy level. The number of electrons in each of the principal energy levels is listed in Table 6.1.

Problem 6.1 How many electrons can exist in the principal energy level designated as N?

Solution The energy level N is the same as the fourth energy level. Since $n = 4$, it follows that the maximum number of electrons in that energy level is:

$$2n^2 = 2 \times (4)^2 = 2 \times 16 = 32$$

Why are electrons located in the lowest energy levels accessible to them?

The electrons in an atom are located in the lowest energy levels accessible to them. For example, the K shell can hold two electrons. Therefore, the electrons in an atom with one or two electrons would be found in this shell and not in a higher energy shell. Only when an atom has more electrons than a lower energy level can hold are electrons found in a higher energy level. Using a representation similar to a planetary model, we may then represent $^{4}_{2}He$ as follows:

shell
1 or K

$2p$
$2n$
$2)e^-$

$^{4}_{2}He$

The two electrons in helium are located in and fill the lowest energy level.

Elements of atomic numbers 3 through 10 have electrons in the second energy level in addition to the first energy level. The isotope $^{16}_{8}O$ has 8 electrons, more than can be contained in the first energy level. Therefore, 2 electrons occupy the first energy level and 6 are located in the second energy level.

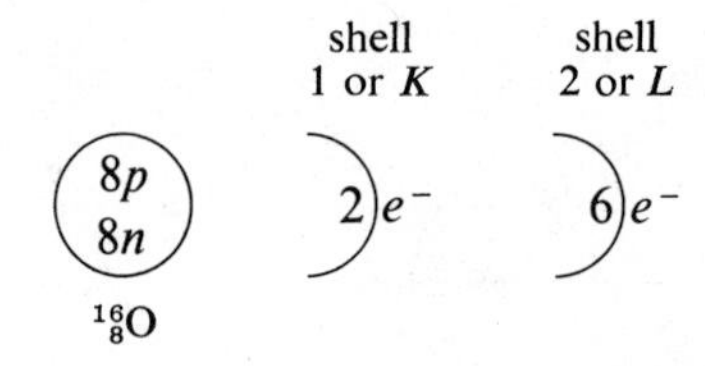

Why doesn't sodium contain 9 electrons in the second energy level and 2 in the first energy level?

When atoms of an element contain more than 10 electrons, as in the case of sodium, some electrons occupy the third energy level.

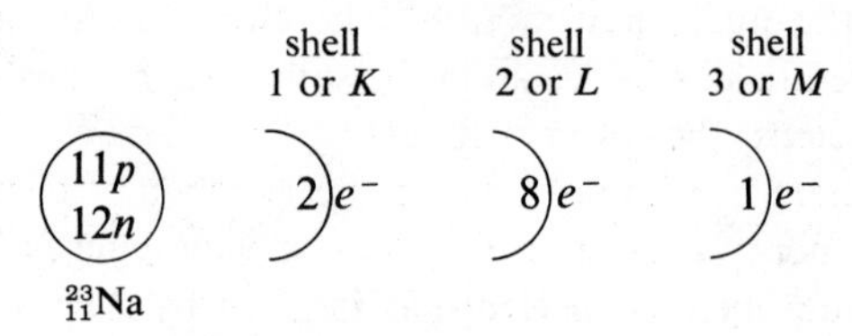

Elements whose atomic numbers are 11 through 18 have electrons in the third energy level in addition to the first and second energy levels.

Problem 6.2 What is the electron arrangement in $^{35}_{17}Cl$?

Solution To do this problem, you need to remember that (1) electrons occupy the lowest energy levels, and (2) the energy levels hold, 2, 8, 18, and so on, electrons in the order of increasing energy. Then you may mentally distribute the 17 electrons according to these rules. Two electrons are located in the first energy level. Eight electrons are in the next highest energy level. The remaining seven electrons must be located in the third energy level.

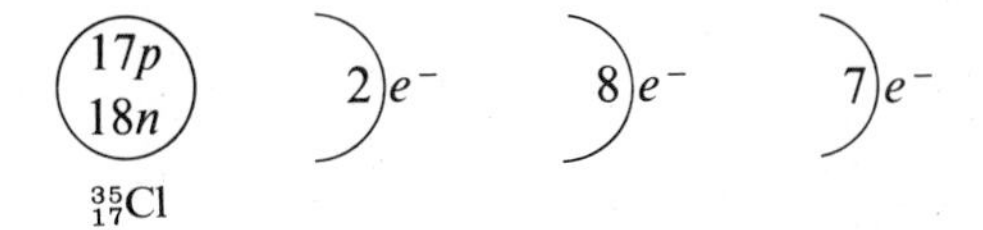

A discussion of the arrangement of electrons in excess of 18 will be postponed until later in this chapter.

6.2 SUBENERGY LEVELS (SUBSHELLS)

The two electrons in the lowest energy level of an atom have the same energy and behave similarly. However, this is not always true for the different electrons in the higher energy levels. Within these energy levels there are subenergy levels or subshells. **Electrons within different subenergy levels of the same energy level differ in both energy and behavior.** Since the higher energy levels have more electrons in them, the higher the energy level, the more subenergy levels it has. Of all the known atoms, only four types of subenergy levels exist. These have been labeled by the lower case letters *s*, *p*, *d*, and *f*. For clarity, the subenergy levels in an atom are represented by symbols such as 2*s*, 4*d*, 5*p*, and so on. **The number represents the energy level while the letter represents the subenergy level within that energy level.**

To make this point a little clearer, let's return to our corporate analogy. Electrons in subenergy levels within an energy level would be like employees doing different jobs within a given corporate level of responsibility (Figure 6.1). Of course, since they are doing different jobs they earn slightly different salaries. In our corporation, the first level of responsibility is the simplest and involves only one type of employee—salespeople who earn \$9000. In other levels of the corporate structure, there are employees other than salespeople. The second level has both salespeople and production personnel. The salespeople earn \$12,000 while the production personnel earn \$14,000. Both of these classes of employees have similar salaries which are clearly higher than the salary of salespeople in the first level of responsibility. Thus it is

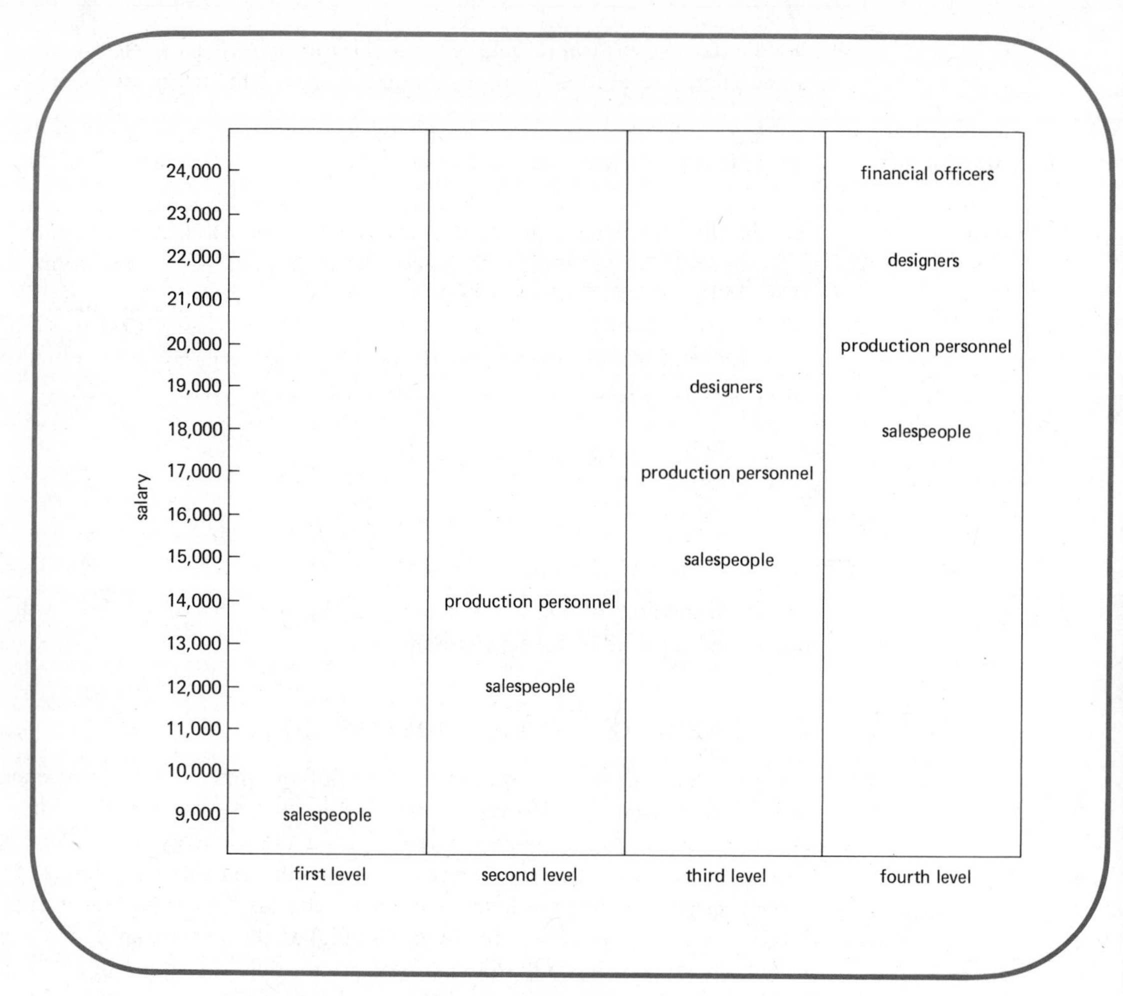

Figure 6.1 The structure of a corporation.

convenient to classify both the $12,000 salespeople and the $14,000 production personnel as members of an average $13,000 salary bracket. However, while their level of responsibility and salaries are similar, they are distinguishable by the jobs which they perform.

To carry our analogy further, let's assume that there are more job classifications at the third level of responsibility which, of course, command higher salaries. The average salary of this level is $17,000. Within this level there are salespeople who earn $15,000 because they have more responsibility than salespeople in other lower levels who earn $9000 and $12,000. Similarly, there are production personnel in the third level who earn $17,000 as they are above the production personnel in the second level who earn only $14,000. In addition to salespeople and production personnel, there are designers in the third level who are even more important to the corporation and earn $19,000. Thus the average salary of the third level is $17,000 as it reflects the salaries of $15,000, $17,000, and $19,000 for salespeople, production personnel, and designers, respectively.

Finally, our top-heavy corporate structure has a fourth level of responsibility which commands an average salary of $21,000. At this level, salespeople earn $18,000, production personnel earn $20,000, and designers earn $22,000. In addition there is a fourth job for financial officers who are more important and earn $24,000. Thus within this level of responsibility, salespeople still earn the least followed by higher salaried production personnel, designers, and financial officers. The order of compensation remains the same at all lower levels. However, at lower levels there are fewer job classifications.

A careful analysis of Figure 6.1 reveals that our corporation has levels of responsibility and salaries that intermix at the higher levels. The salespeople at the fourth level earn more than salespeople in the lower levels but actually earn less than the designers of the third level. Still, they are salespeople because of their job, and they are in the fourth level because of their level of responsibility in the area of sales. Our corporation has simply decided that designers of the third level are to be higher salaried than the fourth level salesperson. Similar situations exist within many job classifications in our society.

Our corporation might also set up its structure so that the number of people within any job description is limited. Furthermore, the number of people in each job need not be the same.

Getting back to the atom, it is known that there are restrictions on the number of electrons that each of the subenergy levels can hold. **The maximum number of electrons in each of the subenergy levels is as follows: 2 in the *s* subenergy level, 6 in the *p* subenergy level, 10 in the *d* subenergy level, and 14 in the *f* subenergy level.** The order of energy of the subenergy levels is $s < p < d < f$. Thus within an energy level, electrons will be located first within the *s* subenergy level, next in the *p* subenergy level, and so on. In our analogous corporation, this is equivalent to the previous observation that salespeople within a level earn less than the production staff, the production staff earn less than the designers, and the designers earn less than the financial officers.

Since the first energy level can hold only 2 electrons, it consists of only an *s* subshell. The second energy level can hold 8 electrons. It consists of an *s* and a *p* subenergy level. The third energy level has a potential occupancy of 18 electrons and has an *s*, *p*, and *d* subenergy level. An accounting of energy levels, subenergy levels and the electrons they may hold is given in Table 6.2.

How many subenergy levels are there in the third energy level?

Since the electrons in an atom are known to be in the lowest energy levels available to them, it should be easy to determine the arrangement of electrons in a particular atom. This task, however, is not quite as straightforward as might be imagined because of the energy spread of the subenergy levels, as illustrated in Figure 6.2. The subenergy level within a high energy level may be of lower energy than the subenergy level within a lower energy level. For example, the 4*s* subenergy level is of lower energy than the 3*d* subshell. In our corporate analogy, this would mean that a salesperson in the fourth level may actually earn less than designers in the third level.

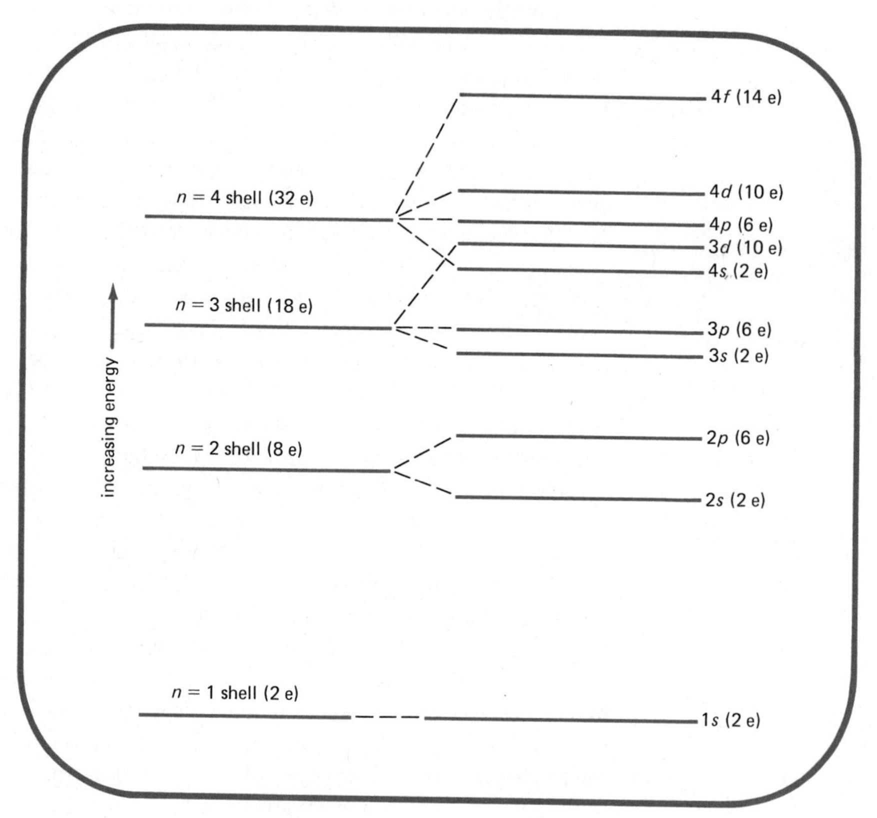

Figure 6.2 Energy of subshells.

The ordering of energy levels and their subenergy levels is $1s < 2s < 2p < 3s < 3p < 4s < 3d < 4p < 5s < 4d < 5p < 6s < (4f < 5d) < 6p < 7s < (5f < 6d)$. The subenergy levels designated within parentheses are quite close, and in some elements there are apparent inconsistencies in the arrangement of the electrons. Note that the *d* subenergy level of a given *n* energy level is always of higher energy than the *s* subenergy level of the $n + 1$ energy level and of lower energy than the *p* subenergy level of the $n + 1$ energy level.

$4s < 3d < 4p$
$5s < 4d < 5p$
$6s < 5d < 6p$

In addition, the *f* subenergy level of the *n* energy level is of higher energy than the s subenergy level of the $n + 2$ energy level and of lower energy than the d subshell of the $n + 1$ energy level.

$6s < 4f < 5d$
$7s < 5f < 6d$

A complete tabulation of the electrons which may be placed within each energy level and subenergy level is given in Table 6.2. Some subenergy levels other than s, p, d, and f are designated. However, there are no known elements with sufficient numbers of electrons to occupy these sublevels. A method of memorizing the order of energies of the sublevels is given in Figure 6.3.

TABLE 6.2

electrons and subenergy levels

Energy level	Sublevel	Electrons in sublevel	Total electrons possible in energy level
1	*s*	2	2
2	*s*	2	
	p	6	8
3	*s*	2	
	p	6	
	d	10	18
4	*s*	2	
	p	6	
	d	10	
	f	14	32
5	*s*	2	
	p	6	
	d	10	
	f	14	
	g	18	50
6	*s*	2	
	p	6	
	d	10	
	f	14	
	g	18	
	h	22	72
7	*s*	2	
	p	6	
	d	10	
	f	14	
	g	18	
	h	22	
	i	26	98

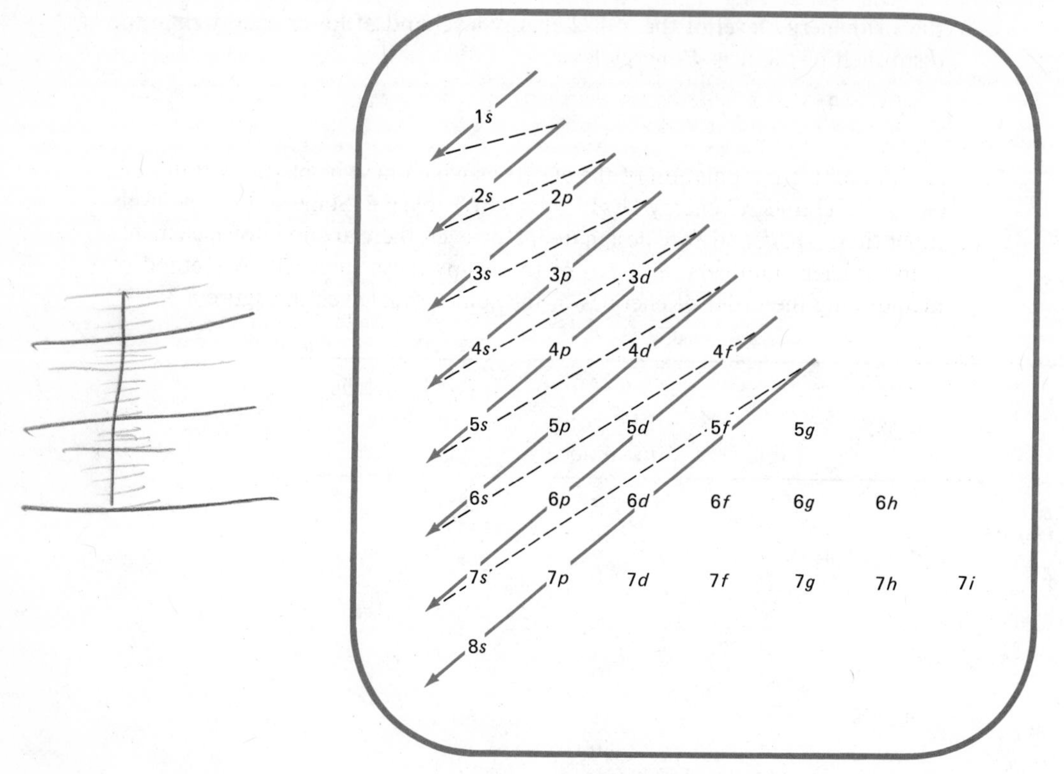

Figure 6.3 Order of filling subenergy levels.

Start from the 1*s* subenergy level and follow the solid line from the start of the arrow line to the head of the arrow. Then proceed via the dotted line to the start of the next solid-line arrow. This procedure will reproduce the order of energies previously given. (In Chapter 8 you will learn another more satisfying procedure using a list called the periodic table.) With the order of energy level available to you, it is now possible to describe the electronic arrangement of atoms containing more than 18 electrons. Some of these will be presented in the next section.

6.3 ELECTRONIC ARRANGEMENT OF ATOMS

A convenient way of designating energy levels, subenergy levels, and the number of electrons which they contain is shown below.

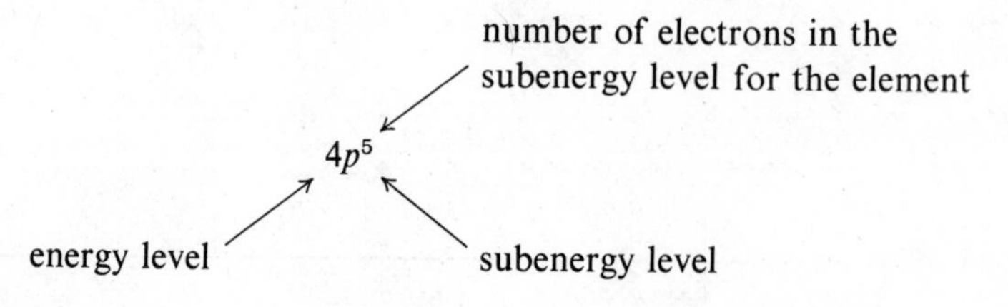

The shorthand method of showing the electrons in an atom is the number of the energy level followed by the letter of the subenergy level and a superscript for the number of electrons within that subenergy level.

When writing a complete electronic description of an atom, the sequence of symbols for subenergy levels is written from left to right in order of increasing energy. The number of electrons contained in each sublevel is designated by a superscript to the right of the letter. Several examples should give you a working knowledge of the system.

$^{4}_{2}He$ $1s^2$
$^{9}_{4}Be$ $1s^2 2s^2$
$^{16}_{8}O$ $1s^2 2s^2 2p^4$
$^{20}_{10}Ne$ $1s^2 2s^2 2p^6$
$^{23}_{11}Na$ $1s^2 2s^2 2p^6 3s^1$
$^{35}_{17}Cl$ $1s^2 2s^2 2p^6 3s^2 3p^5$
$^{40}_{20}Ca$ $1s^2 2s^2 2p^6 3s^2 3p^6 4s^2$
$^{58}_{28}Ni$ $1s^2 2s^2 2p^6 3s^2 3p^6 4s^2 3d^8$

Does an element have the electronic configuration $1s^2 2s^2 2p^7$? Why?

A complete listing of the electronic arrangements or configurations of the elements is given in Appendix 2. You may note that there are a number of exceptions to the orderly picture presented here. These exceptions are understood in terms of additional factors which will not be presented here. A number of these exceptions are due to the closeness of some subenergy levels given as ($4f < 5d$) and ($5f < 6d$).

Problem 6.3 Write the electronic arrangement of $^{88}_{38}Sr$.

Solution The electrons are located in the lowest energy subenergy levels accessible to them. The ordering of subenergy levels is:

$$1s < 2s < 2p < 3s < 3p < 4s < 3d < 4p < 5s$$

Electrons up to a total of 38 are then placed within each subenergy level to obtain:

$$1s^2 2s^2 2p^6 3s^2 3p^6 4s^2 3d^{10} 4p^6 5s^2$$

Problem 6.4 What is the arrangement of electrons in $^{59}_{27}Co$?

Solution Write down the order of energies of the subenergy levels, filling each in order:

$$1s^2 2s^2 2p^6 3s^2 3p^6 4s^2$$

The above arrangement accounts for 20 electrons. The remaining 7 electrons are placed in an incomplete $3d$ subenergy level to give:

$$1s^2 2s^2 2p^6 3s^2 3p^6 4s^2 3d^7$$

6.4 ELECTRON DOT SYMBOLS

The last principal energy level of an atom containing electrons in *s* and *p* subshells is called the valence energy level or valence shell. The electrons in this level are known as valence electrons. All of the other electrons together with the nucleus comprise the core. This division is useful because it is the valence electrons which take part in the formation of chemical bonds (Chapter 7) (In general we will not consider the chemistry of the elements in which the *d* electrons are the highest energy. These elements are called transition elements.) Because valence electrons are chemically important, we often visually display them in electron dot formulas.

Electron dot formulas are written according to the following rules:

1. **The symbol of the element is taken to represent the core.**
2. **Dots to represent valence electrons may be placed to either side or on the top or bottom of the symbol.**
3. **Each of the positions about the symbol are equivalent.**
4. **A maximum of two electrons per each side, top, or bottom of the symbol is allowed. Thus only eight valence electrons can be placed about the symbol.**
5. **The valence electrons are placed one to each of the four possible positions about the symbol until a maximum of four is reached.**
6. **Additional valence electrons beyond four are paired at each position until four pairs result.**
7. **Hydrogen and helium are exceptions to the maximum of eight electrons. They have a maximum of two electrons.**

$\cdot\ddot{N}\cdot$ Dots represent valence shell electrons.

N represents the nucleus and electrons other than the valence shell for the element nitrogen.

These rules are derived to obtain symbols which are used as models to explain chemical observations. These observations and the related theory will be explained later. Several examples in Figure 6.4 illustrate the use of electron dot symbols.

When we study bonding in Chapter 7, electron dot symbols will be quite useful. You should learn to recognize them and understand which electrons are valence electrons.

6.5 ORBITALS

In Section 6.1, we drew an analogy between the planetary system and the electrons about the nucleus. That analogy is not very accurate when the atom is considered in detail. The electrons do not circle the nucleus in planet-like orbits. However, they do occupy a region of space called an orbital. **An orbital is a region of space having a characteristic shape and in which two electrons of a specific energy may be located.** There are different orbital shapes corresponding to the *s*, *p*, *d*, and *f* sublevels. The *s* orbitals are spherical, as shown in Figure 6.5.

The size of the orbital increases with increasing principal energy levels. The 2*s* orbital is a bigger sphere than the 1*s* orbital, and the 3*s* orbital is larger

^{4}He	$1s^2$	He: or He or :He or He
$^{11}_{5}B$	$1s^2 2s^2 2p^1$	·B· or ·B· or ·B or B·
$^{12}_{6}C$	$1s^2 2s^2 2p^2$	·C·
$^{14}_{7}N$	$1s^2 2s^2 2p^3$	·N· or ·N: or ·N· or :N·
$^{16}_{8}O$	$1s^2 2s^2 2p^4$	·O: or ·O· or :O· etc.
$^{19}_{9}F$	$1s^2 2s^2 2p^5$	:F· or :F: or ·F: or :F:
$^{24}_{12}Mg$	$1s^2 2s^2 2p^6 3s^2$	·Mg or ·Mg· or Mg· etc.
$^{32}_{16}S$	$1s^2 2s^2 2p^6 3s^2 3p^4$	:S· or ·S· etc.

Figure 6.4 Electron dot symbols.

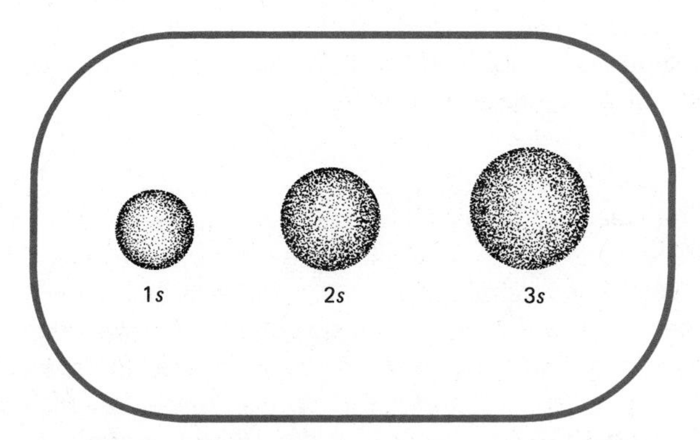

Figure 6.5 The shape of *s* orbitals.

than the 2*s* orbital. However, each *s* orbital can still contain only two electrons. The two electrons may be located any place within the orbital.

There are three *p* orbitals within each *p* sublevel because the subenergy level can hold six electrons, but each orbital can only be occupied by two electrons. The shapes of the three *p* orbitals are identical but are located at right angles to one another (Figure 6.6). Note that each orbital is dumbbell shaped and consists of two teardrop shapes that meet at the nucleus. The two electrons within a *p* orbital can occupy any of the area shown as a *p* orbital. The electrons do not distribute themselves one each to the two teardrop shapes. The orbital is merely a shape within which two electrons may exist.

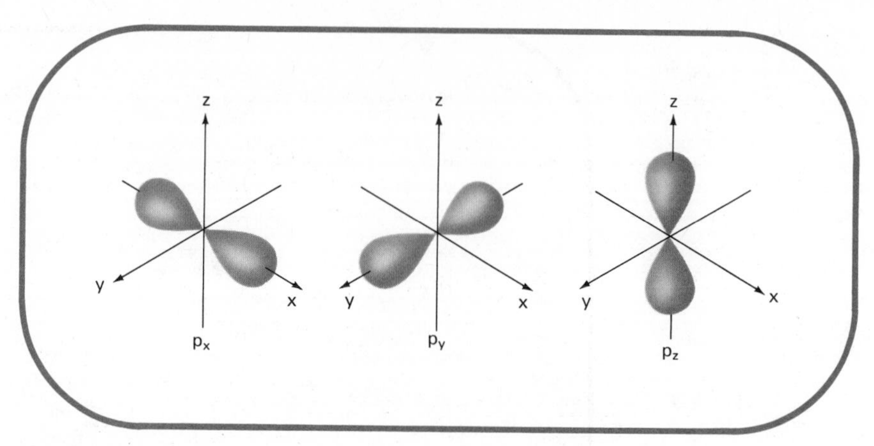

Figure 6.6 The shape of *p* orbitals.

The shapes of *d* and *f* orbitals will not be described in this text. They are more complex and the electrons in these subshells are not part of the valence shell.

We do not concern ourselves with exactly what the electrons are doing in terms of motion within an orbital. Indeed, we do not have any physical means of checking that closely on the electron. We are satisfied to know that there is a region of space within which electrons may be located. It is these regions of space with their associated electrons that are responsible for chemical reactions and bonding.

SUMMARY

The electrons about the nucleus are located in ***energy levels*** *which increase in energy with distance from the nucleus. The number of electrons contained in an energy level is* $2n^2$*, where* n *is the number of the energy level. Energy levels consist of* ***subenergy levels*** *designated as* s, p, d, *and* f. *The order of the energy of the subenergy levels is* s < p < d < f, *and the electrons which they may contain are* 2, 6, 10, *and* 14, *respectively. The ordering of energy levels and their sublevels is* 1s < 2s < 2p < 3s < 3p < 4s < 3d < 4p < 5s < 4d < 5p < 6s < (4f < 5d) < 6p < 7s < (5f < 6d).

Electrons occupy levels in order of increasing energy. Electron descriptions involve a sequence of numbers and letters describing levels and sublevels written from left to right in order of increasing energy. The number of electrons contained in each sublevel is written as a superscript to the right of the symbol for the sublevel.

The highest energy s *and* p *electrons are called* ***valence electrons. Electron dot symbols*** *are used to represent the number of valence electrons by placing dots about the symbol for the element.*

Electrons are contained in regions of space called ***orbitals.*** *Each orbital can contain a maximum of two electrons.*

QUESTIONS AND PROBLEMS

Definitions

1. Define each of the following.
(**a**) orbital
(**b**) valence electrons
(**c**) electron dot symbol
(**d**) energy level
(**e**) subenergy level
(**f**) core
(**g**) *s* subshell
(**h**) *p* subshell
(**i**) *d* subshell
(**j**) *f* subshell

2. Distinguish between the following terms.
(**a**) core and valence electrons
(**b**) energy level and subenergy level
(**c**) orbital and electron dot symbol

3. Explain the meaning of each of the following symbols.
(**a**) $1s^2$
(**b**) $3p^4$
(**c**) $4d^5$
(**d**) $5f^1$

4. Describe the arrangement of electrons about atoms.

Electrons and symbols

5. For each of the following, indicate how many electrons are contained in one atom.
(**a**) ${}^{16}_{8}O$
(**b**) ${}^{36}_{18}Ar$
(**c**) ${}^{235}_{92}U$
(**d**) ${}^{9}_{4}Be$
(**e**) ${}^{13}_{6}C$
(**f**) ${}^{42}_{20}Ca$
(**g**) ${}^{31}_{15}P$
(**h**) ${}^{55}_{25}Mn$

6. How many electrons are contained in each of the following ions?
(**a**) ${}^{23}_{11}Na^{+}$
(**b**) ${}^{16}_{8}O^{2-}$
(**c**) ${}^{31}_{15}P^{3-}$
(**d**) ${}^{42}_{20}Ca^{2+}$
(**e**) ${}^{19}_{9}F^{-}$
(**f**) ${}^{35}_{17}Cl^{-}$
(**g**) ${}^{7}_{3}Li^{+}$
(**h**) ${}^{27}_{13}Al^{3+}$

Principal energy levels

7. List the number of electrons which may be located in each of the four lowest energy levels.

8. Why aren't there more than seven energy levels discussed in this chapter?

9. List the number of electrons contained in each of the following by principal energy levels.
(**a**) ${}^{19}_{9}F$
(**b**) ${}^{39}_{19}K$
(**c**) ${}^{42}_{20}Ca$
(**d**) ${}^{32}_{16}S$
(**e**) ${}^{14}_{7}N$
(**f**) ${}^{24}_{12}Mg$
(**g**) ${}^{35}_{17}Cl$
(**h**) ${}^{16}_{8}O$
(**i**) ${}^{10}_{4}Be$
(**j**) ${}^{28}_{14}Si$

10. How many electrons are contained in the valence shell of each of the following?

(a) $^{23}_{11}\text{Na}$ **(b)** $^{7}_{3}\text{Li}$
(c) $^{19}_{9}\text{F}$ **(d)** $^{35}_{17}\text{Cl}$
(e) $^{31}_{15}\text{P}$ **(f)** $^{14}_{7}\text{N}$
(g) $^{27}_{13}\text{Al}$ **(h)** $^{24}_{12}\text{Mg}$
(i) $^{12}_{6}\text{C}$ **(j)** $^{28}_{14}\text{Si}$

11. Is it necessary to list the mass number in order to ascertain the electronic arrangement of an atom? Explain.

Subenergy levels

12. Write out the entire electronic arrangement for each of the following.

(a) $^{1}_{1}\text{H}$ **(b)** $^{7}_{3}\text{Li}$
(c) $^{20}_{10}\text{Ne}$ **(d)** $^{23}_{11}\text{Na}$
(e) $^{32}_{16}\text{S}$ **(f)** $^{37}_{17}\text{Cl}$
(g) $^{16}_{8}\text{O}$ **(h)** $^{27}_{13}\text{Al}$
(i) $^{79}_{35}\text{Br}$ **(j)** $^{59}_{27}\text{Co}$
(k) $^{91}_{40}\text{Zr}$ **(l)** $^{118}_{50}\text{Sn}$

13. How many electrons are contained in the indicated valence shell of each of the following? What subenergy levels are occupied with electrons in the valence shell?

(a) $^{1}_{1}\text{H}$ **(b)** $^{23}_{11}\text{Na}$
(c) $^{37}_{17}\text{Cl}$ **(d)** $^{7}_{3}\text{Li}$
(e) $^{19}_{9}\text{F}$ **(f)** $^{14}_{7}\text{N}$
(g) $^{16}_{8}\text{O}$ **(h)** $^{39}_{19}\text{K}$
(i) $^{32}_{16}\text{S}$ **(j)** $^{31}_{15}\text{P}$
(k) $^{27}_{13}\text{Al}$ **(l)** $^{28}_{14}\text{Si}$
(m) $^{24}_{12}\text{Mg}$ **(n)** $^{11}_{5}\text{B}$

Electron dot symbols

14. Depict each of the following using electron dot symbols.

(a) $^{16}_{8}\text{O}$ **(b)** $^{23}_{11}\text{Na}$
(c) $^{19}_{9}\text{F}$ **(d)** $^{27}_{13}\text{Al}$
(e) $^{14}_{7}\text{N}$ **(f)** $^{31}_{15}\text{P}$
(g) $^{11}_{5}\text{B}$ **(h)** $^{35}_{17}\text{Cl}$
(i) $^{7}_{3}\text{Li}$ **(j)** $^{39}_{19}\text{K}$
(k) $^{12}_{6}\text{C}$ **(l)** $^{28}_{14}\text{Si}$
(m) $^{32}_{16}\text{S}$ **(n)** $^{9}_{4}\text{Be}$

15. What similarities are noted in the electron dot symbols of each of the following pairs of elements?

(a) $^{19}_{9}\text{F}$ and $^{35}_{17}\text{Cl}$ **(b)** $^{23}_{11}\text{Na}$ and $^{39}_{19}\text{K}$
(c) $^{16}_{8}\text{O}$ and $^{32}_{16}\text{S}$ **(d)** $^{14}_{7}\text{N}$ and $^{31}_{15}\text{P}$
(e) $^{24}_{12}\text{Mg}$ and $^{40}_{20}\text{Ca}$ **(f)** $^{12}_{6}\text{C}$ and $^{28}_{14}\text{Si}$
(g) $^{11}_{5}\text{B}$ and $^{27}_{13}\text{Al}$

7 CHEMICAL BONDS

LEARNING OBJECTIVES

When you finish this chapter you should be able to:

(a) explain why atoms combine using the concept of electronic arrangements within the atom.

(b) classify all bonds into two major categories, ionic and covalent.

(c) distinguish between the types of covalent bonds.

(d) formulate the electron transfers necessary between atoms to form ionic bonds.

(e) arrange the shared electrons in electron dot representations of molecules to depict covalent bonds.

(f) use electronegativity values to predict which bonds will be polar.

(g) express the rules of oxidation numbers.

(h) match the following terms with their corresponding definitions.

bond
coordinate covalent bond
covalent bond
electronegativity
ionic bond
Lewis octet rule
oxidation number
polar covalent bond

Most substances, even the elements, are not composed of individual atoms. Instead, they are composed of atoms in combination with one another. Some exceptions are the noble gases, such as argon which constitutes one percent of the atmosphere. The remainder of the atmosphere, the water of the earth, and the earth itself consist predominantly of atoms combined with each other to form molecules. The molecules of some substances consist of very few atoms. Water is such a substance. The molecules of other substances, such as DNA, contain many hundreds of atoms in a complex array.

From the small number of elements there are derived millions of known compounds, and there are an infinite number possible. This is because there are so many possible ways in which atoms may combine to form molecules. The combinations of atoms into molecules are similar to the combinations of the small number of letters of an alphabet to form the words of a language. The study of the twenty-six letters of our alphabet is a limited subject. But clearly the formation of words is a subject of both immense scope and subtle variety. In the formation of words, there are rules of the language which govern the types of possible combinations. In the case of atoms, the rules are based on the number and type of electrons which the atoms possess.

In this chapter we shall study how atoms and their electrons are involved in bonding each other into molecules. Fortunately, there are only a few general types of bonds, each of which encompasses many bonds with similar electronic characteristics. These characteristics determine why salt is a crystalline solid and water is a liquid at room temperature.

7.1 THE HUMAN BOND

People are bound to each other by their human qualities which allow them to form interpersonal relationships. We talk of the bonds of friendship, the bonds of marriage, the bonds of the family, the bonds of solidarity. In each case the reference is to how individuals group themselves into social units. Thus we can break down human relationships into groups of types of interactions. Each of the groups can then be analyzed as to characteristics and variety.

If we were to compare the bonds of marriage to the bonds of friendship, it should be possible to list characteristics which distinguish one from the other although it must be granted that they each have many characteristics in common. Once a common thread of the bond of marriage, such as the sharing of self, is defined, we would find that there are individual differences among the various married couples. Of course this is to be expected because individuals differ in personality and ability to share oneself with another person. Furthermore, the degree to which a mutual sharing occurs depends not on one person but on both persons. Thus the marriage of Ann with Bill will, of necessity, involve a different extent of sharing on Ann's part than would occur if she were married to Charles. Similarly, Bill will share to a different extent with Ann than with an alternate spouse, Doris. In each case, although bonds of marriage are involved, there are differences in degree but not in kind.

7.2 THE CHEMICAL BOND

The chemical bond is every bit as varied as the human bond. It depends on a more precisely definable quantity, that is, the individual electronic characteristics of the atoms. **The chemical bond results from a change in the electronic structure of two atoms as they associate with each other.** Thus the type of bond that is formed depends on the atoms involved in the bond.

Could you explain to a sixth grade student the concept of a chemical bond?

As indicated in the prologue of this chapter, there is an immense number of possible combinations of atoms to form molecules. Thus there are many individual bonds which are known. In order to organize the many bonds into a more manageable subject for study, it is convenient to define a limited number of classes of bonds. Within each of these classes are bonds which have similar characteristics but vary in degree.

While chemists could divide up all of the known bonds into any number of classes, **the simplest approach which has historically been found convenient is to describe two main bonding classes, the ionic bonds and the covalent bonds.** In subsequent sections, the electronic characteristics of each class will be described. **The class of covalent bonds can be further subdivided into pure covalent, polar covalent, and coordinate covalent.**

7.3 THE CONCEPT OF VALENCE

The beginning of our modern theory of bonding can be traced to the concept of *valence* introduced in 1850. **Each element was said to have a valence equal to its combining capacity. The number of hydrogen or chlorine atoms with which another atom combines is called its combining capacity. The valence of these two reference elements was set as 1.** Therefore, oxygen, which reacts with hydrogen to produce H_2O, was said to have a valence of 2. By using such definitions it was possible to predict the formulas of some compounds. For example, it would be expected that oxygen, with a valence of 2, and chlorine, with a valence of 1, would combine to produce OCl_2, which actually is a known molecule.

As a starting point the concept of valence was extremely useful, although some elements have more than one valence. Oxygen also combines with hydrogen to form another compound, H_2O_2, called hydrogen peroxide. Nitrogen combines with hydrogen to form either ammonia, NH_3, or hydrazine, N_2H_4. Under such circumstances, the combination of several elements

What is the valence of nitrogen in NH_3?

having multiple valences can lead to many compounds, and the predictive aspect of the valence concept is lost. The series of compounds containing nitrogen and oxygen alone illustrate the point, because N_2O, NO, NO_2, N_2O_3, N_2O_4, and N_2O_5 are all known oxides of nitrogen.

7.4 WHY DO ATOMS COMBINE?

To ask why atoms combine is much like asking why any physical phenomenon occurs. Invariably it is necessary to return to fundamental truths that must be accepted as natural facts of life. One of these is that **all natural systems tend to lose potential energy and become more stable.** Other things being equal, a system that has stored potential energy is less stable than a system that has none. The term *stable* refers to systems that do not easily undergo spontaneous change. Unstable systems either change spontaneously to stable systems or do so under conducive conditions. For example, water on the middle of a slope has a higher potential energy than water at the bottom of a hill. If there are no obstructions, the water will flow downhill to the more stable position, releasing energy in the process. In a similar manner, elements have stored potential energy and may, under the proper conditions, release this energy and form compounds.

Because chemical bonding is an electronic phenomenon, it is instructive to consider the electronic structure of atoms and their chemical reactivity. The noble gases are noted for their lack of chemical reactivity. There are no known compounds of helium, neon, and argon. Why are these elements so unreactive toward other elements? All these gases have electronic structures that consist of filled valence shells. Except for helium, whose electron configuration is $1s^2$, the s and p subshells of the highest energy level contain a total of eight electrons. The particular stability of the filled level suggests that under the proper circumstances atoms of other elements might gain or lose electrons to achieve a noble gas electronic configuration. **The tendency of atoms to achieve the noble gas electronic configuration, ns^2np^6, is referred to as the *Lewis octet rule*.**

Oxygen, fluorine, sodium, and magnesium can achieve the noble gas configuration of neon by, respectively, gaining two electrons, gaining one electron, losing one electron, and losing two electrons:

Is Na^+ an anion or a cation?

$$2e^- + \text{O}\ (1s^22s^22p^4) \longrightarrow \text{O}^{2-}\ (1s^22s^22p^6)$$
$$1e^- + \text{F}\ (1s^22s^22p^5) \longrightarrow \text{F}^-\ (1s^22s^22p^6)$$
$$\text{Na}\ (1s^22s^22p^63s^1) \longrightarrow \text{Na}^+\ (1s^22s^22p^6) + 1e^-$$
$$\text{Mg}\ (1s^22s^22p^63s^2) \longrightarrow \text{Mg}^{2+}\ (1s^22s^22p^6) + 2e^-$$

In subsequent sections dealing with ionic and covalent bonds, it will be shown how the achievement of a noble gas electronic configuration controls the bonds formed between atoms.

7.5 IONIC BONDS

Ionic bonds are formed between two or more atoms as the result of the transfer of one or more electrons between the atoms. This electron transfer results in the formation of anions and cations. Recall from Chapter 4 that anions are negatively charged ions whereas cations are positively charged ions. The oppositely charged ions are attracted to each other and a bond between them is formed. **The bond existing between ions is called an ionic or electrovalent bond. Compounds containing ionic bonds are called ionic compounds.**

A common example of an ionic compound is sodium chloride, NaCl. The free sodium atom has one valence electron ($3s^1$) whereas the chlorine atom has seven valence electrons ($3s^23p^5$). In forming an ionic bond, the sodium loses its valence electron to chlorine. As a result, sodium achieves the noble gas electron configuration of neon ($1s^12s^22p^6$) and becomes a positive ion. Chlorine achieves the noble gas configuration of argon ($1s^22s^22p^63s^23p^6$) and acquires a negative charge (Figure 7.1). The attraction between the sodium ion and the chloride ion is an ionic bond.

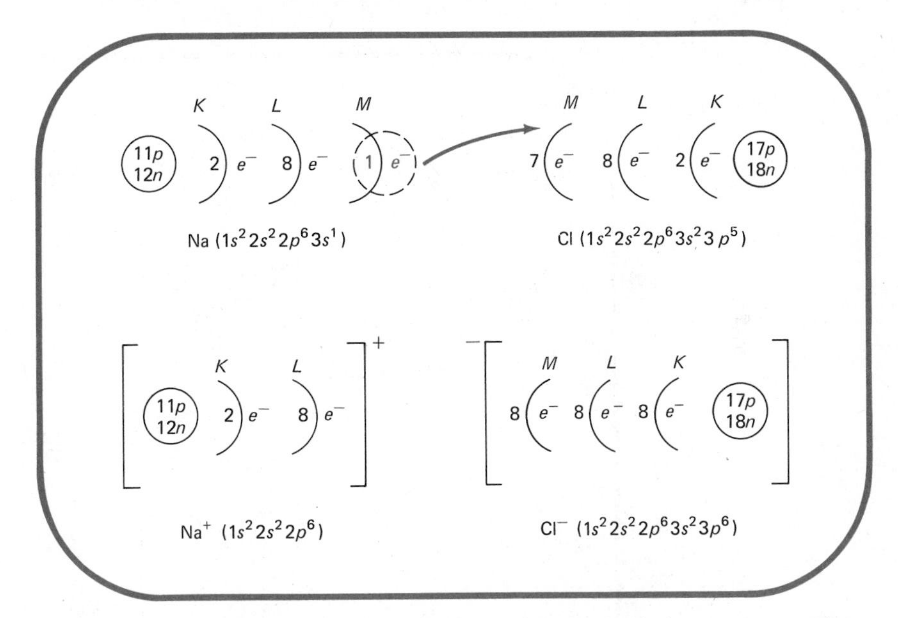

Figure 7.1 The transfer of an electron between sodium and chlorine atoms to produce sodium chloride.

A shorthand way of showing the formation of sodium chloride from the sodium and chlorine atoms involves electron dot symbols. These symbols are sufficient to represent the electronic changes because only the valence electrons are involved.

$$\text{Na}\,(\cdot) + \cdot\ddot{\underset{..}{\text{Cl}}}: \rightarrow \text{Na}^+ + :\ddot{\underset{..}{\text{Cl}}}:^-$$

Figure 7.2 The ionic accommodation.

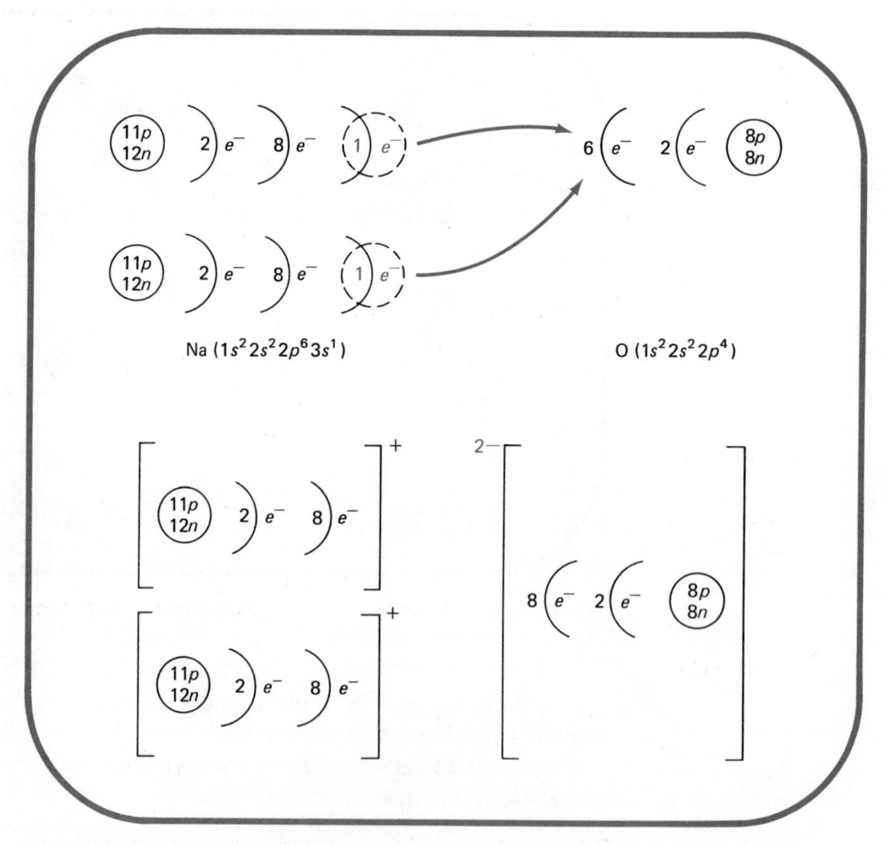

Figure 7.3 The formation of sodium oxide.

A representation of two atoms achieving the stability of an ionic bond is given in Figure 7.2. Harmony is achieved by allowing the valence electron represented by the blanket to be possessed by one atom at the expense of another atom.

NaCl
Na_2O
$CaCl_2$

Sodium combines with oxygen to form the ionic compound sodium oxide, Na_2O. In this compound, two sodium atoms each lose their valence electron and are converted into sodium ions with a neon-like electron configuration. Oxygen, which has six valence electrons, gains two electrons to achieve the neon-like electron configuration. Thus, in total, two sodium atoms each transfer their single valence electron to oxygen (Figure 7.3).

The shorthand method of showing the formation of sodium oxide from the sodium and oxygen atoms involves electron dot symbols. Only the valence electrons which participate in bonding are given in the electron dot symbols.

$$2\,\mathrm{Na}\cdot + \cdot\ddot{\underset{\cdot}{\mathrm{O}}}: \rightarrow 2\mathrm{Na}^+ + :\ddot{\underset{\cdot\cdot}{\mathrm{O}}}:^{2-}$$

In all ionic compounds, the electrons lost by one or more atoms in achieving the noble gas electronic configuration of a cation are accepted by the appropriate number of companion atoms to achieve the noble gas electronic configuration of the anions. Thus the product of the number of electrons lost by a type of atom and the number of such atoms must equal the product of the number of electrons gained by the companion atom and the number of such atoms.

$$\begin{pmatrix}\text{number of}\\\text{cations formed}\end{pmatrix} \times \begin{pmatrix}\text{number of}\\\text{electrons lost}\\\text{in forming one}\\\text{cation}\end{pmatrix} = \begin{pmatrix}\text{number of}\\\text{anions formed}\end{pmatrix} \times \begin{pmatrix}\text{number of}\\\text{electrons gained}\\\text{in forming one}\\\text{anion}\end{pmatrix}$$

2 1 1 2

Problem 7.1 Describe the process by which calcium and chloride reach electronic accommodation in forming calcium chloride.

Solution The electronic configuration of the valence shell electrons of calcium and chlorine are $4s^2$ and $3s^23p^5$, respectively. Therefore, calcium need lose only two electrons to achieve a noble gas electron configuration of argon as the Ca^{2+} cation. Chlorine can gain one electron to achieve the noble gas electron configuration of argon as the Cl^- anion.

$$\mathrm{Ca}\ (4s^2) \rightarrow \mathrm{Ca}^{2+} + 2e^-$$
$$1e^- + \mathrm{Cl}\ (3s^23p^5) \rightarrow \mathrm{Cl}^-\ (3s^23p^6)$$

In order to balance their mutual electronic requirements, two chlorine atoms each accept one electron from calcium and form $CaCl_2$. In terms of electron dot symbols, the process is:

$$\dot{\underset{\cdot}{Ca}} + 2:\ddot{\underset{\cdot}{Cl}}: \longrightarrow Ca^{2+} + 2:\ddot{\underset{\cdot\cdot}{Cl}}:^{-}$$

7.6 COVALENT BONDING

The second major class of bonds cannot be explained by a complete electron transfer process. In the case of simple molecules, such as H_2 and F_2, the bonds hold together identical atoms. Because there is no difference in the atoms, there is no reason for electron transfer to occur.

The hydrogen molecule is composed of two hydrogen atoms, each with one electron in a 1*s* orbital. In order to achieve the electron configuration of the noble gas helium, each hydrogen would have to accept one electron. Since both hydrogen atoms need one electron, neither is likely to give up the one it has. In such a case, the two atoms join so that their two electrons are shared. Each of the two atoms possess both electrons as long as they are together. When atoms are associated by means of one or more pairs of electrons, the atoms are covalently bound. **Covalent bonds are shared electron pairs.** The bonding of two hydrogen atoms to form a hydrogen molecule is pictured below, using the electron dot notation.

Covalent bonds are shared electron pairs.

$$H\cdot + \cdot H \longrightarrow H:H$$

The fluorine molecule is pictured as two fluorine atoms joined by a mutually shared pair of electrons. Each fluorine atom requires one electron to fill the second energy level. Because there is no difference between the electron donating or accepting properties of the two atoms, each must share one electron.

$$:\ddot{\underset{\cdot\cdot}{F}}\cdot + \cdot\ddot{\underset{\cdot\cdot}{F}}: \longrightarrow :\ddot{\underset{\cdot\cdot}{F}}:\ddot{\underset{\cdot\cdot}{F}}:$$

In representing covalently bound molecules, it is convenient to write a pair of electrons as a dash. The hydrogen and fluorine molecules are more easily written using this convention.

$$H{-}H \qquad |\overline{\underline{F}}{-}\overline{\underline{F}}|$$

The concept of a shared pair of electrons in a covalent bond requires that the electrons occupy a region of space common to the bonding atoms. With such a requirement, polyatomic molecules must exist in geometric arrays that satisfy the bonding requirements of all atoms.

A hydrogen molecule can be visualized as two hydrogen atoms with interpenetrating or overlapping 1*s* orbitals (Figure 7.4). The resulting region of high electron probability is located between the two atoms. The strength

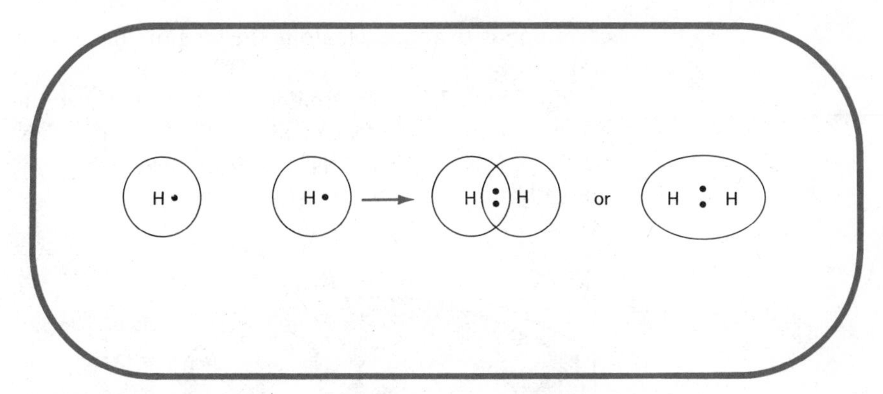

Figure 7.4 The overlap of 1s orbitals in forming the hydrogen molecule.

of the covalent bond is the result of the attraction of the positively charged nuclei for the electron pair. Because the 1s orbital is spherically symmetrical, the same molecule would result regardless of the direction of approach of the two atoms.

In fluorine, the shared electron pair consists of one electron from each fluorine 2p orbital. Owing to the shape of the 2p orbital of fluorine, which contains the single electron, the orbitals overlap in the direction corresponding to the highest probability of finding an electron. This direction

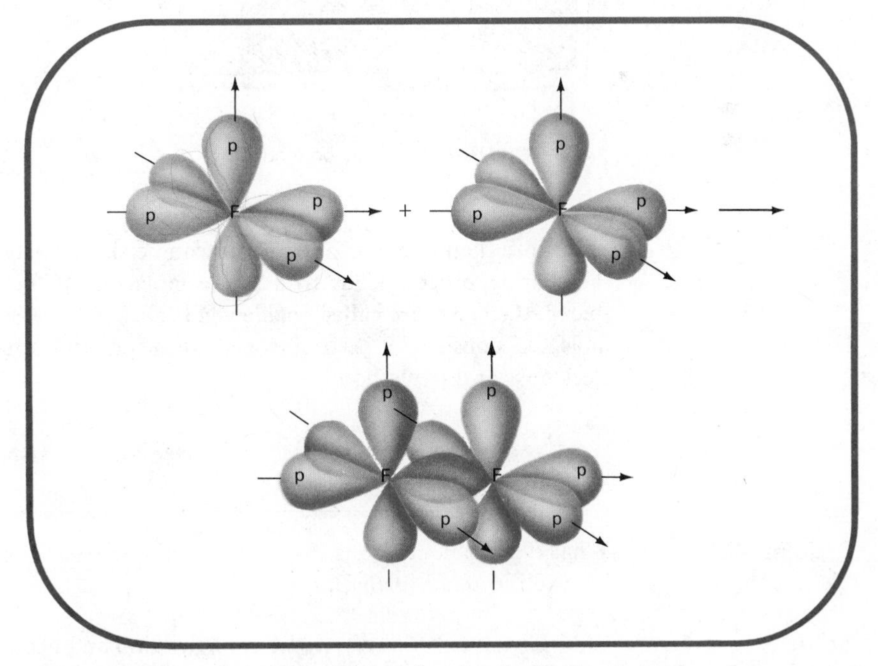

Figure 7.5 The overlap of 2p orbitals in forming the fluorine molecule.

corresponds to approach along the longitudinal axis of the 2*p* orbital (Figure 7.5).

In Figure 7.6 is represented the electronic accommodation of the hydrogen atoms in the hydrogen molecule. Although they do not have enough electrons (blanket) to satisfy each of their needs, they share the electrons mutually.

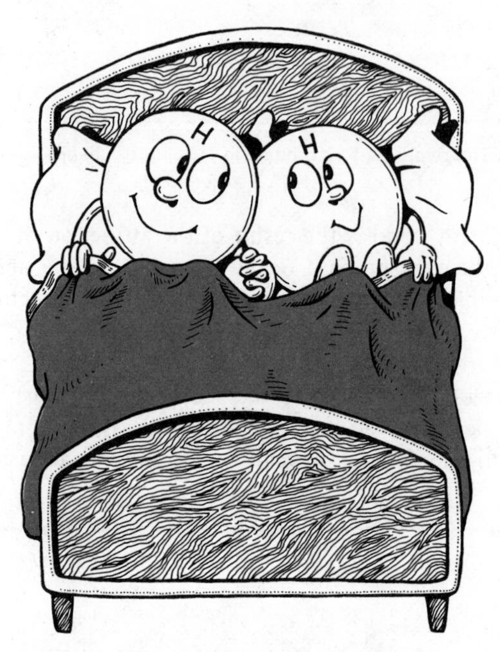

Figure 7.6 The covalent accommodation.

More than one pair of electrons can be shared between pairs of atoms if the sharing process leads to a stable molecule. **If four or six electrons are shared, the bonds are called double and triple bonds, respectively.** The nitrogen molecule consists of two nitrogen atoms bound together by six shared electrons, or a triple bond.

$$\cdot\underset{\cdot}{\ddot{N}}\cdot + \cdot\underset{\cdot}{\ddot{N}}\cdot \longrightarrow :N::N: \quad \text{or} \quad |\,N{\equiv}N\,|$$

Problem 7.2 What type of bond exists in Cl_2? Describe the electrons and orbitals involved in bond formation.

Solution Chlorine has a $3s^23p^5$ electronic configuration and needs only one electron to achieve a noble gas configuration. However, since both chlorine atoms have identical needs, the two must each share one electron in common and form a single covalent bond.

$$:\ddot{\underset{..}{Cl}}\cdot + \cdot\ddot{\underset{..}{Cl}}: \longrightarrow :\ddot{\underset{..}{Cl}}-\ddot{\underset{..}{Cl}}:$$

7.7 POLAR COVALENT BONDING

In addition to molecules containing identical atoms, there are many molecules that contain nonidentical atoms bound together by a covalent bond. The molecule hydrogen fluoride can be pictured as consisting of two atoms, each of which needs one electron for a noble gas configuration. Fluorine has a higher attraction for electrons than does hydrogen. However, fluorine's attraction for hydrogen's electron is not strong enough for the hydrogen to lose its electron to fluorine and form an ionic bond. Consequently, the necessary electrons are shared. The sharing is unequal, however, as only in the case of identical atoms can electrons be shared equally. **A covalent bond in which the electrons are shared unequally between two nonidentical atoms is called a polar covalent bond.** Hydrogen fluoride contains a polar covalent bond. The molecule is represented by the conventional electron dash notation, even though the shared electron pair must be associated to a somewhat larger extent with fluorine.

Unequal sharing of electrons results in a polar covalent bond.

$$^{\delta+}H-\overline{\underline{F}}|^{\delta-}$$

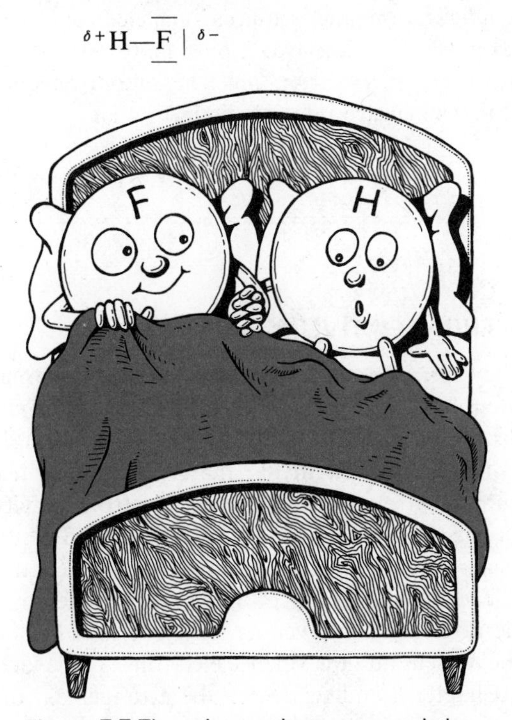

Figure 7.7 The polar covalent accommodation.

The result of this unequal sharing is that the fluorine end of the molecule has acquired a partial negative charge whereas the hydrogen end has acquired a partial positive charge. The symbol δ (the Greek lowercase letter delta) is used to denote the fractional charge located at a site within a molecule. The HF molecule is said to possess a *dipole* (that is, two poles). It should be understood that only a displacement of electrons toward fluorine occurs in HF. There is no transfer of electrons as in NaCl. Thus there are no ions in HF, and the entire molecule is electrically neutral even though there are partial charges on each atom.

To continue the analogy of bonds to atoms in a shared bed, consider Figure 7.7. The hydrogen and fluorine atoms do share their blanket but in an unequal manner. This condition is very common in compounds and is quite prevalent among humans as well. Sharing is seldom perfectly equal because the individuals sharing an item are invariably slightly different.

How well do you share things?

Problem 7.3

What type of bond exists in the molecule HCl? Draw an electron dot representation of the molecule.

Solution

The hydrogen atom has a $1s^1$ electron configuration and requires one electron to achieve the inert gas configuration $1s^2$. The chlorine atom has a $3s^23p^5$ electron configuration and requires one electron to achieve an inert gas configuration $3s^23p^6$. If the two atoms share one electron, a covalent bond results. However, the electrons must be shared unequally as the atoms are nonidentical. The electron dot configuration is:

$$\mathrm{H{:}\overset{\cdot\cdot}{\underset{\cdot\cdot}{Cl}}{:}}$$

7.8 ELECTRONEGATIVITY

The tendency for an atom to attract a pair of electrons when bonded to a different atom in a polar covalent bond is given by the electronegativity of the atom. A listing of electronegativity values is given in Table 7.1. **The larger the value of the electronegativity, the greater is the tendency to attract an electron pair.** The series of decreasing electronegativity of some common elements is F > O > Cl = N > Br > I = C = S > P = H > B > Si.

What element is the most electronegative?

The electronegativity of an atom depends on (1) the number of protons in the nucleus, (2) the distance of the valence electrons from the nucleus, and (3) the number of intervening electrons in lower energy levels that are between the nucleus and the valence electrons. These factors will be discussed further in Chapter 8 and related to the arrangement of the elements in the periodic table.

TABLE 7.1 **electronegativity values of atoms**

Element	Electronegativity
F	4.0
Cl	3.0
Br	2.8
I	2.5
O	3.5
S	2.5
N	3.0
P	2.1
C	2.5
Si	1.8
B	2.0
H	2.1

7.9 COORDINATE COVALENT BOND

In both the pure covalent bond and the polar covalent bond, each of the bonded atoms contributes to the pair of electrons they share between them. **The coordinate covalent bond is one formed in which both electrons of the electron pair shared by the two atoms are derived from one atom.** This bond is sometimes designated by an arrow pointing away from the atom donating the pair of electrons. However, once formed, a coordinate covalent bond is indistinguishable from any other covalent bond. All types of covalent bonds have in common a shared electron pair between atoms.

One example of the formation of a coordinate covalent bond is the reaction of ammonia with a hydrogen cation (proton) to form the ammonium ion.

$$
\begin{array}{ccccccc}
 & \mathrm{H} & & & & \mathrm{H} & \\
 & | & & & & | & \\
\mathrm{H}- & \mathrm{N}: & + \ \mathrm{H}^{+} & \longrightarrow & \mathrm{H}- & \overset{+}{\mathrm{N}} & -\mathrm{H} \\
 & | & & & & | & \\
 & \mathrm{H} & & & & \mathrm{H} &
\end{array}
$$

In the electron dot representation of ammonia, nitrogen shares three of its five valence electrons with hydrogen. The remaining two electrons are not involved in bonding. This pair of electrons, called an unshared pair of electrons, is used in forming the bond to the proton for the ammonium ion. The proton has no electrons in its valence shell and needs two electrons to fill this shell. Since the nitrogen in ammonia has two electrons it can share, the proton forms a coordinate covalent bond with nitrogen. However once the coordinate covalent bond is formed, it is indistinguishable from the other three covalent bonds. The positive charge of the ammonium ion is dispersed over the entire ion.

7.10 OXIDATION NUMBER

The oxidation number of an atom is a positive or negative integer assigned to describe how an atom will combine as a free atom, an ion, or in a molecule. As an ion or in an ionic compound, the charge of a cation or an anion results from the transfer of electrons away from or to the neutral atom. The positive charge on the ion indicates the number of electrons lost while a negative charge indicates the number of electrons gained relative to the neutral atom.

As a reference point, **the oxidation number of an element, regardless of whether it is monatomic, diatomic, and so on, is set at zero.** When an electron is removed from an atom as Na to yield Na^+, the oxidation number of the cation is $+1$. In the case of gaining an electron, such as Cl to yield Cl^-, the oxidation number is -1. Thus **the charge of a monatomic ion is equal to its oxidation number.** These oxidation numbers provide information about the compounds which may be formed. One sodium ion whose oxidation number is $+1$ combines with one chloride ion whose oxidation number is -1. **In any ionic compound, the sum of the oxidation numbers must be numerically equal to zero since the compound is electrically neutral overall.**

The oxidation number of a monatomic ion is equal to its charge.

In covalent compounds, electrons are not completely transferred but rather are shared between atoms. The pure covalent bond as exists in the elements such as H_2, F_2, P_4, and S_8 is the result of equal sharing. No single atom has gained or lost an electron and therefore the oxidation number of each atom in the molecule is zero. However, the sharing of electrons in polar

TABLE 7.2 oxidation number rules

1. **The oxidation number of an element in its uncombined state is zero.**
2. **The algebraic sum of the oxidation numbers in an ionic or covalent compound is zero.**
3. **The oxidation number of a monatomic ion is the same as the charge of the ion.**
4. **The algebraic sum of the oxidation numbers in a complex ion equals the charge of the ion.**
5. **Metals generally have positive oxidation numbers in the combined state.**
6. **Negative oxidation numbers in covalent compounds of two unlike atoms are assigned to the more electronegative atom.**
7. **Most hydrogen compounds contain hydrogen with a $+1$ oxidation number. Hydrides of very active metals, such as Li, Na, K, Mg, and Ca, contain hydrogen with a -1 oxidation number.**
8. **In most oxygen compounds, the oxidation number is -2. Exceptions occur in peroxides whose oxidation number is -1.**
9. **The oxidation number of F, Cl, Br, and I is -1 except when combined with a more electronegative element.**
10. **Sulfides have an oxidation number of -2.**

covalent bonds is unequal. As a result, one atom gains a partial negative charge at the expense of another atom which then has a partial positive charge. For the sake of convenience, the oxidation number of atoms bonded by polar covalent bonds is based on an assumed complete electron transfer.

The most electronegative element in the polar covalent bond is assigned the negative oxidation number as it has the greatest attraction for electrons. Thus in carbon tetrachloride, CCl_4, each electronegative chlorine atom is assigned a -1 oxidation number based on its tendency to gain an electron in its unequal sharing with carbon. Carbon is assigned a $+4$ oxidation number because it has a tendency to lose four electrons in bonding to four chlorine atoms. Since the number of electrons partially gained or lost must be equivalent, **the sum of the oxidation numbers of atoms in a molecule must be equal to zero.**

Why must the sum of the oxidation numbers of atoms in a molecule be equal to zero?

In order to facilitate the assignment of oxidation numbers, the rules listed in Table 7.2 are useful.

Problem 7.4 What is the oxidation number of sulfur in SO_3?

Solution The sum of the oxidation numbers of a neutral compound must be zero (rule 2).

$$\begin{pmatrix}\text{oxidation number}\\ \text{of sulfur}\end{pmatrix} + 3\begin{pmatrix}\text{oxidation number}\\ \text{of oxygen}\end{pmatrix} = 0$$

Since the oxidation number of oxygen is -2 (rule 8), the oxidation number of sulfur may be algebraically calculated.

$$\begin{pmatrix}\text{oxidation number}\\ \text{of sulfur}\end{pmatrix} + 3(-2) = 0$$

$$\begin{pmatrix}\text{oxidation number}\\ \text{of sulfur}\end{pmatrix} - 6 = 0$$

$$\text{oxidation number of sulfur} = +6$$

Problem 7.5 What is the oxidation number of nitrogen in $NaNO_2$?

Solution The sum of the oxidation numbers of sodium, nitrogen, and oxygen must be equal to zero (rule 2).

$$\begin{pmatrix}\text{oxidation number}\\ \text{of sodium}\end{pmatrix} + \begin{pmatrix}\text{oxidation number}\\ \text{of nitrogen}\end{pmatrix} + 2\begin{pmatrix}\text{oxidation number}\\ \text{of oxygen}\end{pmatrix} = 0$$

The oxidation number of oxygen is -2 (rule 8). The oxidation number of sodium is positive since it is a metal (rule 5). Sodium is known to exist as only the $+1$ ion and its oxidation number is equal to that charge (rule 3).

$$(+1) + \begin{pmatrix}\text{oxidation number}\\ \text{of nitrogen}\end{pmatrix} + 2(-2) = 0$$

$$\begin{pmatrix}\text{oxidation number}\\ \text{of nitrogen}\end{pmatrix} + 1 - 4 = 0$$

$$\text{oxidation number of nitrogen} = +3$$

Problem 7.6

What is the oxidation number of manganese in the permanganate ion MnO_4^-?

Solution

The sum of the oxidation numbers of a complex ion is equal to the charge of the ion (rule 4).

$$\begin{pmatrix}\text{oxidation number}\\ \text{of manganese}\end{pmatrix} + 4\begin{pmatrix}\text{oxidation number}\\ \text{of oxygen}\end{pmatrix} = -1$$

Since the oxidation number of oxygen is -2 (rule 8), the oxidation number of manganese can be calculated:

$$\begin{pmatrix}\text{oxidation number}\\ \text{of manganese}\end{pmatrix} + 4(-2) = -1$$

$$\begin{pmatrix}\text{oxidation number}\\ \text{of manganese}\end{pmatrix} - 8 = -1$$

$$\begin{pmatrix}\text{oxidation number}\\ \text{of manganese}\end{pmatrix} = +7$$

7.11 ELECTRON DOT REPRESENTATIONS OF MOLECULES AND IONS

Throughout this chapter, electron dot representations have been used to illustrate bonds formed in molecules and ions. With the exception of hydrogen, which requires only two electrons to fill its subshell, the majority of the common substances you will encounter are governed by the octet rule. The substances which do not conform to this rule will not be covered in this text.

In order to progress in the study of chemistry, it is important that you are able to write electron dot representations of molecules easily. The following guide lines are beneficial in depicting molecules:

1 The electron dot symbol of the atoms that occur in the molecule should be written.

2 The number of excess electrons in the case of negative complex ions should be written.
3 Each atom should be examined to determine how many electrons are required to achieve an octet.
4 The atoms are mutually bonded so as to achieve the desired octets.

Problem 7.7 Draw the electron dot representation of H_2O.

Solution The electron dot symbols of hydrogen and oxygen are as follows:

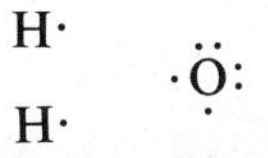

H·
·Ö:
H·

No extra electrons are added as the substance is electrically neutral. The oxygen atom requires two electrons to achieve an octet. Each hydrogen atom requires one electron. Thus by arranging the hydrogen atoms about the oxygen atom, the requisite number of covalent bonds are formed.

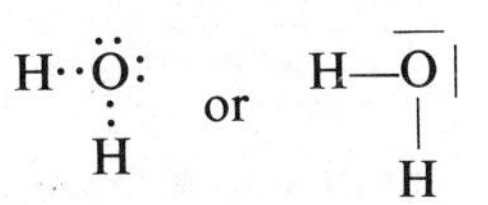

```
H··Ö:        _
   :   or  H—O|
   H         |
             H
```

Problem 7.8 Draw the electron dot representation of CCl_4.

Solution The electron dot symbols of carbon and chlorine are as follows:

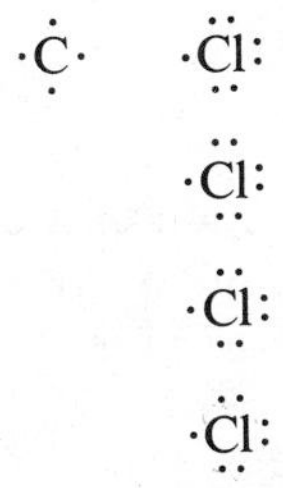

·Ċ· ·C̤̈l:

·C̤̈l:

·C̤̈l:

·C̤̈l:

Each chlorine atom needs to gain one electron. The carbon atom requires four electrons or may donate four electrons to achieve a noble gas electron configuration. In order to achieve a balance between carbon and chlorine, the carbon must use all four of its valence shell electrons.

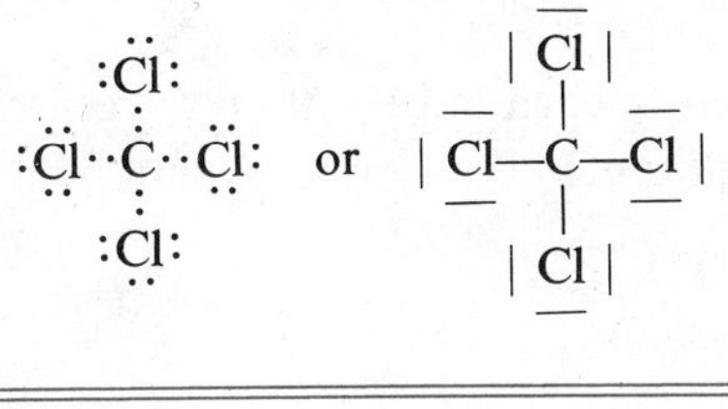

```
     ..                _
    :Cl:             | Cl |
     ..             _  |  _
 ..  ..  ..  ..    | Cl—C—Cl |
:Cl··C··Cl:   or    ‾  |  ‾
 ..  ..  ..          | Cl |
    :Cl:               ‾
     ..
```

Problem 7.9 Draw the electron dot representation of the hydroxide ion OH^-.

Solution The electron dot symbols of hydrogen and oxygen are as follows:

H· ·Ö:

Since the ion bears a negative charge, an extra electron must be added. For clarity of visualization, a small circle, ∘, will be used for this electron which is placed on the more electronegative element.

H· ·Ö:$^-$

Now it can be seen that oxygen with its extra electron needs only one more electron for an octet. The hydrogen and oxygen share one electron each.

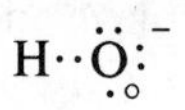

H··Ö:$^-$

Problem 7.10 Draw the electron dot representation of the sulfite ion $SO_3{}^{2-}$.

Solution The electron dot symbols of sulfur and oxygen are as follows:

:S̈· ·Ö:

·Ö:

·Ö:

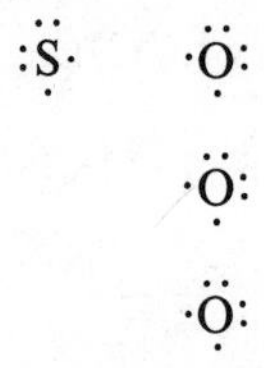

Two electrons are added to the collection of atoms to achieve the negative charge. They are added one to each of two oxygen atoms which are more electronegative than sulfur.

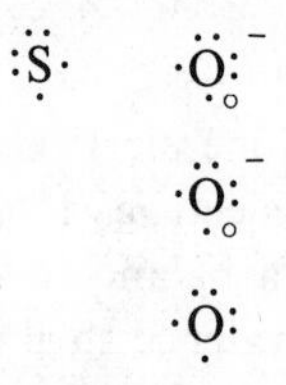

:S̈· ·Ö:$^-$

·Ö:$^-$

·Ö:

Sulfur requires only two electrons to achieve an octet. Two of the oxygens require one electron each but the third oxygen requires two electrons. First the oxygens requiring only one electron are bonded to sulfur.

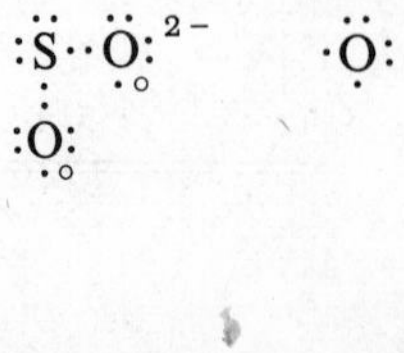

:S̈··Ö:$^{2-}$ ·Ö:

:Ö:

The remaining oxygen atom's electrons are rearranged to produce three pairs, and it then bonds to one of sulfur's unshared pairs of electrons.

:Ö: ⟵ :S̈··Ö: or |O—S—O|
:O: |O|

Once the coordinate covalent bond is formed, it is no different from the other two sulfur-oxygen bonds.

Problem 7.11 Write the electron dot representation of carbon bisulfide CS_2.

Solution The electron dot symbols of the elements are:

·Ċ· ·S̈:

·S̈:

Each sulfur requires two electrons and the carbon requires four electrons. If only one electron were shared between carbon and each sulfur, neither element would achieve an octet of electrons.

·Ċ··S̈:

:S·

The solution is to have each sulfur share two electrons with carbon to form double covalent bonds. The electron dots are rearranged to achieve the necessary bonds.

:S̈::C::S̈: or |S=C=S|

SUMMARY

Most elements and all compounds consist of atoms held together in units by ***chemical bonds****. The types of bonds are classified according to the nature of the electron involvement into* ***ionic*** *and* ***covalent****. The class of covalent bonds is further subdivided into* ***pure covalent****,* ***polar covalent****, and* ***coordinate covalent****.*

Atoms have a tendency to achieve a noble gas electronic configuration according to the ***Lewis octet rule****. By using this rule, the structure of molecules may be depicted using* ***electron dot representations****.*

Ionic bonds exist between oppositely charged ions. Covalent bonds are shared electron pairs. Polar covalent bonds result from unequal sharing of electron pairs due to the differences in the ***electronegativity*** *of the atoms. Coordinate covalent bonds result from both electrons in an electron pair being provided by a single atom.*

The ***oxidation number*** *of an atom is a positive or negative number assigned to describe the way an atom may combine as a free atom, an ion, or in a molecule.*

QUESTIONS AND PROBLEMS

Definitions and use of terms

1. Define the following terms.

(a) anion
(b) cation
(c) combining capacity
(d) valence
(e) Lewis octet rule
(f) ionic bond
(g) covalent bond
(h) polar covalent bond
(i) coordinate covalent bond
(j) dipole
(k) single bond
(l) triple bond
(m) oxidation number
(n) electronegativity

2. Compare the use of the terms valence and bond.

3. If two elements combine in a one-to-one ratio, does that necessarily mean that their valences are both one?

4. What is the difference between each of the following?

(a) anion and cation
(b) ionic bond and covalent bond
(c) covalent bond and polar covalent bond
(d) a single bond and a double bond
(e) a positive and a negative oxidation number
(f) a bonded pair and an unshared pair of electrons

Electronic configuration of ions

5. Write out the entire electronic configuration of each of the following ions.

(a) H^+
(b) H^-
(c) Li^+
(d) Na^+
(e) K^+
(f) Mg^{2+}
(g) Ca^{2+}
(h) F^-
(i) Cl^-
(j) Br^-
(k) I^-
(l) O^{2-}
(m) S^{2-}
(n) N^{3-}
(o) P^{3-}

6. Write only the electronic configuration of the valence shell electrons of the ions in Question 5.

7. Write the electron dot symbols of the ions in Question 5.

Ionic bonds

8. Write the formula for the ionic compound which results from the combination of the following ions.

(a) Na^+ and Cl^- **(b)** Ca^{2+} and Cl^-
(c) Mg^{2+} and O^{2-} **(d)** Mg^{2+} and Br^-
(e) Ba^{2+} and H^- **(f)** Ba^{2+} and P^{3-}
(g) Li^+ and N^{3-} **(h)** Al^{3+} and Cl^-

9. Write the formula for the ionic compound that results from the reaction of atoms of the following elements.

(a) Li and F **(b)** Mg and Br
(c) Li and O **(d)** Mg and S
(e) Al and F **(f)** Na and Se
(g) Ca and I **(h)** Na and N

Covalent bonds

10. Why can true covalent bonds only result from interaction of identical atoms?

11. Write electron dot representations of the following covalent molecules.

(a) H_2 **(b)** Cl_2
(c) I_2 **(d)** N_2
(e) HF **(f)** HI

12. Write electron dot representations of the following covalent molecules.

(a) H_2S **(b)** PH_3
(c) CS_2 **(d)** CF_4

13. Write electron dot representations of the following ions.

(a) SH^- **(b)** CN^-
(c) PH_4^+ **(d)** SO_4^{2-}

Polar covalent bonds

14. How can one predict the direction of the polarity of a bond using electronegativity values?

15. How might the order of the polarity of a series of bonds be predicted?

16. Indicate which element will be the positive end of the dipole when the following pairs of elements are bonded to each other.

(a) H and Br **(b)** Br and Cl
(c) H and O **(d)** O and F
(e) O and Cl **(f)** Si and F
(g) N and H **(h)** N and F
(i) P and Cl **(j)** P and F

Oxidation numbers

17. Calculate the oxidation of the indicated element in each of the following.

(a) Cl in $NaClO_3$ (b) S in H_2SO_3
(c) N in NO_2^- (d) I in IO_4^-
(e) S in SO_4^{2-} (f) P in PO_4^{3-}
(g) Mn in $KMnO_4$ (h) S in H_2S
(i) S in $S_2O_3^{2-}$ (j) Cr in $Cr_2O_7^{2-}$

18. In what way does oxidation number differ from valence?

19. How could hydrogen exist with a negative oxidation number?

8 THE PERIODIC TABLE

LEARNING OBJECTIVES

When you finish this chapter you should be able to:

(a) list the attempts to classify elements in chronological order.

(b) illustrate the periodic law for the elements.

(c) distinguish between periods and groups.

(d) interpret the properties of the elements using the periodic table.

(e) outline the periodic table in terms of the subenergy levels of the electrons in the atom.

(f) distinguish between representative elements and transition elements.

(g) locate the metals, nonmetals, and semimetals within the periodic table.

(h) employ the periodic table to calculate maximum and minimum oxidation states of the elements.

(i) identify trends in atomic radii, ionization potential, and electronegativity within the periodic table.

(j) formulate the types of compounds which may be formed from the elements based on their position in the periodic table.

(k) match the following terms with their proper definitions.

atomic radii
electronegativity
group
ionization potential
period
periodic table

At this point in your study of chemistry, the accumulation of facts concerning atomic symbols, atomic weights, subatomic particles, electronic configuration, and types of bonds in compounds may be beginning to overwhelm you. How does the chemist handle all of this information and much more? The answer is that he has the equivalent of a program at a college football game. Such a program consists of an orderly arrangement of the names of the players, their positions, numbers, vital physical statistics, personal data, accomplishments, and any other information designed to present an interesting picture of the individual and his relationship to the rest of the team. As the vendor may say, "You need a program to know the players." Chemists have as their program an arrangement of the elements called the periodic table. The periodic table provides a wealth of information interrelating the chemical elements and thus reduces the burden of memorizing an excessive amount of information.

In this chapter you will learn about the origin of the periodic table, its relationship to electronic configuration, and its use in predicting chemical properties of elements and compounds.

8.1 THE TRIADS OF DOBEREINER

Early in chemical history it was recognized that many elements have similar sets of properties, and these were used for the purpose of classification. The earliest division entailed only two large groups of elements designated as metals and nonmetals. The German chemist Johann Dobereiner in 1829 made the first serious attempt to classify the elements into smaller and simpler subgroups than metals and nonmetals. He observed that **elements with similar physical and chemical properties fall into groups of three. Because these groups contained three related elements, he called them triads.** One such triad includes the elements chlorine, bromine, and iodine. These elements all form compounds of the same general molecular formula. Thus the acids HCl, HBr, and HI are all compounds in which one atom of the triad is combined with one atom of hydrogen. Similarly, the compounds NaCl, NaBr, and NaI are all related by the combination of one atom of the triad with one atom of sodium.

In light of present-day knowledge, the members of a triad of elements are related by their atomic weights. The atomic weight of one element of the triad is approximately the average of the atomic weights of the other two elements. The approximate atomic weights of chlorine and iodine are 35.5 amu and 126.6 amu, respectively. The average of these two numbers is 81 amu, a value close to the atomic weight of bromine, which is 79.9 amu.

$\left.\begin{matrix} Cl_2 \\ Br_2 \\ I_2 \end{matrix}\right\}$ a triad

A list of other Dobereiner triads is given in Table 8.1. In the triad of the elements sulfur, selenium, and tellurium, the atomic weight of selenium is the average of the other two atomic weights. In addition, it is found that the density and the melting point of the middle member of the triad is intermediate between the values for the other two members of the triad. The compounds of hydrogen and these elements have similar molecular formulas —H_2S, H_2Se, and H_2Te—and all smell vile.

TABLE 8.1

Dobereiner's triads

Element	Atomic weight	Averaged atomic weight	Density g/ml	Melting point °C
chlorine	35.5		1.6	−101
bromine	79.9	81.2	3.1	−7
iodine	126.9		4.9	113.5
sulfur	32.1		2.1	95.5
selenium	79.0	79.8	4.8	217
tellurium	127.6		6.2	452
calcium	40.1		1.6	842
strontium	87.6	88.7	2.5	769
barium	137.3		3.5	725

A similar ordering of atomic weights, densities, and melting points is observed for the triad calcium, strontium, and barium. These elements also form similar compounds such as CaO, SrO, and BaO.

There are other triads of chemically similar elements. However, the grouping together of only three elements is not significant because there were only a limited number of elements to classify when Dobereiner made his observations. Nevertheless, such observations provided the background and incentive to seek further classification schemes for the elements.

8.2 NEWLANDS' OCTAVES

It was not until 1866, when the English chemist John Newlands examined the then known elements, that an order of great significance was realized. Newlands arranged the elements in an increasing order of their atomic weight. He noted that chemically similar elements occur at regular intervals in the list (Figure 8.1). Every eighth element has similar properties. Lithium, the second in the list, is comparable to sodium, the ninth in the list, and to potassium, the sixteenth in the list. In like manner, beryllium, magnesium, and calcium are third, tenth, and seventeenth in the list. The similarity of this periodic reoccurrence of the properties of elements to the octave on the musical scale prompted Newlands to postulate a **law of octaves. Chemically similar elements reoccur in octaves when arranged in order of increasing atomic weight.**

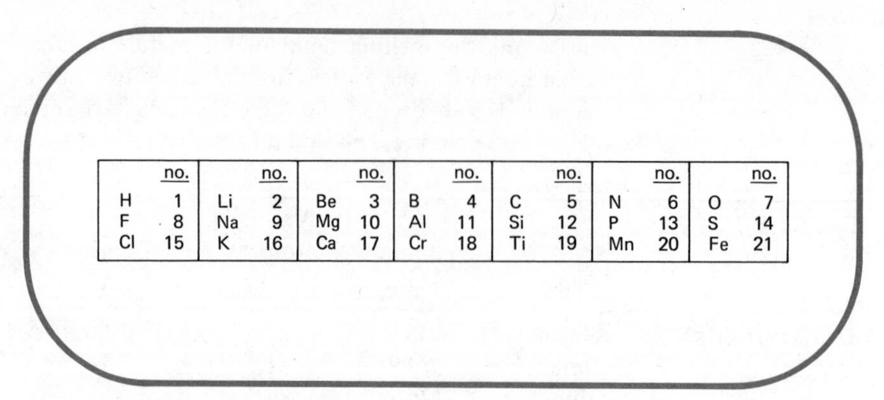

	no.		no.		no.		no.		no.		no.		no.
H	1	Li	2	Be	3	B	4	C	5	N	6	O	7
F	8	Na	9	Mg	10	Al	11	Si	12	P	13	S	14
Cl	15	K	16	Ca	17	Cr	18	Ti	19	Mn	20	Fe	21

Figure 8.1 Newlands' arrangement of elements in octaves.

Ideas ahead of their time are often unjustifiably criticized.

When Newlands first presented his classification of the elements at the Chemical Society of London, he was severely criticized and his paper was rejected as unsuitable for publication. However, in 1887 the Royal Society awarded him the Davy medal for his contribution to the development of the classification of the elements.

There are some valid criticisms of Newlands' octaves. Several elements do not fit well in his scheme. Their properties are not similar to the others in the previous octave. For example, chromium fits poorly under aluminum although both are metals. The metals manganese and iron do not resemble the nonmetals phosphorus and sulfur. Unfortunately, when Newlands listed the elements by increasing atomic weight, there were two things he did not consider. One, maybe not all the elements had been discovered, and two, maybe some of the atomic weights were inaccurate. Gallium, which resembles aluminum, and germanium, which resembles silicon, were not known in Newlands' time. Furthermore, titanium is now known to be of lower atomic weight than chromium.

The most important conclusion that can be derived from Newlands' work is that the regularity of similar properties of elements appears to be more than a coincidence. An ordering of the elements more extensive than the triads of Dobereiner was indicated.

8.3 MENDELEEV'S AND MEYER'S PERIODIC TABLES

In 1869, two chemists working independently proposed similar classification schemes for the elements. The Russian chemist Dimitri Mendeleev, who had no knowledge of Newlands' law of octaves, published a classification in which he arranged all of the elements in a table according to their atomic weight and physical and chemical properties (Figure 8.2). The German chemist Lothar Meyer independently proposed a similar table at the same time, but Mendeleev is mainly credited with the concept. Mendeleev did more than just arrange the elements according to the atomic weights available—he

Row	Group I	Group II	Group III	Group IV	Group V	Group VI	Group VII	Group VIII
1	H = 1							
2	Li = 7	Be = 9.4	B = 11	C = 12	N = 14	O = 16	F = 19	
3	Na = 23	Mg = 24	Al = 27.3	Si = 28	P = 31	S = 32	Cl = 35.5	
4	K = 39	Ca = 40	? = 44	Ti = 48	V = 51	Cr = 52	Mn = 55	Fe = 56, Co = 59, Ni = 59
5	Cu = 63	Zn = 63	? = 68	? = 72	As = 75	Se = 78	Br = 80	
6	Rb = 85	Sr = 87	Y = 88	Zr = 90	Nb = 94	Mo = 96	? = 100	Ru = 104, Rh = 104 Pd = 106
7	Ag = 108	Cd = 112	In = 113	Sn = 118	Sb = 122	Te = 125	I = 127	
8	Cs = 133	Ba = 137	Dy = 138	Ce = 140				
9								
10			Er = 178	La = 180	Ta = 182	W = 184		Os = 195, Ir = 197, Pt = 198
11	Au = 199	Hg = 200	Tl = 204	Pb = 207	Bi = 208			
12				Th = 231	U = 240			

Figure 8.2 Mendeleev's periodic table.

also contributed valuable experimental data on atomic weights. For example, he found the atomic weight of chromium was 52.0 amu not 43.4 amu. Although there was a place for chromium between calcium and titanium with its faulty value of 43.4 amu, the properties of chromium did not seem to fit with this placement. The faulty value was responsible for one of the improper placements in Newlands' octaves. Thus Mendeleev focused on chemical and physical properties to guide him to proper placement of the elements in the table.

As another example, the atomic weight of indium was taken as 77 amu before the time of Mendeleev's development of a periodic table of elements. This value required that it be placed between arsenic and selenium. However, the properties of these two elements indicated they belonged to the adjacent Groups V and VI, respectively. In addition, the properties of indium suggested that it was related to other elements in Group III of the table. A reevaluation of the molecular formulas of compounds of indium revealed that the atomic weight should be 114.8 amu.

Mendeleev left gaps in his table where there were no known elements whose properties could fit there. The next element of higher atomic weight than calcium known at the time of Mendeleev was titanium. If titanium were placed after calcium, it would fall under aluminum in Group III of the table. However, titanium forms compounds in a manner more similar to the elements of Group IV. Accordingly, Mendeleev placed titanium in Group IV, and left a space for an undiscovered element in Group III.

The classification of the elements attracted considerable attention; an order of nature had been unveiled, and the several gaps in the table suggested that new elements could be found to fill them. In addition, the location of these gaps enabled Mendeleev and other chemists to make predictions of the chemical and physical properties of the unknown elements. These predictions guided the search for new elements, because it is easier to find something that can be partially described as compared to something that is not even known to exist. The prediction of Mendeleev probably also stimulated scientists to prove that Mendeleev's gaps in the table were wrong. In 1871, Mendeleev suggested that elements similar to aluminum and silicon should exist. Gallium (similar to aluminum) was discovered in 1875 by Lecog de Boisbaudran, and germanium was discovered in 1886 by Clemens Alexander Winkler.

How would you describe a person who states that there is an unknown quantity to be discovered and then describes what it should be?

The predictions of chemical and physical properties by Mendeleev were based on deduction from the properties of those elements surrounding the space in his table. Thus the predicted properties of ekasilicon listed in Table 8.2 are seen to be close to those found for germaniun. *Eka* (Greek) means *first beyond* or *first after*.

TABLE 8.2

ekasilicon and germanium

	Ekasilicon	Germanium
atomic weight	72	72.32
specific gravity	5.5	5.47
color	dark grey	greyish white
formula of oxide	EsO_2	GeO_2
specific gravity of oxide	4.7	4.70
formula of chloride	$EsCl_4$	$GeCl_4$
specific gravity of chloride	1.9	1.887
boiling point of chloride	below 100°C	83°C

Mendeleev's table did include some exceptions to the relationship between properties and atomic mass. Today, in the modern periodic table shown in Figure 8.3, cobalt (58.93 amu) is placed before nickel (58.71 amu), and tellurium (127.60 amu) is placed before iodine (126.90 amu). In addition, argon, which was not discovered until many years later, is placed before potassium in spite of its larger atomic weight. More exceptions existed in Mendeleev's time, but this was because the atomic weights were incorrect. In spite of the three exceptions of nickel, iodine, and potassium, which are not placed according to atomic weight in the periodic table the arrangement of the other elements does allow the general statement of a **periodic law:**

The chemical and physical properties of the elements are periodic functions of their atomic weight.

Now it is known that the atomic weights are averages of the various isotopes of an element and different numbers of neutrons may be present and not affect the chemical properties of the element. There is little significance to the correlation between atomic weight and the properties of the elements. In 1913, H. G. J. Moseley observed that the proper correlation between the properties of the elements involves the atomic number. Thus the periodic law is now stated: **The properties of the elements are periodic functions of their atomic numbers.**

Have you ever encountered periodic behavior in your life?

Problem 8.1

Silicon and tin form chlorides of formulas $SiCl_4$ and $SnCl_4$ with boiling points 58°C and 114°C, respectively. Predict the boiling point of germanium chloride, $GeCl_4$.

Solution

The atomic weights of silicon, germanium, and tin are 28.06, 72.59, and 118.69, respectively. Because the atomic weight of germanium is intermediate between those of silicon and tin, it could be regarded as reasonable that the boiling point of $GeCl_4$ ought to lie between that of $SiCl_4$ and $SnCl_4$. The average of 58 and 114 is 86. The predicted boiling point is 86°C, which compares very favorably with the observed value of 83°C. Note that Mendeleev only predicted that the boiling point would be below 100°C. Even this great chemist must have had some scientific conservatism in him.

8.4 PERIODS AND GROUPS

The modern periodic table shown in Figure 8.3 represents a considerable refinement of Mendeleev's table of 63 elements. It displays the periodic chemical relationships that are now known to exist between the 106 known elements. **The periodic table is arranged in 7 horizontal rows called periods and 18 vertical columns called groups or families.** There are no gaps remaining to be filled with unknown elements. However, additional elements of atomic numbers larger than 106 are actively being sought in nuclear reactions. These man-made elements will find a place at the end of the periodic table.

G
R
O
U
PERIOD

The first period contains only the two elements hydrogen and helium of atomic numbers 1 and 2, respectively. Period 2 consists of eight elements from lithium through neon with atomic numbers 3 through 10. Similarly, period 3 also contains eight elements; these are sodium to argon of atomic numbers 11 through 18.

In the fourth period there are 18 elements, and for the first time a continuous progression of elements from left to right occurs. The elements potassium and calcium appear on the left of the period in positions below sodium and magnesium of the third period. Six elements, gallium of atomic

			transition elements															
period	IA	IIA	IIIB	IVB	VB	VIB	VIIB	VIII			IB	IIB	IIIA	IVA	VA	VIA	VIIA	0
1	1 H 1.008																1 H 1.008	2 He 4.003
2	3 Li 6.939	4 Be 9.012											5 B 10.81	6 C 12.011	7 N 14.007	8 O 15.999	9 F 19.00	10 Ne 20.183
3	11 Na 22.990	12 Mg 24.31											13 Al 26.98	14 Si 28.086	15 P 30.974	16 S 32.064	17 Cl 35.453	18 Ar 39.948
4	19 K 39.102	20 Ca 40.08	21 Sc 44.956	22 Ti 47.90	23 V 50.94	24 Cr 51.996	25 Mn 54.94	26 Fe 55.85	27 Co 58.93	28 Ni 58.71	29 Cu 63.54	30 Zn 65.37	31 Ga 69.72	32 Ge 72.59	33 As 74.92	34 Se 78.96	35 Br 79.904	36 Kr 83.80
5	37 Rb 85.47	38 Sr 87.62	39 Y 88.91	40 Zr 91.22	41 Nb 92.91	42 Mo 95.94	43 Tc 99	44 Ru 101.07	45 Rh 102.91	46 Pd 106.4	47 Ag 107.870	48 Cd 112.40	49 In 114.82	50 Sn 118.69	51 Sb 121.75	52 Te 127.60	53 I 126.90	54 Xe 131.30
6	55 Cs 132.91	56 Ba 137.34	57 La 138.91	72 Hf 178.49	73 Ta 180.95	74 W 183.85	75 Re 186.2	76 Os 190.2	77 Ir 192.2	78 Pt 195.09	79 Au 196.97	80 Hg 200.59	81 Tl 204.37	82 Pb 207.19	83 Bi 208.98	84 Po 210	85 At 210	86 Rn 222
7	87 Fr 223	88 Ra 226.05	89 Ac 227	104 Rf	105 Ha	106												

lanthanides	58 Ce 140.12	59 Pr 140.91	60 Nd 144.24	61 Pm 145	62 Sm 150.35	63 Eu 151.96	64 Gd 157.25	65 Tb 158.92	66 Dy 162.50	67 Ho 164.93	68 Er 167.26	69 Tm 168.93	70 Yb 173.04	71 Lu 174.97
actinides	90 Th 232.04	91 Pa 231	92 U 238.03	93 Np 237	94 Pu 242	95 Am 243	96 Cm 247	97 Bk 249	98 Cf 251	99 Es 254	100 Fm 253	101 Md 256	102 No 253	103 Lw 257

Figure 8.3 The modern periodic table.

number 31 through krypton of atomic number 36, appear beneath aluminum through argon of the third period. However, the 10 elements scandium through zinc have no counterparts in the previous period. **These 10 elements are called the transition elements, whereas those occurring to the right and left of the transition elements as well as those in the three previous periods are called the representative elements.** The fifth period contains 18 elements; there are 8 representative elements and 10 transition elements. The 10 transition elements yttrium through cadmium occupy positions directly beneath the transition elements of period 4.

ELEMENTS
Representative Transition

An interruption of the sequence of atomic numbers occurs in the sixth period. The 14 missing **elements of atomic numbers 58 through 71 are located outside the main body of the periodic table and are known as the lanthanides.** As shown in Figure 8.4, the inclusion of the lanthanides in the main body of the period table would make the table rather unwieldy. Although the groups of 10 transition elements are contained in the main body of the periodic table, the addition of 14 lanthanides would make the sixth period 32 elements long.

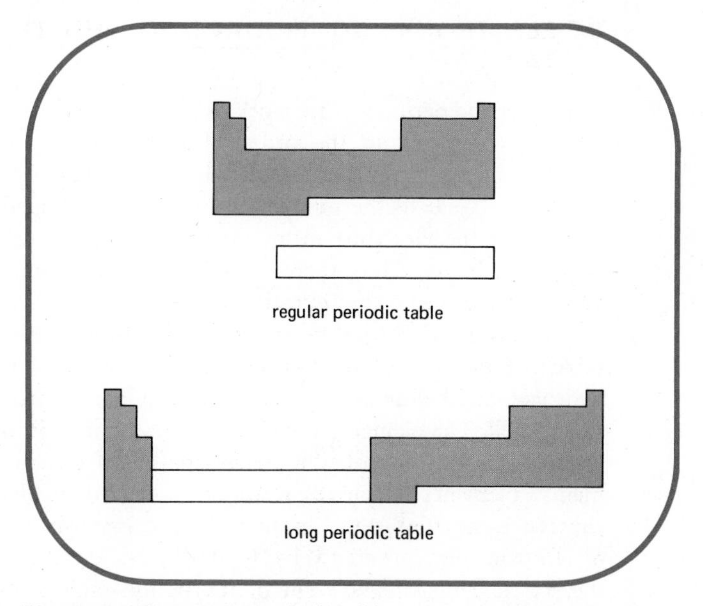

Figure 8.4 The regular periodic table and the long periodic table.

The seventh period contains only 2 representative elements and the start of a group of transition elements. In addition, **a series of 14 elements of atomic numbers 90 through 103 appear outside the main body of the periodic table and are known as the actinides.** The elements rutherfordium, hahnium, and an as yet unnamed element of atomic number 106 are the most recent of the man-made elements produced by nuclear reactions. The elements constitute the start of a fourth series of transition elements.

Most of the 18 vertical columns of the periodic table are labeled with a Roman numeral in combination with an A or a B. The exceptions are the extreme right column, which is labeled 0, and 3 interior columns, which are labeled as a unit as VIII without any letter. **With the exception of Group 0, the representative elements are designated by Roman numerals I through VII in combination with the letter A.** Group 0 consists of the unreactive gases which were discovered 30 years after the time of Mendeleev's proposed periodic table. **The transition metals are in IB through VIIB Groups and in 3 columns designated by VIII. The lanthanides and actinides are classed as Group IIIB.**

Each of the elements in a family or group possesses similar properties. Thus several of the families have been given names as a class. **Group IA elements, excluding hydrogen, are known as the alkali metals,** while **Group IIA is called the alkaline earth metals.** On the other side of the periodic table, **the Group VIIA is known as the halogens** and **Group 0 as the noble gases.**

8.5 ELECTRONIC CONFIGURATION AND THE PERIODIC TABLE

Why does the periodic table work? Consider that the periodic table is systematically arranged in the order of increasing atomic number such that elements of similar properties periodically reoccur. Furthermore, the atomic number is equal to the number of electrons contained in an atom, and we know that the electrons in an atom are systematically located in specific subenergy levels. Thus there must be a connection between the periodic table and electronic configuration of atoms.

The order of filling the energy levels of atoms is shown by the periodic table in Figure 8.5. In the first period there are only two elements. These are hydrogen and helium, which have one and two electrons, respectively, in the $1s$ subshell. The elements lithium (atomic number 3) and beryllium (4) have electrons in the $2s$ subshell and are separated from boron and five other elements. Elements boron (5) through neon (10) involve the addition of six successive electrons to the $2p$ subshell. The $3s$ subshell is filled by the addition of electrons eleven and twelve to produce sodium and magnesium. Again, the six elements on the right of the period table involve the filling of a p subshell ($3p$). Elements 19 and 20 are in the same region in which s subshells were filled in earlier periods, which in this case is the $4s$ subshell. Now for the first time a center region of the periodic table contains elements. The electronic configurations of the ten elements scandium (21) through zinc (30) are successively pictured by the addition of electrons to the $3d$ subshell. Then elements 31 through 36 appear in the same region where p subshells have been filled previously, In this case it is the $4p$ subshell, as it is in the fourth period. The remainder of the periodic table follows in an orderly fashion in the case of s and p subshells. The d subshells are always

Could you tell by inspection that an element of electronic configuration $1s^22s^22p^63s^23p^64s^23d^2$ is a transition element?

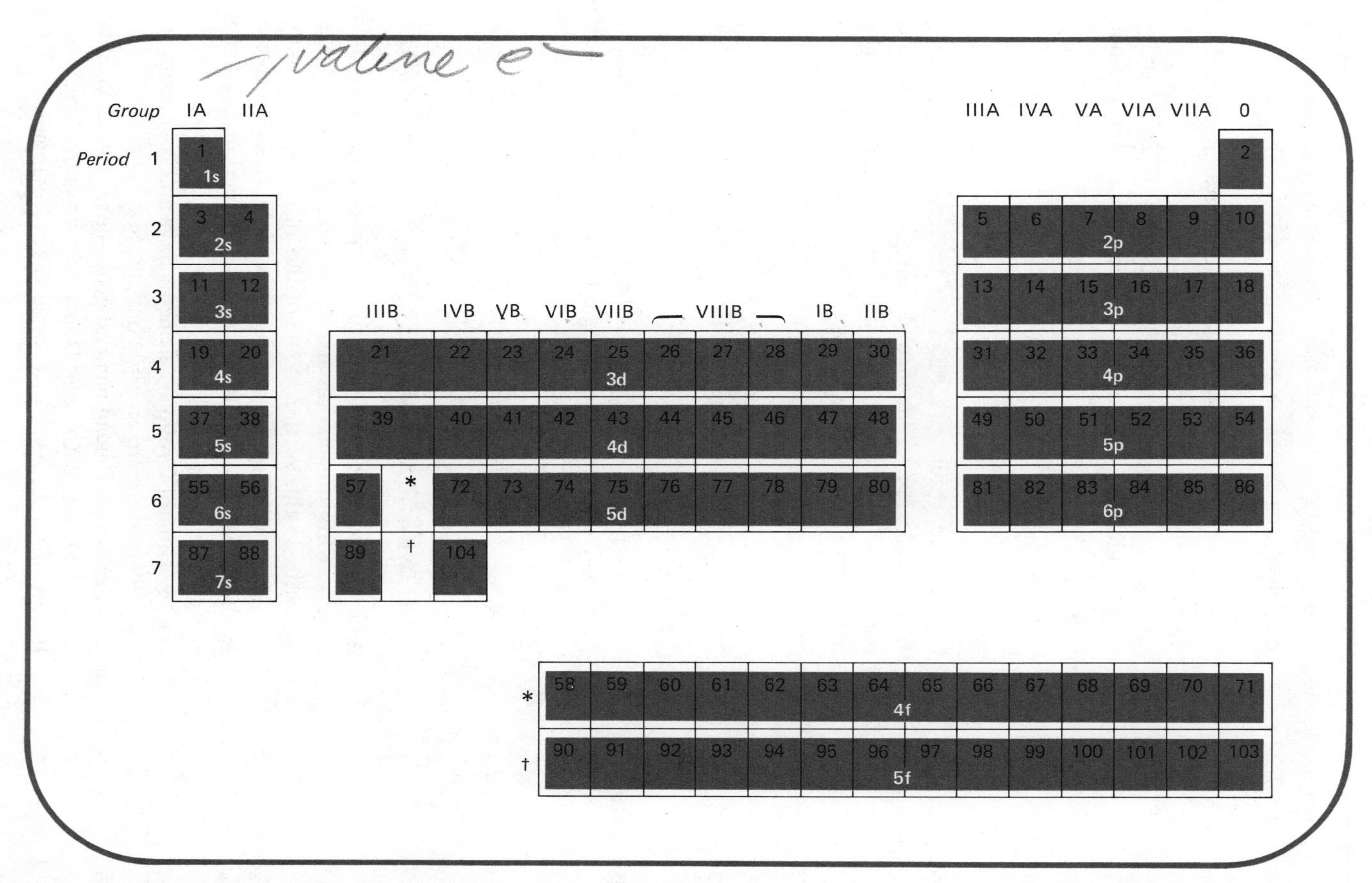

Figure 8.5 The relation between the periodic table and electronic configuration.

one behind the period number. Outside the general body of the periodic table are located two groups of fourteen elements that represent the filling of the $4f$ and $5f$ subshells.

Problem 8.2 What is the electronic configuration of silicon?

Solution The electronic configuration of silicon, atomic number 14, can be written down by examining the periodic table and going through all regions until silicon is reached. The configuration is $1s^2 2s^2 2p^6 3s^2 3p^2$, with the superscripts indicating the number of electrons in each subshell.

Problem 8.3 What is the electron configuration of zirconium (40)?

Solution In order to reach zirconium by reading through the periodic table, the $1s$, $2s$, $2p$, $3s$, $3p$, $4s$, $3d$, $4p$, and $5s$ regions must be traversed. Then one has to proceed by two steps into the $4d$ region. The entire electronic configuration is $1s^2 2s^2 2p^6 3s^2 3p^6 4s^2 3d^{10} 4p^6 5s^2 4d^2$.

Problem 8.4 How many electrons are contained in the highest energy subshell of rhenium (75)?

Solution Rhenium is in the sixth period of the periodic table. It is in the d region and is located five in from the $6s$ region. The d regions are one behind the period number. Therefore, the highest energy subshell is $5d$ and it contains five electrons.

8.6 GENERAL INFORMATION DERIVED FROM GROUPS

As indicated in the prologue to this chapter, the periodic table provides the chemist with information about the chemical and physical properties of the elements. In this section, some of this information is outlined.

The number of valence electrons of the Group A elements is given directly by the Group Roman numeral. Since the period number is the same as the energy level being filled, one knows the valence electrons configuration at a glance. Thus all Group IA elements have an s^1 configuration, while all Group VIIA elements have an s^2p^5 configuration.

The Group Roman numeral gives the maximum positive oxidation number for both A and B Groups. Hence the maximum positive oxidation number for tin of Group IVA is $+4$ and that of manganese of Group VIIB is $+7$.

What is the maximum positive oxidation number of Group VA elements? What is the maximum negative oxidation number of the same elements?

However, the most common oxidation number may be less than these maximum values. **The maximum negative oxidation number can be calculated by subtracting 8 from the Group Roman number.** For chlorine of Group VIIA the maximum negative oxidation number is $7 - 8 = -1$. The maximum negative oxidation number for sulfur is -2. Oxidation numbers of the elements in compounds may be at either of the two maximum values or values in between. For chlorine the oxidation numbers of ClO_4^-, ClO_3^-, ClO_2^-, ClO^-, and Cl^- are $+7$, $+5$, $+3$, $+1$, and -1, respectively.

Elements within groups have similar chemical properties due to their similarity of electronic configurations. Thus for the alkali metals, the s^1 configuration allows all members to lose one electron and form singly charged cations. Therefore, similar compounds such as LiCl, NaCl, KCl, RbCl, CsCl, and FrCl can be formed.

Within an A Group, the metallic properties increase with increasing atomic number and the nonmetallic properties decrease. For Group VA, the most metallic element is bismuth. Antimony and arsenic have borderline

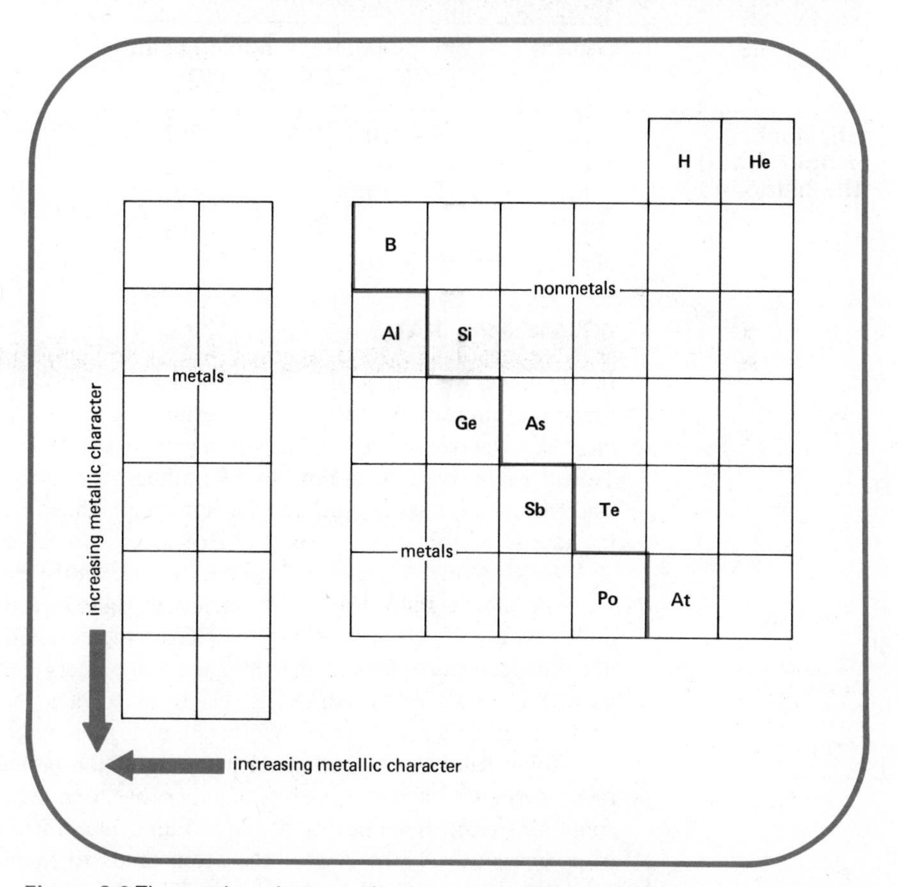

Figure 8.6 The metals and nonmetals of the representative elements.

properties and are semimetals, whereas phosphorus and nitrogen are nonmetals. Metallic properties within periods increase toward the left and nonmetallic properties increase toward the right. Thus within a group, the dividing line between metal and nonmetals occurs at higher atomic numbers for groups on the right than groups on the left. For Group IVA, lead and tin are metals and the semimetals are germanium and silicon. The boundary, then, for Group IVA occurs between period 3 and period 4, whereas for Group VA the boundary is between period 4 and period 5. The stairstep line shown in Figure 8.6 separates the metals and nonmetals. Elements to the right of the line are nonmetals and elements to the left of the line are metals. The semimetals lie immediately to either side of the line.

Many of the semimetals located along the stairstep line in the periodic table are used in the manufacture of transistors.

Within a group there is a uniform change in the physical and chemical properties of the elements and their derived compounds with similar structures. The properties of O_2 and S_8 cannot be compared, nor can N_2 and P_4 because their molecular formulas differ. However, for F_2, Cl_2, Br_2, and I_2, the melting points and boiling points (Table 8.3) show a uniform change.

TABLE 8.3
physical properties of the halogens

Element	Melting point (°C)	Boiling point (°C)
F_2	−220	−187
Cl_2	−102	−35
Br_2	−7	+59
I_2	+113	+183

8.7 ATOMIC RADII

In Chapter 6 we learned that the electrons within an atom do not all possess the same amount of energy. Furthermore, as the energy of the electron increases, the size of the orbital (volume of space) which it occupies increases. Thus the size of an atom reflects the number of electrons and their energies. The net overall shape of an atom is spherical. Thus its size may be given by the radius of the sphere. The radii of atoms are given in angstrom units ($1\ Å = 10^{-8}$ cm) in Figure 8.7 for the representative elements.

The atomic radii decrease from left to right in a period of the periodic table. Within a period, the nuclear charge increases in the same direction, but the electrons are located within the same energy level. This increase in nuclear charge tends to draw the electrons toward the nucleus; the probability of finding an electron then becomes higher within a smaller volume.

The atomic radii of atoms are an indication of the balance between electron-nucleus attraction and the energy of the electrons.

Going from top to bottom in a family of the periodic table, the atomic radii increase. Each successive member has one additional energy level containing electrons. Because the size of an orbital increases with the number of its energy level, the size of the atom tends to increase. The increase in orbital size is partially balanced by the increase in the nuclear charge.

1 H **0.37**							2 He
3 Li **1.23**	4 Be **0.89**	5 B **0.80**	6 C **0.77**	7 N **0.74**	8 O 0.74	9 F **0.72**	10 Ne
11 Na **1.57**	12 Mg **1.36**	13 Al **1.25**	14 Si **1.17**	15 P **1.10**	16 S **1.04**	17 Cl **0.99**	18 Ar
19 K **2.03**	20 Ca **1.74**	31 Ga **1.25**	32 Ge **1.22**	33 As **1.21**	34 Se **1.17**	35 Br **1.14**	36 Kr
37 Rb **2.16**	38 Sr **1.91**	49 In **1.50**	50 Sn **1.41**	51 Sb **1.41**	52 Te **1.37**	53 I **1.33**	54 Xe
55 Cs **2.35**	56 Ba **1.98**	81 Tl **1.55**	82 Pb **1.54**	83 Bi **1.52**	84 Po **1.53**	85 At	86 Rn
87 Fr	88 Ra						

increasing radii (down) / increasing radii (left)

Figure 8.7 Atomic radii of the representative elements.

8.8 IONIZATION POTENTIAL

If sufficient energy is added to an atom of a gas, an electron may be removed to form a positive ion, or cation. **The process of electron removal is called *ionization*, and the amount of energy required to accomplish the removal is called the *ionization potential*. Ionization potentials are expressed in electron volts per atom.** The ionization potential of gaseous lithium, Li(g), is 5.4 electron volts (eV) per atom:

$$\text{Li(g)} \longrightarrow \text{Li}^+\text{(g)} + 1e^-$$

The lithium cation has a superscript plus sign to indicate its unit positive charge. Because the ionization potential is a measure of the energy required to remove an electron from an element, it indicates the energy of the electron in the element. Higher energy electrons require less additional energy (lower ionization potential) to dissociate them from the element; conversely, low energy electrons are closely bound to the atom and require more energy for ionization. The ionization potentials of the representative elements are given in Figure 8.8.

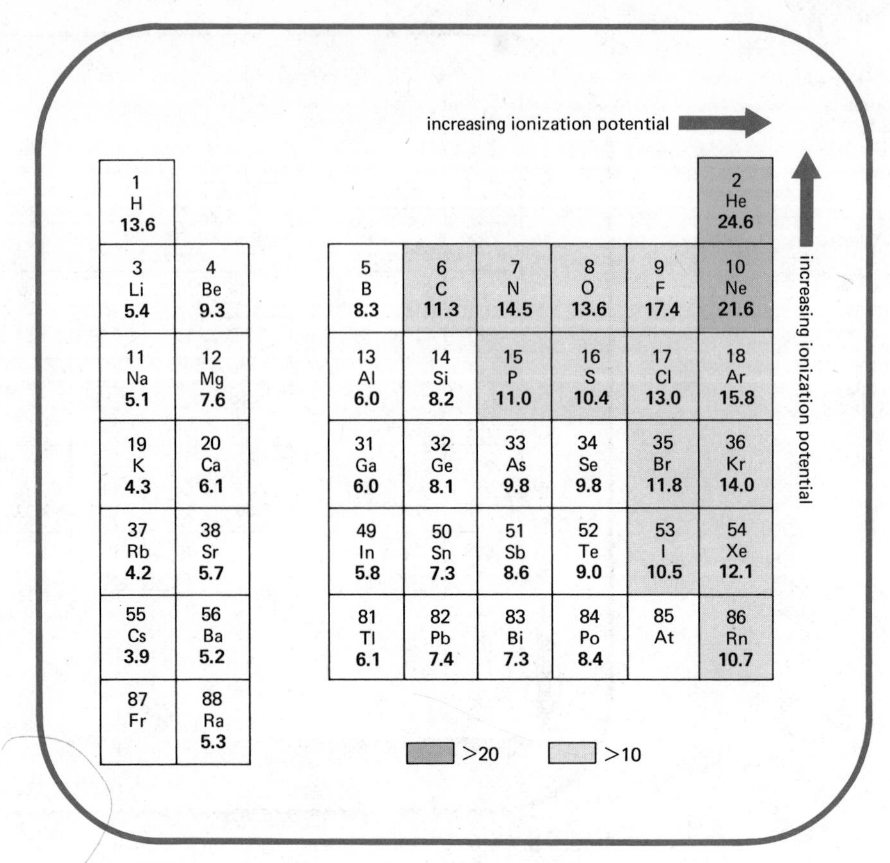

Figure 8.8 The ionization potentials of the representative elements.

The ionization potentials of the representative elements increase from left to right within a period. This trend is due to the increase in the nuclear charge which causes an increase in the electrostatic attraction for the electron. The ionization potential for the fluorine atom is 17.4 eV/atom, a value much higher than for lithium at the start of the period. Metals have low ionization potentials; nonmetals have higher ionization potentials.

Which area of the periodic table contains the elements with the lowest ionization potentials?

Within a group there is a trend of decreasing ionization potential that is in agreement with the increasing metallic character of elements of higher atomic number. Thus the ionization potential of cesium is 3.9 eV/atom, a value less than the 5.4 eV/atom of lithium.

8.9 ELECTRONEGATIVITY

Electronegativity is a measure of the ability of an atom to attract electrons in the presence of a second atom. The electronegativity values of the representative elements are given in Figure 8.9. The larger the electronegativity value,

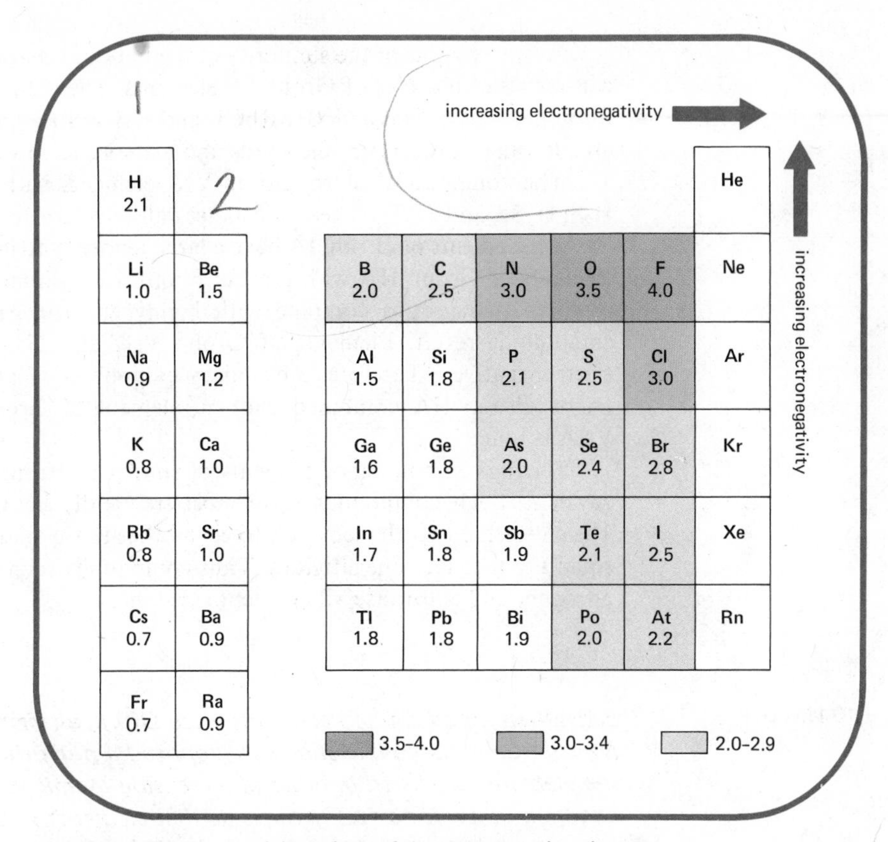

Figure 8.9 Electronegativity values of the representative elements.

the greater the tendency to attract electrons. Within a period, the electronegativities increase from left to right. Within a group, the electronegativities decrease with increasing atomic weight. Thus nonmetals have higher electronegativity values than metals. **Metals are often referred to as being electropositive.**

The most electronegative elements are located in the upper right region of the periodic table.

The electronegativities of atoms determine the types of chemical bonds which form between them. If the electronegativities are very different, electron transfer will occur and an ionic bond result. As the electronegativities become more similar, electrons are shared in a polar covalent bond. The negative end of the bond dipole is the more electronegative atom and the positive end is the more electropositive.

8.10 FORMULAS AND BONDING INFORMATION FROM THE PERIODIC TABLE

The outer electrons of all the atoms within a group are in the same type of subshell. Therefore, the oxidation numbers of these elements are the same and they bond similarly to other elements.

As an example of the similarity of the bonding characteristics of a group, consider the chlorides of Group IA elements. They all have similar chemical formulas—LiCl, NaCl, KCl, RbCl, and CsCl. In addition, the substances are all ionic white crystalline solids and are soluble in water.

The compounds of the Group VIA elements with hydrogen are H_2O, H_2S, H_2Se, and H_2Te. These substances all have covalent bonds.

All elements of Group IA have a large tendency to lose electrons as do all elements of Group IIA with the exception of beryllium. When the elements of these two groups combine with highly electronegative elements, ionic compounds result. Elements of Group VIA and Group VIIA are highly electronegative. Therefore, any binary compound of an element of Group IA or Group IIA combined with an element of Group VIA and Group VIIA is ionic.

Binary compounds of elements of similar electronegativities tend to be covalent. Such compounds form most frequently between two nonmetals. The closer the two elements are to each other in the periodic table, the more equal are the electronegativities. Thus compounds such as oxides of carbon, nitrogen, and sulfur are all covalent substances.

SUMMARY

Early attempts to classify elements were ***Dobereiner's triads*** *and* ***Newlands' octaves****. Mendeleev proposed a* ***periodic table*** *in which the elements are listed in order of increasing atomic weight and are arranged in* ***periods*** *and* ***groups*** *so that elements of similar chemical properties are arranged to reflect their similarities.*

The periodic table reflects the systematic arrangement of electrons within atoms. Within the periodic table, areas are identified with specific subenergy levels in which the highest energy electrons are located.

The periodic table yields information about the maximum and minimum oxidation numbers of the elements. In addition, the ordering of elements in terms of ***atomic radii, ionization potential****, and* ***electronegativity*** *may be obtained from the periodic table.*

QUESTIONS AND PROBLEMS

Definitions

1. Define each of the following terms.

(a) triad **(b)** octave
(c) periodic **(d)** representative element
(e) transition element **(f)** group
(g) period **(h)** lanthanide
(i) actinide **(j)** alkali metal
(k) noble gas **(l)** alkaline earth metal
(m) halogen

2. Indicate the difference between each of the following.
(a) a triad and an octave
(b) a group and a period
(c) a representative and a transition element
(d) alkali metal and alkaline earth metal
(e) lanthanide and actinide

The periodic table

3. How are the triads related in the modern periodic table?

4. Why did Dobereiner only discover triad relationships when there are more elements within a family of the modern periodic table?

5. What relationship is there between Newlands' octaves and the modern periodic table?

6. What did Mendeleev do to develop his periodic table that may be considered a bold proposition?

7. Why are A and B used in the periodic table?

8. How does the modern periodic law differ from Mendeleev's periodic law?

9. Indicate the location of each of the following elements by period number and group designation.
(a) ${}_{7}N$ **(b)** ${}_{20}Ca$
(c) ${}_{48}Cd$ **(d)** ${}_{35}Br$
(e) ${}_{3}Li$ **(f)** ${}_{13}Al$
(g) ${}_{50}Sn$ **(h)** ${}_{16}S$
(i) ${}_{54}Xe$ **(j)** ${}_{26}Fe$
(k) ${}_{24}Cr$ **(l)** ${}_{29}Cu$

10. What elements occupy the indicated locations in the periodic table?
(a) period 2, Group IA **(b)** period 3, Group IIIA
(c) period 6, Group IIA **(d)** period 4, Group IVA
(e) period 5, Group VII **(f)** period 1, Group 0
(g) period 4, Group VIIB **(h)** period 5, Group IB
(i) period 6, Group VIB **(j)** period 3, Group IVB

Metals and nonmetals

11. Using the periodic table, classify each of the following as a metal or a nonmetal.
(a) carbon **(b)** calcium
(c) mercury **(d)** gallium
(e) iodine **(f)** phosphorus
(g) strontium **(h)** cesium
(i) selenium **(j)** vanadium

12. Using the periodic table, indicate which member of the following pairs is the most metallic.
(a) magnesium and silicon (b) germanium and bromine
(c) sulfur and selenium (d) silicon and tin
(e) phosphorus and antimony (f) oxygen and tellurium
(g) indium and tin (h) sodium and cesium

Oxidation numbers

13. What is the maximum positive oxidation number for each of the following metals?
(a) $_{24}Cr$ (b) $_{38}Sr$
(c) $_{82}Pb$ (d) $_{49}In$
(e) $_{80}Hg$ (f) $_{37}Rb$
(g) $_{25}Mn$ (h) $_{40}Zr$

14. Determine both the maximum positive oxidation number and maximum negative oxidation number of the following nonmetals.
(a) $_{16}S$ (b) $_{15}P$
(c) $_{53}I$ (d) $_{6}C$
(e) $_{5}B$ (f) $_{33}As$

Electron configuration

15. Indicate the number of valence electrons for each of the following elements.
(a) silicon (b) selenium
(c) phosphorus (d) krypton
(e) francium (f) arsenic
(g) calcium (h) aluminum

16. In what period and group do the elements having the following electronic configuration belong?
(a) $1s^2 2s^2 2p^3$
(b) $1s^2 2s^2 2p^6 3s^1$
(c) $1s^2 2s^2 2p^6 3s^2 3p^5$
(d) $1s^2 2s^2 2p^6 3s^2 3p^6 4s^2 3d^5$

Predictions of physical properties

17. In each of the following series, estimate the missing property of the element or compound.

(a)

ELEMENT	RADIUS (Å)
F	0.72
Cl	?
Br	1.14

(b)

ELEMENT	DENSITY (g/cm^3)
Ca	1.54
Sr	2.60
Ba	?

(c)

ELEMENT	MELTING POINT (°C)
Na	?
K	63
Rb	39

(d)

COMPOUND	BOILING POINT (°C)
$SiCl_4$	58
$GeCl_4$	83
$SnCl_4$	?

(e)

COMPOUND	MELTING POINT (°C)
H_2S	−86
H_2Se	?
H_2Te	−49

(f)

COMPOUND	DENSITY (g/cm^3)
B_2S_3	1.6
Al_2S_3	2.4
Ga_2S_3	?

Prediction of formulas

18. Using the periodic table, write the formulas of each of the following compounds.

(a) rubidium nitrate **(b)** barium sulfate
(c) aluminum sulfide **(d)** calcium selenide
(e) sodium bromate **(f)** magnesium selenate

Prediction of bonding

19. Classify each of the following as ionic or covalent compounds.

(a) CsI **(b)** SO_2
(c) TeO_2 **(d)** OCl_2
(e) $SiCl_4$ **(f)** BaF_2
(g) CS_2 **(h)** NaBr
(i) MgO **(j)** CO_2

Prediction of man-made elements

20. Elements of atomic numbers 93 to 106 have been made by nuclear reactions. Additional elements will probably continue to be produced. Predict the following features of elements 112, 115, 118, and 120.

(a) What group will the element be located in?
(b) How many valence electrons will the element have?
(c) Will the element be a metal or a nonmetal?

(d) What element will it resemble?
(e) What will be its maximum positive oxidation number?

Prediction of atomic radii

21. Indicate which member of each of the following pairs of elements has the largest radius.
(a) F or Cl (b) Si or S
(c) Li or K (d) C or F
(e) Se or Br (f) Ge or Pb

Prediction of electronegativity

22. Indicate which member of each of the following pairs of elements has the highest electronegativity.
(a) O or F (b) S or Se
(c) Li or K (d) C or N
(e) Cl or Br (f) P or S

9 CHEMICAL NAMES

LEARNING OBJECTIVES

When you finish this chapter you should be able to:

(a) explain the advantages of systematic names over common names.

(b) express the name of a binary compound given its formula; write the formula of a binary compound given its name.

(c) illustrate the use of *-ous* and *-ic* suffixes for metals of variable oxidation numbers and contrast their use with the Stock system.

(d) construct the name of compounds containing only nonmetals.

(e) review the names of complex ions and name ternary compounds.

(f) name acids given the formula and write the formula given the name.

(g) name bases given their formulas and write formulas of bases given their names.

(h) differentiate between the following prefixes and suffixes and employ them in naming compounds.

per- and hypo-	**-ate and -ite**
-ous and -ic	

(i) match the following terms with their corresponding definitions.

acid	**IUPAC**
acid salt	**mixed salt**
base	**salt**
basic salt	**Stock system**
binary compound	**ternary compound**

When one travels to a foreign country, a knowledge of the language of that country is necessary to really understand its people and how they behave and feel. Without any knowledge of the language and in the absence of a willing translator, one would be at a complete loss to understand what was occurring. In a sense we are surrounded by the foreign languages of the various sciences even in our native land. The names of chemicals reported in the news as having spilled from a tank car of a derailed train or dumped into a local stream by an industry have little meaning to many people. There are chemical names on containers of food and on products such as mouthwash, mosquito repellent, soap, deodorant, oven cleaner, and drain cleaner. As greater and stricter labeling of products is required by our government, chemical names will become a more significant part of the language of our modern society. Unless we are to become foreigners in our own country, our education must include some fundamentals of the chemical language.

In this chapter the first step is taken toward understanding the language of chemistry. The step will be a small one; it will be analogous to learning the simplest phrases of a foreign language, such as, "What time is it?" "How much does it cost?" "Where is the rest room?" "Please," "Hello," and "You're welcome." Clearly a crash program will not allow you to understand the subtleties of a foreign language. Similarly, this chapter cannot provide you with the ability to recognize and understand the names of all compounds. This chapter, then, is devoted to the naming of simple inorganic compounds.

9.1 COMMON OR TRIVIAL NAMES

Arbitrary names of compounds unrelated to chemical composition are called common or trivial names. Any chemical could be assigned an arbitrary name which all of us would then have to relate to that substance. Water is such a name; it refers to a widely distributed and well-recognized substance. If every known substance were assigned such an arbitrary name, then we would have to learn an immense number of one-to-one relationships between name and substance. Since the number of known substances is in the millions, the human mind could not cope with such a system.

To give you an example of the difficulties of dealing with common names, consider the millions of people living in your city or state. Each has a somewhat arbitrary name. If you are introduced to a large number of people over a period of time, you might not be able to recall the name of a particular individual at a later date; or if you heard the name, you might not be able to put a face to the name. However, if someone were to describe the features of that person, such as hair color, height, shape, and so on, you might recognize the person readily. Thus if everyone had names based specifically on their constituent characteristics, then we might be able to relate the identity of the person to the name more easily and vice versa.

While common chemical names have distinct limitations, they are prevalent in chemistry. Many substances were named very early in history before their composition was understood. Water and ammonia are such examples. Even after the composition of such substances was established, the common names lingered on as part of the language.

Common names are often used by industry and by nonchemical specialists because the exact systematic chemical name is too long or technical for convenient usage. Such an example is the *hypo* used by photographers which is really sodium thiosulfate. Even the modern chemist often resorts to similar invented names for everyday usage because the systematic names may be too long to say easily. For example, the compound dichlorodiphenyltrichloroethane is known by an abbreviation DDT. A list of some substances in common use along with their common names is given in Table 9.1.

TABLE 9.1
common substances and names

Name	Formula
water	H_2O
ammonia	NH_3
salt	$NaCl$
lye	$NaOH$
saltpeter	$NaNO_3$
baking soda	$NaHCO_3$
hypo	$Na_2S_2O_3$
muriatic acid	HCl
sal ammoniac	NH_4Cl
carbon tet	CCl_4
lime	CaO
marble	$CaCO_3$
gypsum	$CaSO_4 \cdot 2H_2O$
plaster of paris	$CaSO_4 \cdot \frac{1}{2}H_2O$
galena	PbS
litharge	PbO
milk of magnesia	$Mg(OH)_2$
epsom salts	$MgSO_4 \cdot 7H_2O$
wood alcohol	CH_3OH
grain alcohol	C_2H_5OH
cane sugar	$C_{12}H_{22}O_{11}$

9.2 SYSTEMATIC NAMES

The common name is not satisfactory to the chemist for more reasons than the burden of using and learning many names. A chemist wants to have a

systematic name which indicates the composition of the substance. With such a systematic name it is possible to glean much desirable information from merely glancing at the words even if the name is encountered for the first time.

IUPAC

The Commission on the Nomenclature of Inorganic Chemistry of the International Union of Pure and Applied Chemistry (IUPAC) first met in 1921 and continues to meet to constantly update its work. What is its work? The name of the organization tells you. Nomenclature means naming. Therefore, its name indicates that its work is to devise names of inorganic chemicals to be used internationally.

Why is chloride sodium an incorrect name?

The systematic names of inorganic compounds are arranged so that every substance can be named from its formula and each formula has a name specific for that formula. **In the systematic name, the more positive portion is named first.** This may be a metal (such as sodium), a positive polyatomic ion (such as ammonium, NH_4^+), or the least electronegative of the nonmetals present in the substance. **The more negative portion is then named second.** The negative portion may be an electronegative element (such as chlorine) or a negatively charged ion (such as nitrate, NO_3^-). It is not necessary to specify directly in the name whether the substance is ionic or covalent. This information can be derived from the knowledge of the constituent elements in the compound and their electronegativities.

In order to present the rules of nomenclature in an orderly fashion, the compounds will be divided into binary (two elements) compounds, ternary (three elements) compounds, and acids, bases, and salts. Within some of these categories, even further subdivisions will be discussed.

9.3 BINARY COMPOUNDS CONTAINING METALS OF FIXED OXIDATION NUMBER

A number of metals exist in compounds with only one oxidation number. For example, sodium has an oxidation number of +1 and calcium has an oxidation number of +2. Since this information becomes general knowledge as one studies chemistry, there is no need to indicate the oxidation number in the name. In naming the compound, **the metal is named first and the name of the nonmetal with an ending-*ide* follows.** Examples are listed in Table 9.2. Note that compounds may contain more than one atom of the same element, but as long as they contain only two elements, they are considered binary. The number of atoms of a given element is not indicated in the name.

The majority of the metals having fixed oxidation numbers are members of Group IA and Group IIA, and the compounds which they form with nonmetals are ionic. The Group IA elements are electropositive and have a +1 oxidation number while those elements of Group IIA have a +2 oxidation number. Binary compounds of hydrogen with metals contain hydrogen as the hydride ion (H^-) and are ionic compounds.

It is a simple matter to name binary compounds containing a metal of a single oxidation number and a nonmetal. The metal is written first and then

TABLE 9.2

binary compounds of fixed oxidation numbers

Formula	Name
$NaCl$	sodium chloride
KCl	potassium chloride
KBr	potassium bromide
$AgBr$	silver bromide
Ag_2O	silver oxide
CaO	calcium oxide
CaS	calcium sulfide
CaF_2	calcium fluoride
NaF	sodium fluoride
Na_2O	sodium oxide
Na_3P	sodium phosphide
Mg_3P_2	magnesium phosphide
Al_2S_3	aluminum sulfide
NaH	sodium hydride
KH	potassium hydride

the nonmetal with the *-ide* ending. However, writing the formula from a name is somewhat more demanding as some learned and stored knowledge is necessary. You must know the oxidation number of the metal cations and the nonmetal anions. Only then can the requisite number of each ion be coupled together to produce a neutral compound. Thus calcium, which exists as Ca^{2+}, requires two bromide ions, which exist as Br^-. Calcium bromide must then be $CaBr_2$. Examples are listed in Table 9.3. In each case you should verify that the formula is correct. Note that **the number of cations multiplied by the oxidation number is equal in magnitude but opposite in sign to the product of the number of anions multiplied by its oxidation number.** For $CaBr_2$, the calculation is:

$$(\text{number of } Ca^{2+})(\text{charge of } Ca^{2+}) = -[(\text{number of } Br^-)(\text{charge of } Br^-)]$$
$$1 \times 2 = -[2 \times (-1)]$$
$$2 = 2$$

TABLE 9.3

formulas derived from names

Name	Cation	Anion	Formula
potassium fluoride	K^+	F^-	KF
calcium bromide	Ca^{2+}	Br^-	$CaBr_2$
barium oxide	Ba^{2+}	O^{2-}	BaO
aluminum chloride	Al^{3+}	Cl^-	$AlCl_3$
aluminum oxide	Al^{3+}	O^{2-}	Al_2O_3
sodium sulfide	Na^+	S^{2-}	Na_2S
sodium nitride	Na^+	N^{3-}	Na_3N
magnesium nitride	Mg^{2+}	N^{3-}	Mg_3N_2

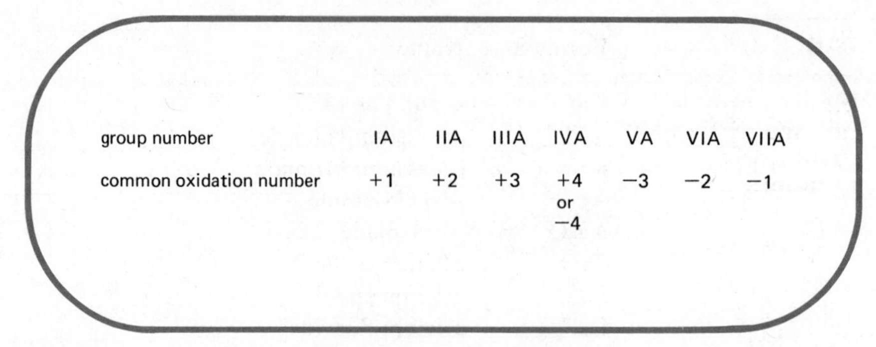

group number	IA	IIA	IIIA	IVA	VA	VIA	VIIA
common oxidation number	+1	+2	+3	+4 or −4	−3	−2	−1

Figure 9.1 Common oxidation numbers and the periodic table.

The oxidation numbers of the metals of the representative elements are predictable from the group number given in the periodic table (Figure 9.1). Remember that the group number of the representative elements indicates the number of electrons in the valence shell. For the metals, the loss of these electrons results then in an ion whose positive charge corresponds to the group number. The nonmetals are electronegative and tend to gain electrons to form anions. The charge of the anion is derived by subtracting 8 from the group number.

9.4 BINARY COMPOUNDS CONTAINING METALS WITH VARIABLE OXIDATION NUMBERS

There are many more metals with variable oxidation numbers than there are with fixed oxidation numbers. In compounds containing such metals, it is vital that the oxidation number be stated because it dictates the number of anions with which the metal combines. Iron may exist as Fe^{2+} and Fe^{3+} and thus can be combined with chlorine as $FeCl_2$ and $FeCl_3$, while titanium may exist with oxidation numbers of +3 or +4 in $TiCl_3$ and $TiCl_4$, respectively.

Two systems are in concurrent use to name compounds in this category. In the older system, metals that had two oxidation numbers were designated by modifying a stem of the metal name with different suffixes. The suffix *-ous* designates the lower oxidation state and *-ic* the higher oxidation state. Thus the chloride of titanium with the formula $TiCl_3$ is titanous chloride, while $TiCl_4$ is titanic chloride. Some metals are named using a stem derived from the same word for the element which was used to form the chemical symbol. Thus cuprum (Cu), ferrum (Fe), and stannum (Sn) are used to give the stems cupr, ferr, and stann, respectively. The compound $FeCl_2$ is ferrous chloride while $FeCl_3$ is ferric chloride. Examples of the usage of the stem and the *-ous* and *-ic* suffixes are given in Table 9.4.

Why is $SnCl_4$ stannic chloride and $FeCl_3$ ferric chloride when the substances do not contain the same number of chlorine atoms?

The IUPAC recommends the use of the Stock system of nomenclature in which the oxidation state of the metal is indicated by a Roman numeral in parentheses immediately following the name of the metal. The two chlorides of

TABLE 9.4 ions of variable oxidation numbers

Ion	Old name	Stock name
Cu^{+} or $(Cu_2)^{2+}$	cuprous	copper(I)
Cu^{2+}	cupric	copper(II)
Fe^{2+}	ferrous	iron(II)
Fe^{3+}	ferric	iron(III)
Hg^{+} or $(Hg_2)^{2+}$	mercurous	mercury(I)
Hg^{2+}	mercuric	mercury(II)
Sn^{2+}	stannous	tin(II)
Sn^{4+}	stannic	tin(IV)

iron are iron(II) chloride and iron(III) chloride. Note that the Roman numeral does not indicate the number of units of the nonmetal combined with the metal but only the oxidation state of the metal. The oxides of iron, FeO and Fe_2O_3, are iron(II) oxide and iron(III) oxide, respectively. Examples of both the older name and the Stock name are given in Table 9.4. Note that the endings *-ous* and *-ic* do not directly indicate to the reader the oxidation state without prior learning of the values. The term *-ous* means $+1$ for copper and mercury but $+2$ for iron and tin. The term *-ic* means $+2$ for copper, $+3$ for iron, and $+4$ for tin. The Stock system gives you this information directly.

The relationships between chemical names and formulas are listed in Table 9.5.

TABLE 9.5 names of compounds containing metals of variable oxidation numbers

Formula	Older name	Stock name
FeO	ferrous oxide	iron(II) oxide
Fe_2O_3	ferric oxide	iron(III) oxide
$SnCl_2$	stannous chloride	tin(II) chloride
$SnCl_4$	stannic chloride	tin(IV) chloride
$HgCl_2$	mercuric chloride	mercury(II) chloride
Cu_2S	cuprous sulfide	copper(I) sulfide
PbO_2	plumbic oxide	lead(IV) oxide
Cu_3P_2	cupric phosphide	copper(II) phosphide

9.5 BINARY COMPOUNDS OF NONMETALS

With only few exceptions, the names of binary compounds containing only two nonmetals consist of the names of the two elements with the second listed element having an *-ide* ending. Compounds such as water and ammonia are the most obvious exceptions. The most electropositive element is written first and is treated in the same manner as the metal in earlier sections. However, it should be noted that in binary compounds of two nonmetals, there are no positive or negative ions but rather the compounds are covalent substances. The sequence of preferred first names is B > Si > C > P > N > H > S > I > Br > Cl > O > F. The elements to the left are named first when combined with an element to the right in the series. For example, in a compound of sulfur and oxygen, sulfur would precede oxygen in the name.

mono- 1
di- 2
tri- 3
tetra- 4
penta- 5

The number of atoms of each element in a molecule of the compound is indicated by a Greek prefix preceding that element. The prefixes are *mono-*, *di-*, *tri-*, *tetra-*, *penta-*, *hexa-*, *hepta*, *octa-*, *nona-*, and *deca-* for the numbers one through ten. The prefix *mono-* is rarely used except in cases where an ambiguous name would result. Examples of these names are given in Table 9.6.

TABLE 9.6

binary compounds of nonmetals

Name	Formula
carbon monoxide	CO
carbon dioxide	CO_2
sulfur dioxide	SO_2
sulfur trioxide	SO_3
carbon tetrachloride	CCl_4
phosphorus pentachloride	PCl_5
nitrogen oxide	NO
dinitrogen oxide	N_2O
dinitrogen trioxide	N_2O_3
dinitrogen pentoxide	N_2O_5

9.6 TERNARY COMPOUNDS

Ternary compounds contain three elements. This occurs most frequently when a metal or hydrogen with an oxidation number of +1 is combined with a negatively charged complex ion containing two elements. Many of these complex ions contain oxygen and are listed in Table 9.7. In order to name these ternary compounds it is necessary to know the names of the complex ions.

-ate
-ite

The complex ions containing oxygen commonly have the suffixes *-ate* and *-ite*. While the suffix does not indicate the absolute number of oxygen atoms in the complex ion, **the ion with the *-ate* suffix contains more oxygen than that with the *-ite* suffix.** Thus while nitrate and nitrite are NO_3^- and NO_2^-, respectively, sulfate and sulfite are SO_4^{2-} and SO_3^{2-}, respectively. It is for this reason that you should learn the symbols corresponding to the names of the complex ions and vice versa.

There are several combinations of two elements which give rise to more than two complex ions. Therefore, more endings than *-ate* and *-ite* are required. For example, Cl and O actually form four different complex ions. In Table 9.7 are listed perchlorate (ClO_4^-), chlorate (ClO_3^-), chlorite (ClO_2^-).

TABLE 9.7 common complex ions

Formula	Name
NO_3^-	nitrate
NO_2^-	nitrite
SO_4^{2-}	sulfate
SO_3^{2-}	sulfite
HSO_4^-	bisulfate
HSO_3^-	bisulfite
PO_4^{3-}	phosphate
CO_3^{2-}	carbonate
HCO_3^-	bicarbonate
OH^-	hydroxide
CN^-	cyanide
CrO_4^{2-}	chromate
MnO_4^-	permanganate
ClO_4^-	perchlorate
ClO_3^-	chlorate
ClO_2^-	chlorite
ClO^-	hypochlorite
NH_4^+	ammonium
PH_4^+	phosphonium

and hypochlorite (ClO^-). **The prefix *per-* is used to mean over.** Thus perchlorate has one oxygen more than the number of oxygen in chlorate. **The prefix *hypo-* means under.** Thus hypochlorite is related to chlorite by having one less oxygen atom. These prefixes are also used with other similar ions such as hypoiodite for IO^- and perbromate for BrO_4^-.

Only two, commonly occurring, negatively charged complex ions do not use the *-ate* or *-ite* ending. These are hydroxide for OH^- and cyanide for CN^- which employ the *-ide* ending similar to the negative ion of a binary compound.

The ammonium ion, NH_4^+, is the most frequently encountered positive complex ion. A closely related ion, PH_4^+, is called phosphonium in a similar manner but is far less frequently observed in compounds.

In naming ternary compounds the same procedure is followed as for binary compounds. The only difference is that the name of the complex ion replaces the name of the monatomic ion. Either the older system or Stock convention may be used for the positive part of the compound. Several examples are listed in Table 9.8. You should check to verify that the formulas are correct. Remember that the product of the number of positive ions and their charge must be equal in magnitude but opposite in sign to the product of the number of negative ions and their charge. Thus cupric phosphate contains three doubly charged cupric ions which are balanced by two triply charged phosphate ions.

TABLE 9.8 names of ternary compounds

Formula	Name
KNO_2	potassium nitrite
$NaNO_3$	sodium nitrate
$Ca(NO_3)_2$	calcium nitrate
$LiClO$	lithium hypochlorite
$NaClO_2$	sodium chlorite
$Ba(ClO_3)_2$	barium chlorate
$AgClO_4$	silver perchlorate
$Ca_3(PO_4)_2$	calcium phosphate
$KMnO_4$	potassium permanganate
$PbCO_3$	lead(II) carbonate
$CuCN$	copper(I) cyanide
$Fe_2(CrO_4)_3$	iron(III) chromate
$Al_2(SO_4)_3$	aluminum sulfate
$BaSO_3$	barium sulfite
$NaOH$	sodium hydroxide
KCN	potassium cyanide
$Ba(OH)_2$	barium hydroxide
NH_4Cl	ammonium chloride
$(NH_4)_2CO_3$	ammonium carbonate
$(NH_4)_3PO_4$	ammonium phosphate

9.7 ACIDS

Hydrogen is less electropositive than many metals but is more electropositive than the simple anions and negative complex ions. **Compounds in which hydrogen is the more electropositive element are called acids.** The properties of acids are completely different from those of compounds containing metals. For example, HCl is a gaseous compound whereas NaCl is a solid.

In the gaseous or liquid state, the acids are named as hydrogen compounds. Thus HCl is hydrogen chloride, H_2S is hydrogen sulfide, and H_2SO_4 is hydrogen sulfate. In water, the acids undergo a change which is called ionization. The compounds separate into positive and negative ions as the hydrogen atom gives up its electron to the more negative element. The H^+ ion, a bare proton, combines with water to form the hydronium ion, H_3O^+. The chemistry of acids will be presented in Chapter 18. In water solution, the binary acids are named by the prefix *hydro-* followed by the anion name in which *-ide* has been replaced by *-ic* acid. Thus hydrogen chloride in water is hydrochloric acid. For the ternary acid hydrogen cyanide, HCN, a similar process gives hydrocyanic acid.

Most ternary acids do not include the *hydro-* prefix and all reference to hydrogen is dropped. The name of the polyatomic negative ion is used and the suffixes *-ate* and *-ite* are replaced by *-ic* acid and *-ous* acid, respectively.

Thus HNO_3 and HNO_2 in aqueous solution are nitric acid and nitrous acid, respectively. For cases where more than two complex ions are known, the prefixes *per-* and *hypo-* are used. Examples of the names of acids are listed in Table 9.9.

TABLE 9.9
names of acids

Formula	Name	Name in aqueous solution
HCl	hydrogen chloride	hydrochloric acid
HBr	hydrogen bromide	hydrobromic acid
HI	hydrogen iodide	hydroiodic acid
H_2S	hydrogen sulfide	hydrosulfuric acid[a]
HCN	hydrogen cyanide	hydrocyanic acid
HNO_3	hydrogen nitrate	nitric acid
HNO_2	hydrogen nitrite	nitrous acid
H_2SO_4	hydrogen sulfate	sulfuric acid[a]
H_2SO_3	hydrogen sulfite	sulfurous acid[a]
H_3PO_4	hydrogen phosphate	phosphoric acid
$HClO_4$	hydrogen perchlorate	perchloric acid
$HClO_3$	hydrogen chlorate	chloric acid
$HClO_2$	hydrogen chlorite	chlorous acid
$HClO$	hydrogen hypochlorite	hypochlorous acid

[a] In the acids containing sulfur, the *ur* is not dropped in the name of the aqueous solution.

9.8 BASES

Inorganic bases contain the hydroxide ion in combination with a metal ion. The properties of these compounds will be presented in Chapter 18. Although these bases are not binary compounds, they are named using the *-ide* ending. Several examples of bases are listed in Table 9.10.

In cases where the metal can exist in a variety of oxidation states, the name must indicate which oxidation state is present.

TABLE 9.10
names of bases

Formula	Name
$LiOH$	lithium hydroxide
$NaOH$	sodium hydroxide
KOH	potassium hydroxide
$Ca(OH)_2$	calcium hydroxide
$Ba(OH)_2$	barium hydroxide
$Mg(OH)_2$	magnesium hydroxide
$Fe(OH)_3$	iron(III) hydroxide
$Al(OH)_3$	aluminum hydroxide

9.9 SALTS

A salt is formed when one or more of the hydrogen ions of an acid reacts with one or more hydroxide ions of a base. In the case of hydrogen chloride and sodium hydroxide, the salt formed is sodium chloride.

$$NaOH + HCl \longrightarrow NaCl + H_2O$$

The reaction of an acid and a base to form a salt and water is called neutralization. Further amplification will be given in Chapter 18.

All of the binary and ternary compounds containing a metal and a nonmetal or complex ion which have been discussed in previous sections are salts. In each case they could be produced from the reaction of an acid and a base. The metal is derived from the base and the nonmetal or complex ion from the acid. Thus potassium bromide can be formed from potassium hydroxide and hydrogen bromide. Similarly, calcium nitrate is formed from calcium hydroxide and nitric acid, and sodium sulfate is formed from sodium hydroxide and sulfuric acid.

$$KOH + HBr \longrightarrow KBr + H_2O$$
$$Ca(OH)_2 + 2HNO_3 \longrightarrow Ca(NO_3)_2 + 2H_2O$$
$$2NaOH + H_2SO_4 \longrightarrow Na_2SO_4 + 2H_2O$$

Salts which contain one or more hydrogen atoms which were originally a part of the acid are called acid salts. For example, while sulfuric acid may react with sodium hydroxide to produce sodium sulfate when both hydrogens react as indicated above, it is also possible that only one hydrogen may ionize to produce sodium hydrogen sulfate (or bisulfate). The acid salt contains the HSO_4^- anion.

$$NaOH + H_2SO_4 \longrightarrow NaHSO_4 + H_2O$$

Other anions which also contain hydrogen are listed in Table 9.11. Note that both phosphoric acid (H_3PO_4) and phosphorous acid (H_3PO_3) contain three hydrogen atoms, so there are two possible acid salts that may be formed if not all the hydrogen atoms ionize. To name these anions and the salts

TABLE 9.11 anions containing hydrogen

Formula	Name
HSO_4^-	hydrogen sulfate, bisulfate
HSO_3^-	hydrogen sulfite, bisulfite
HCO_3^-	hydrogen carbonate, bicarbonate
$H_2PO_4^-$	dihydrogen phosphate
HPO_4^{2-}	hydrogen phosphate
$H_2PO_3^-$	dihydrogen phosphite
HPO_3^{2-}	hydrogen phosphite

which contain them, a Greek prefix is used to denote the number of hydrogen atoms present.

Salts which contain one or more hydroxide ions which were originally part of a base are called basic salts or hydroxy salts. Calcium hydroxide may react with nitric acid to yield calcium nitrate if both hydroxide ions react. However, reaction of only one hydroxide ion produces calcium hydroxynitrate.

$$Ca(OH)_2 + HNO_3 \longrightarrow Ca(OH)NO_3 + H_2O$$

There is no complex ion known as hydroxynitrate in this compound but rather two distinctly different ions, the hydroxide ion and the nitrate ion, are present. Note that one of the anions is put inside parentheses to indicate that the hydroxide ion and the nitrate ion, are present. Note that one of the anions is put inside parentheses to indicate that the hydroxide ion and nitrate ion are separate ions.

Salts which contain two or more different cations are called mixed salts. These salts are named by naming each of the cations and then the nonmetal anion or negative polyatomic ion. Greek prefixes are used if more than one of the same cation is present. Thus $NaKSO_4$ is sodium potassium sulfate and $K_2NH_4PO_4$ is dipotassium ammonium phosphate.

SUMMARY

__Common names__ of compounds are unrelated to chemical composition, whereas __systematic names__ indicate the elemental composition. Systematic names are prepared by the __IUPAC__.

In naming __binary__ compounds, the metal or electropositive portion of the substance is named first and then the name of the nonmetal or more electronegative element follows. If the metal is one with a variable oxidation number, the oxidation number must be indicated by use of an __-ous__ or __-ic__ suffix or by using the Stock system.

__Ternary__ compounds frequently contain a complex anion which is named as a unit. The prefixes __per-__ and __hypo-__ and the suffixes __-ate__ and __-ite__ are used to name complex ions of an element and variable amounts of oxygen.

Compounds in which hydrogen is the more electropositive element are called __acids__. Acids are named according to whether they are pure compounds or in aqueous solution.

__Bases__ contain an hydroxide ion in combination with a metal ion.

__Salts__ are formed when one or more hydrogen ions of an acid react with one or more hydroxide ions of a base.

QUESTIONS AND PROBLEMS

Definitions and terms

1. Indicate the usage of each of the following prefixes or suffixes and give one name using the term.

(a) *-ate* **(b)** *-ite*
(c) *-ous* **(d)** *-ic*

(e) *-ide* (f) *-ic* acid
(g) *-ous* acid (h) *hydro-*
(i) *hypo-* (j) *per-*

2. Define each of the following terms.
(a) salt (b) acid
(c) acid salt (d) base
(e) basic salt (f) mixed salt
(g) binary compound (h) ternary compound
(i) *-ate* and *-ite* names (j) *-ous* and *-ic* names
(k) Stock system (l) IUPAC

Names of ions

3. Name each of the following ions.
(a) Cl^- (b) F^-
(c) Br^- (d) I^-
(e) OH^- (f) S^{2-}
(g) H^+ (h) CN^-
(i) NH_4^+ (j) Na^+
(k) Li^+ (l) Ag^+
(m) Mg^{2+} (n) Ca^{2+}
(o) K^+ (p) Al^{3+}

4. Name each of the following anions.
(a) ClO^- (b) MnO_4^-
(c) BrO_4^- (d) ClO_2^-
(e) SO_4^{2-} (f) SO_3^{2-}
(g) PO_4^{3-} (h) ClO_3^-
(i) CO_3^{2-} (j) ClO_4^-
(k) IO^-

5. Name each of the following cations.
(a) Cu^+ (b) Fe^{3+}
(c) Hg^{2+} (d) Hg_2^{2+}
(e) Fe^{2+} (f) Cu^{2+}
(g) Sn^{2+} (h) Pb^{4+}
(i) Pb^{2+} (j) Sn^{4+}

6. Name each of the following anions.
(a) HSO_4^- (b) HCO_3^-
(c) $H_2PO_4^-$ (d) HSO_3^-
(e) $H_2PO_3^-$ (f) HPO_4^{2-}

Formulas of ions

7. Write the symbols of each of the following cations.
(a) sodium ion (b) cuprous ion
(c) ferric ion (d) potassium ion

(e) stannic ion (f) ferrous ion
(g) cupric ion (h) mercuric ion
(i) calcium ion (j) aluminum ion
(k) barium ion (l) stannous ion

8. Write the symbols of each of the following ions.
(a) bromide (b) sulfite
(c) chlorate (d) bicarbonate
(e) nitrate (f) perchlorate
(g) bromite (h) nitrite
(i) sulfate (j) hypoiodite
(k) hydrogen sulfate (l) cyanide
(m) hydroxide (n) bisulfite

Naming compounds

9. Write the systematic name for each of the following.
(a) NH_4Cl (b) $Ba(NO_3)_2$
(c) $Al_2(SO_4)_3$ (d) $NaHCO_3$
(e) $KClO_4$ (f) PbI_2
(g) $Hg(CN)_2$ (h) $LiHSO_3$
(i) $NaNO_2$ (j) $NaMnO_4$
(k) $Al(OH)_3$ (l) $Cd(ClO_3)_2$
(m) $Ca(HSO_4)_2$ (n) $(NH_4)_2CO_3$
(o) SO_3 (p) $SnCl_4$
(q) $LiIO$ (r) K_2SO_3
(s) Fe_2O_3 (t) $NaBrO_3$
(u) PCl_3 (v) H_2SO_4
(w) $Al(OH)Cl_2$ (x) CO
(y) KOH (z) N_2O_3

10. Some metals can exist in three or more oxidation states. What advantage does the Stock system have over the common names for compounds containing these metals?

Formulas of compounds

11. Write the formula for each of the following compounds.
(a) calcium hydroxide (b) cuprous chloride
(c) oxygen difluoride (d) sodium sulfate
(e) hydrogen chloride (f) nitrous acid
(g) lead(II) fluoride (h) stannous carbonate
(i) aluminum oxide (j) copper(II) cyanide
(k) perchloric acid (l) ferrous oxide
(m) dinitrogen pentoxide (n) sodium potassium sulfate
(o) cupric nitrate (p) mercuric bromide
(q) potassium bisulfate (r) iron(III) oxide

(s) hydrogen sulfide
(t) lithium hydroxide
(u) lead(II) phosphate
(v) hydrochloric acid
(w) sulfur dioxide
(x) phosphorous pentachloride
(y) magnesium sulfite
(z) nitric acid

Incorrect names and formulas

12. Indicate why each of the following names is improper.
(a) chloric
(b) mononitrogen monooxide
(c) lead chloride
(d) sulfur oxide
(e) iron oxide
(f) nitrate acid
(g) sodium bicarbonide
(h) copper(IV) oxide

13. Indicate why each of the following formulas is incorrect.
(a) $NaCl_2$
(b) Fe_3O_2
(c) $Cu(NO_4)_2$
(d) H_2ClO_4
(e) H_2SO_2
(f) $Ba_2(PO_4)_3$
(g) $NaHPO_4$
(h) $Al(CN)_2$
(i) $Ca(HSO_4)_3$
(j) CO_3

Acids, bases, and salts

14. Classify each of the following as an acid, base, acid salt, basic salt, or mixed salt.
(a) $NaOH$
(b) $Fe(OH)Cl_2$
(c) $(NH_4)_2HPO_4$
(d) $Ca(OH)_2$
(e) $HClO_3$
(f) HBr
(g) H_3PO_4
(h) $NaHSO_3$

10 COMPOSITION OF COMPOUNDS

LEARNING OBJECTIVES

When you finish this chapter you should be able to:

(a) illustrate the difference between an empirical formula and a molecular formula.

(b) express the molecular weight or formula weight of a substance in atomic mass units given the formula of the substance.

(c) use the mole concept in describing quantities of matter in terms of grams or the number of particles which a substance contains.

(d) calculate the percent composition of a substance given its formula.

(e) formulate the empirical formula of a substance given its percent composition.

(f) use the empirical formula and the molecular weight to determine the molecular formula of a substance.

(g) match the following terms with their corresponding definitions.

Avogadro's number
empirical formula
formula weight
mole
molecular formula
molecular weight
percent composition

In the previous chapter we learned how to name compounds from the formula and how to write formulas from a chemical name. However, the question that must be raised is, How do we know the formula in the first place? Compounds found in nature or produced in a reaction in a chemical laboratory do not carry labels to identify themselves for our convenience. Thus if a new medicinal drug were found in a South American plant and it were desirable to synthesize the substance in a chemical laboratory, it would be necessary to know its formula. Similarly, any new substance discovered or produced by chemists must have its formula established in order to properly identify it.

The chemist needs to know quantitatively the elemental composition of a compound before a formula can be established. Thus both the identity of the elements (a qualitative determination) and their amounts contained in the compound (a quantitative determination) must be established.

The identification of a compound in quantitative terms might be regarded as analogous to identifying a criminal to the police if you were an observer of the crime. The police would want you to identify qualitatively the features of the individual. Thus was the suspect bald or did he have hair? Was he short or tall? Was he white, black, yellow, or red? In addition, in order to draw a likeness, a police artist would like to know the suspect's features more quantitatively. Thus, if he had hair, how much hair? If he was tall, how tall?

In this chapter a systematic approach is given which can be used to determine the formula of any substance. The method depends on knowing the identity of the elements and their percent composition in the compound.

10.1 EMPIRICAL FORMULA VERSUS MOLECULAR FORMULA

The empirical formula of a compound, or simplest formula, gives only the smallest ratio of the atoms that are present in a compound. This contrasts with the **molecular formula, the true formula, which gives the total number of atoms of each type of element that are present in one molecule of the compound.** In the case of water, the empirical formula and the molecular formula are the same. The formula H_2O gives the simplest ratio of two hydrogen atoms for every oxygen atom in the molecule. Thus H_2O serves as the empirical formula. However, each molecule of water contains two hydrogen atoms and one oxygen atom, and therefore, the formula H_2O is also the molecular formula. Hydrogen peroxide has the molecular formula H_2O_2. The molecular formula does not correspond to the simplest formula as the ratio of the atoms is one hydrogen atom to one oxygen atom. In this case the empirical formula is HO.

H_2O_2 molecular formula
HO empirical formula

A molecular formula is always a multiple of the empirical formula (Table 10.1), but the factor may be unity, as in the case of water. The empirical formula of the gas used in welding torches, acetylene, is CH. This empirical formula indicates that there is one carbon atom for every hydrogen atom in the molecule. However, this does not necessarily mean that there is only one carbon atom and one hydrogen atom in one molecule of acetylene. The acetylene molecule actually contains two carbon atoms and two hydrogen atoms, and the molecular formula is C_2H_2. The subscripts of the empirical formula are multiplied by 2 to obtain the molecular formula. Benzene, an important major product of the chemical industry which is used for production of polystyrene and nylon, is a liquid and has an empirical formula of CH. Again, as in the case of acetylene, the empirical formula only indicates that there is a one-to-one ratio of carbon and hydrogen atoms. The molecular formula of benzene is C_6H_6. One molecule of benzene contains exactly six atoms of carbon and six atoms of hydrogen. The subscripts of the empirical formula of benzene must be multiplied by 6 to obtain the molecular formula.

TABLE 10.1
the empirical formula and the molecular formula

Substance	Molecular formula	Empirical formula
water	H_2O	H_2O
hydrogen peroxide	H_2O_2	HO
acetylene	C_2H_2	CH
benzene	C_6H_6	CH
sucrose	$C_{12}H_{22}O_{11}$	$C_{12}H_{22}O_{11}$
glucose	$C_6H_{12}O_6$	CH_2O

Problem 10.1 What are the empirical formulas and molecular formulas of ammonia, which contains one nitrogen atom and three hydrogen atoms in a molecule, and of hydrazine, which contains two atoms of nitrogen and four atoms of hydrogen in a molecule?

Solution The empirical formula of ammonia is NH_3, as the ratio of nitrogen atoms to hydrogen atoms is one-to-three. The molecular formula for ammonia is identical to the empirical formula as each molecule contains only one atom of nitrogen and three atoms of hydrogen. In the case of hydrazine, the empirical formula and molecular formula are not the same. The empirical formula is NH_2, as the ratio of nitrogen atoms to hydrogen atoms is one-to-two. However, the molecular formula is N_2H_4 as there are two nitrogen atoms and four hydrogen atoms in every molecule.

Problem 10.2 The molecular formula of vitamin C is $C_6H_8O_6$. What is the empirical formula of vitamin C?

Solution

The empirical formula must give only the smallest ratio of atoms present in the substance. The subscripts 6, 8, and 6 are not the smallest ratio as they are all divisible by 2 to give 3, 4, and 3. Therefore, the empirical formula is $C_3H_4O_3$.

10.2 MOLECULAR AND FORMULA WEIGHTS

The molecular weight of water is 18 amu.

In Section 5.9 it was shown that **the weight of a molecule relative to that of its atoms is the sum of the atomic weights of all of the constituent atoms.** This quantity is the molecular weight of the molecule.

Problem 10.3

What is the molecular weight of vitamin C, $C_6H_8O_6$?

Solution

Sum the atomic weights of the constituent atoms, taking into account the number of atoms of each kind present in the molecule.

$$\begin{array}{lll} \text{mass of 6 carbon atoms} & = 6 \times 12.0 \text{ amu} = & 72.0 \text{ amu} \\ \text{mass of 8 hydrogen atoms} & = 8 \times 1.0 \text{ amu} = & 8.0 \text{ amu} \\ \underline{\text{mass of 6 oxygen atoms}} & = \underline{6 \times 16.0 \text{ amu}} = & \underline{96.0 \text{ amu}} \\ \text{mass of 1 } C_6H_8O_6 \text{ molecule} = & & 176.0 \text{ amu} \end{array}$$

The molecular weight is 176.0 amu.

Problem 10.4

What is the molecular weight of dinitrogen pentoxide?

Solution

The name indicates that the molecular formula is N_2O_5. The molecular weight is obtained by summing the atomic weights of the individual atoms multiplied by the proper subscript.

$$\begin{array}{lll} \text{mass of 2 nitrogen atoms} & = 2 \times 14.0 \text{ amu} = & 28.0 \text{ amu} \\ \underline{\text{mass of 5 oxygen atoms}} & = \underline{5 \times 16.0 \text{ amu}} = & \underline{80.0 \text{ amu}} \\ \text{mass of 1 } N_2O_5 \text{ molecule} = & & 108.0 \text{ amu} \end{array}$$

The molecular weight is 108.0 amu.

Ionic substances do not exist as molecules but rather as collections of ions in a definite ratio described by a formula giving the simplest ratio of ions present in the compound. **The formula weight of an ionic compound is the sum of the atomic weights of the atoms indicated by a formula unit of the substance.**

Problem 10.5 What is the formula weight of calcium carbonate $CaCO_3$?

Solution Calcium carbonate is an ionic compound consisting of calcium ions and carbonate ions in a one-to-one ratio. The formula weight is obtained by summing the atomic weights.

$$
\begin{array}{lcccc}
\text{mass of 1 calcium atom} & = & 1 \times 40.1 \text{ amu} & = & 40.1 \text{ amu} \\
\text{mass of 1 carbon atom} & = & 1 \times 12.0 \text{ amu} & = & 12.0 \text{ amu} \\
\underline{\text{mass of 3 oxygen atoms}} & = & \underline{3 \times 16.0 \text{ amu}} & = & \underline{48.0 \text{ amu}} \\
\text{sum} & = & & & 100.1 \text{ amu}
\end{array}
$$

The formula weight is 100.1 amu.

10.3 THE MOLE

The atomic weights of atoms were discussed in Chapter 5 and defined as the relative weights of atoms in reference to a chosen standard. The atomic weights of hydrogen, carbon, and oxygen are 1.0 amu, 12.0 amu, and 16.0 amu, respectively. Thus a hydrogen atom is $\frac{1}{12}$ the weight or mass of a carbon atom. Furthermore, an oxygen atom is $\frac{4}{3}$ the mass of a carbon atom.

Molecular weights are defined in a similar manner. They are obtained by summing the atomic weights. Thus the molecular weight of carbon dioxide is 12.0 amu + 2(16.0) amu = 44.0 amu. The molecular weight of water is 2(1.0) amu + 16.0 amu = 18.0 amu.

In the chemical laboratory, substances are weighed in terms of grams rather than atomic mass units. Recall from Section 5.2 that 1 amu is 1.66×10^{-24} g. This value is incredibly small, and no balance can detect quantities less than 10^{-6} g. Thus the chemist never directly weighs on a balance a single atom of even millions, billions, trillions, or quadrillions of atoms. Approximately 10^{18} atoms would be required to be detected on the most sensitive balance. Accordingly, the chemists have devised a system for comparing quantities of a substance which does not depend on single atoms or molecules but on large collections of atoms or molecules.

In order to put the weights of atoms and molecules on a weight or mass basis which chemists can deal with more conveniently, the mole is defined. **A mole is that amount of material which contains a mass in grams equal to its mass on the atomic weight scale.** Therefore, a mole of carbon atoms is exactly 12.0 g of carbon atoms. A mole of oxygen atoms is exactly 16.0 g of oxygen atoms. In a similar manner, a mole of carbon dioxide molecules has a mass of 44.0 g. A mole of water molecules has a mass of 18.0 g.

Since the mass of a mole of a substance is directly proportional to the mass of a particle of that substance, it follows that **a mole of any substance contains the same number of particles which reflect the properties of the substance (atoms, formula units, or molecules) as a mole of any other sub-**

stance. Thus a mole of carbon atoms (12.0 g) contains the same number of atoms as a mole of oxygen atoms (16.0 g). Similarly, a mole of carbon dioxide molecules has a mass of 44.0 g and has the same number of molecules as a mole of water molecules, which has a mass of 18.0 g. Since a mole of atoms has a mass equal to the atomic weight in grams, it is sometimes referred to as a *gram atomic weight*. Similarly, a mole of molecules is called a *gram molecular weight*.

Mole
6.02×10^{23}
Avogadro's number

The number of particles in a mole of a substance has been determined as 6.02×10^{23} and is known as Avogadro's number. A mole of carbon atoms contains 6.02×10^{23} atoms of carbon. A mole of oxygen atoms contains 6.02×10^{23} atoms of oxygen. A mole of carbon dioxide contains 6.02×10^{23} molecules of carbon dioxide. A mole of water contains 6.02×10^{23} molecules of water.

The subscripts in the formulas of compounds represent the number of atoms of each element present in a molecule of the substance. **The subscripts also indicate the number of moles of the atoms of the elements present in a mole of molecules.** Thus in a mole of carbon dioxide there are one mole of carbon atoms and two moles of oxygen atoms. In a mole of sulfuric acid there are two moles of hydrogen atoms, one mole of sulfur atoms, and four moles of oxygen atoms.

Although there are no discrete units in ionic compounds which correspond to those represented in formulas such as $CaCO_3$, we nevertheless will find it convenient to extend the mole concept to ionic compounds. **A mole of an ionic compound has a mass in grams corresponding to the formula weight of the substance.** The formula weight of $CaCO_3$ is 100.1 amu and the mass of one mole is 100.1 g.

A mole of an ionic compound contains 6.02×10^{23} formula units. Thus one mole of $CaCO_3$ contains 6.02×10^{23} calcium ions and 6.02×10^{23} carbonate ions which account for the 6.02×10^{23} formula units of $CaCO_3$. In one mole of sodium carbonate, Na_2CO_3, there are $2(6.02 \times 10^{23})$ sodium ions and 6.02×10^{23} carbonate ions. In one mole of iron (III) chloride, there are 6.02×10^{23} iron(III) ions and $3(6.02 \times 10^{23})$ chloride ions.

Problem 10.6 How many moles of gold atoms are present in a 30.0-kg ingot of gold?

Solution The atomic weight of gold is 197.0. Therefore, one mole of gold atoms has a mass of 197.0 g. The number of moles in the ingot is calculated as follows:

$$30,\bar{0}00 \text{ g Au} \times \frac{1 \text{ mole Au atoms}}{197 \text{ g Au}} = 152 \text{ moles Au atoms}$$

Problem 10.7 An average adult exhales 1000 grams of carbon dioxide per day. How many moles are exhaled?

Solution

The molecular weight of carbon dioxide, CO_2, is:

carbon	=	1 × 12.0 amu
oxygen	=	2 × 16.0 amu
carbon dioxide	=	44.0 amu

Therefore, the molecular weight of carbon dioxide is 44.0 amu. The number of moles in 1000 g is:

$$1000 \not{g}\ \not{CO_2} \times \frac{1 \text{ mole } CO_2}{44.0 \not{g}\ \not{CO_2}} = 22.7 \text{ moles } CO_2$$

Problem 10.8

How many moles of calcium carbonate are contained in 25 g of the compound?

Solution

In Problem 10.5, the formula mass of this ionic compound was determined as 100.1 amu. Thus the mass of one mole of $CaCO_3$ is 100.1 g. The number of moles in the sample is:

$$25 \not{g}\ \not{CaCO_3} \times \frac{1 \text{ mole } CaCO_3}{100.1 \not{g}\ \not{CaCO_3}} = 0.25 \text{ mole } CaCO_3$$

Problem 10.9

Calculate the mass of 0.00100 moles of vitamin C, $C_6H_8O_6$.

Solution

In Problem 10.3, the molecular mass of vitamin C was established as 176.0 amu. Therefore, one mole has a mass of 176.0 g and the mass of 0.00100 mole can be calculated as follows:

$$0.00100 \not{\text{mole}}\ \not{C_6H_8O_6} \times \frac{176.0 \text{ g } C_6H_8O_6}{1 \not{\text{mole}}\ \not{C_6H_8O_6}} = 0.176 \text{ g } C_6H_8O_6$$

Problem 10.10

How many molecules of vitamin C are in a 500-milligram tablet of vitamin C?

Solution

In order to determine the number of molecules, it is first necessary to determine the number of moles present in the tablet which has a mass of 0.500 gram.

$$0.500 \not{g}\ \not{C_6H_8O_6} \times \frac{1 \text{ mole } C_6H_8O_6}{176.0 \not{g}\ \not{C_6H_8O_6}} = 0.00284 \text{ mole } C_6H_8O_6$$

The number of molecules of $C_6H_8O_6$ in a mole is 6.02×10^{23}. In the tablet which contains 2.84×10^{-3} mole, the number of molecules is:

$$2.84 \times 10^{-3} \text{ mole } C_6H_8O_6 \times \frac{6.02 \times 10^{23} \text{ molecules } C_6H_8O_6}{1 \text{ mole } C_6H_8O_6}$$
$$= 17.1 \times 10^{20} \text{ molecules } C_6H_8O_6$$

or 1.71×10^{21} molecules of $C_6H_8O_6$.

10.4 PERCENT COMPOSITION

Percent means parts per hundred. Thus if you scored 20 correct answers on a true-false examination consisting of 25 questions, you had 80% correct answers. This is determined by dividing the number of correct answers by the total number of questions and multiplying the quotient by 100.

$\frac{\text{part}}{\text{whole}} \times 100 = \%$

$$\frac{20 \text{ correct answers}}{25 \text{ questions}} \times 100 = 80\% \text{ correct answers}$$

The percent composition of a compound is calculated and defined in a similar manner. **The percent composition of a given element in a compound is equal to the mass of that element divided by the total mass of all of the elements in the compound with the quotient multiplied by 100.** For example, a 100.0-g sample of water contains 11.1 g of hydrogen and 88.9 g of oxygen. The mass percentage of hydrogen in water is 11.1 and the mass percentage of oxygen is 88.9.

$$\frac{11.1 \text{ g hydrogen}}{100.0 \text{ g water}} \times 100 = 11.1\% \text{ hydrogen}$$

$$\frac{88.9 \text{ g oxygen}}{100.0 \text{ g water}} \times 100 = 88.9\% \text{ oxygen}$$

Problem 10.11 A 250-mg tablet of vitamin C contains 102 mg of carbon, 12 mg of hydrogen, and 136 mg of oxygen. What is the mass percent composition of vitamin C?

Solution To solve this problem, the mass of each individual element must be divided by the total mass and the quotient multiplied by 100.

$$\text{percentage of carbon} = \frac{102 \text{ mg carbon}}{250 \text{ mg vitamin C}} \times 100 = 40.8\%$$

$$\text{percentage of hydrogen} = \frac{12 \text{ mg hydrogen}}{250 \text{ mg vitamin C}} \times 100 = 4.8\%$$

$$\text{percentage of oxygen} = \frac{136 \text{ mg oxygen}}{250 \text{ mg vitamin C}} \times 100 = 54.4\%$$

Problem 10.12 A 0.401-g sample of calcium reacts completely with oxygen to give 0.561 g of calcium oxide. What is the percent composition of calcium oxide?

Solution Since the total mass of the oxide is 0.561 g, the mass of oxygen which has combined with the calcium is 0.160 g. The percent of each element is calculated by dividing the mass of each element by the total mass of the oxide.

$$\frac{0.401 \text{ g calcium}}{0.561 \text{ g CaO}} \times 100 = 71.4\% \text{ calcium}$$

$$\frac{0.160 \text{ g oxygen}}{0.561 \text{ g CaO}} \times 100 = 28.6\% \text{ oxygen}$$

10.5 DETERMINATION OF EMPIRICAL FORMULAS

In order to calculate the empirical formula of a compound, it is first necessary to determine the mass percent composition of the substance. However, before using the mass percent composition in determining the empirical formula, it is instructive to show how a molecular formula can yield the mass percent composition. In the case of water, the molecular formula is H_2O. A mole of this substance is 18.0 g. In this one mole, there are two moles of hydrogen atoms or 2.0 g, and one mole of oxygen atoms or 16 g; the percent composition is given by:

$$\frac{2.0 \text{ g hydrogen}}{18.0 \text{ g water}} \times 100 = 11.1\% \text{ hydrogen}$$

$$\frac{16.0 \text{ g oxygen}}{18.0 \text{ g water}} \times 100 = 88.9\% \text{ oxygen}$$

Problem 10.13 What is the mass percent composition of sulfur in sulfuric acid?

Solution The molecular formula for sulfuric acid is H_2SO_4. The gram molecular weight of sulfuric acid is approximately 2(1.0 g) + 1(32.0 g) + 4(16.0 g) = 98.0 g, using atomic weights for hydrogen, sulfur, and oxygen to the nearest integer. The mass percent composition of sulfur is 32.7.

$$\frac{32.0 \text{ g sulfur}}{98.0 \text{ g sulfuric acid}} \times 100 = 32.7\% \text{ sulfur in sulfuric acid}$$

If the composition of a compound is determined by chemical analysis, the mass percent composition is available. Using the mass percent composition, only three simple arithmetical steps are necessary to arrive at the empirical formula. **In the first step, express the composition of each element in the compound in grams.** Thus, for example, an analysis of vitamin C indicates that it contains 40.9 percent carbon, 4.6 percent hydrogen, and 54.5 percent oxygen. A statement in grams is that every 100.0 grams of vitamin C contain 40.9 grams of carbon, 4.6 grams of hydrogen, and 54.5 grams of oxygen.

The second step involves division of each mass by the mass of a mole of atoms of the element to give the number of moles contained in the constituent sample of the element. The atomic weights of carbon, hydrogen, and oxygen expressed to the nearest 0.1 amu are 12.0, 1.0, and 16.0, respectively.

$$\frac{40.9 \text{ g } \cancel{\text{carbon}}}{12.0 \text{ g } \cancel{\text{carbon}}/(1 \text{ mole carbon})} = 3.4 \text{ moles of carbon}$$

$$\frac{4.6 \text{ g } \cancel{\text{hydrogen}}}{1.0 \text{ g } \cancel{\text{hydrogen}}/(1 \text{ mole hydrogen})} = 4.6 \text{ moles of hydrogen}$$

$$\frac{54.5 \text{ g } \cancel{\text{oxygen}}}{16.0 \text{ g } \cancel{\text{oxygen}}/(1 \text{ mole oxygen})} = 3.4 \text{ moles of oxygen}$$

These divisions yield the number of moles of each element contained in the 100-gram sample.

In the third step, the relative number of atoms contained in a molecule or formula unit is obtained by dividing the number of moles of each element by the smallest of these numbers. This process provides subscripts for the empirical formula but integers may not result. In the example considered, the relative number of atoms are 1 atom of carbon, 1.35 atoms of hydrogen, and 1 atom of oxygen. The empirical formula $C_1H_{1.35}O_1$ is not acceptable, but multiplication by 3 yields $C_3H_{4.05}O_3$ or, more likely, the empirical formula is $C_3H_4O_3$.

Problem 10.14

Vinegar contains acetic acid which is a compound containing carbon, hydrogen, and oxygen. The percent compositions are 40.0 percent carbon, 6.7 percent hydrogen, and 53.3 percent oxygen. What is the empirical formula of acetic acid?

Solution

A 100.0-g sample of acetic acid would contain 40.0 g of carbon, 6.7 g of hydrogen, and 53.3 g of oxygen. Division of each of these quantities by the proper gram atomic weight yields the number of gram atomic weights of each element contained in the sample.

$$\frac{40.0 \text{ g } \cancel{\text{carbon}}}{12.0 \text{ g } \cancel{\text{carbon}}/(1 \text{ mole carbon})} = 3.33 \text{ moles carbon}$$

$$\frac{6.7\ \cancel{g}\ \cancel{\text{hydrogen}}}{1.0\ \cancel{g}\ \cancel{\text{hydrogen}}/(1\ \text{mole hydrogen})} = 6.7\ \text{moles hydrogen}$$

$$\frac{53.3\ \cancel{g}\ \cancel{\text{oxygen}}}{16.0\ \cancel{g}\ \cancel{\text{oxygen}}/(1\ \text{mole oxygen})} = 3.34\ \text{moles oxygen}$$

Division of each number of moles by 3.33 yields the numbers 1, 2, and 1. The empirical formula is CH_2O.

Problem 10.15 Ethyl alcohol contains 34.8% oxygen, 52.2% carbon, and 13.0% hydrogen. What is the empirical formula of ethyl alcohol?

Solution In a 100.0-g sample of ethyl alcohol, there are 34.8 g of oxygen atoms, 52.2 g of carbon atoms, and 13.0 g of hydrogen atoms. The number of moles of each type of element is:

$$\frac{34.8\ \cancel{g}\ \cancel{\text{oxygen}}}{16.0\ \cancel{g}\ \cancel{\text{oxygen}}/(1\ \text{mole oxygen})} = 2.18\ \text{moles oxygen}$$

$$\frac{52.2\ \cancel{g}\ \cancel{\text{carbon}}}{12.0\ \cancel{g}\ \cancel{\text{carbon}}/(1\ \text{mole carbon})} = 4.35\ \text{moles carbon}$$

$$\frac{13.0\ \cancel{g}\ \cancel{\text{hydrogen}}}{1.0\ \cancel{g}\ \cancel{\text{hydrogen}}/(1\ \text{mole hydrogen})} = 13.0\ \text{moles hydrogen}$$

Division of each number of moles by the smallest quantity gives the relative number of moles.

$2.18/2.18 = 1$ for oxygen
$4.35/2.18 = 2$ for carbon
$13.0/2.18 = 6$ for hydrogen

The empirical formula is C_2H_6O.

10.6 MOLECULAR FORMULA DETERMINATION

The molecular formula of a substance may be determined from the empirical formula if the molecular weight of the substance can be obtained. As stated in Section 10.1, the molecular formula is always a simple multiple of the empirical formula. For example, the empirical formula of water is H_2O and the molecular formula must be $(H_2O)_n$, or $H_{2n}O_n$, where n may be 1, 2, 3, and so on. Thus the molecular formula could be H_2O, H_4O_2, H_6O_3, and so on. The various possibilities can be distinguished because their molecular weights, are 18 amu, 36 amu, and 54 amu, respectively. The experimental molecular weight of water is 18 amu and the molecular formula must be H_2O.

Acetylene and benzene both have an empirical formula of CH. The molecular formulas may be $(CH)_n$, or C_nH_n. In order to obtain the molecular formulas, one only needs the molecular weight of each substance. The molecular weights of acetylene and benzene are 26 amu and 78 amu, respectively. Therefore, the molecular formulas are C_2H_2 and C_6H_6.

Problem 10.16 The empirical formula of nicotine is C_5H_7N. The molecular weight is 162 amu. What is the molecular formula?

Solution If the molecular formula were the same as the empirical formula, the molecular weight would be 5(12 amu) + 7(1 amu) + 14 amu = 81 amu. Therefore, the molecular formula must be twice the empirical formula, or $C_{10}H_{14}N_2$.

Problem 10.17 A gaseous compound contains 80.0% carbon and 20.0% hydrogen. The molecular weight is 30.0 amu. Determine the molecular formula of the substance.

Solution The empirical formula must first be determined as in Section 10.5. Consider a 100.0-g sample of the compound. The amount of carbon and hydrogen in this sample must be 80.0 g and 20.0 g, respectively.

$$\frac{80.0\ \cancel{g\ carbon}}{12.0\ \cancel{g\ carbon}/(1\ \text{mole carbon})} = 6.66\ \text{moles carbon}$$

$$\frac{20.0\ \cancel{g\ hydrogen}}{1.0\ \cancel{g\ hydrogen}/(1\ \text{mole hydrogen})} = 20.0\ \text{moles hydrogen}$$

Division of each of these numbers by 6.66 yields the subscripts for the empirical formula CH_3.

6.66/6.66 = 1 for carbon

20.0/6.66 = 3 for hydrogen

If CH_3 were the molecular formula, the molecular weight would be:

12.0 amu + 3(1.0 amu) = 15.0 amu

Since the molecular weight of the substance is 30.0 amu, the molecular formula must be twice the empirical formula, or C_2H_6.

SUMMARY

An ***empirical formula*** *or simplest formula gives the smallest ratio of atoms present in a molecule.* ***The molecular formula*** *is a true formula because it gives the total number of atoms of each type of element that are present in the molecule.*

The ***molecular weight*** *of a covalent substance is calculated by summing the atomic weights of the constituent atoms which the molecule contains.* ***Formula weights*** *of ionic compounds are calculated by summing the atomic weights of the atoms indicated in one formula unit of the substances.*

A ***mole*** *is that amount of material which contains a mass in grams equal to its mass on the atomic weight scale. A mole of any substance contains the same number of particles which have the properties of the the substance (atoms, formula units, or molecules) as a mole of any other substance. The number of particles in a mole is* 6.02×10^{23} *and is known as* ***Avogadro's number****.*

The ***percent composition*** *of an element in a compound is equal to the mass of that element divided by the total mass of all of the elements in the compound with the quotient multiplied by 100. Empirical formulas are calculated by (1) determining the percent composition of the compound, (2) dividing the mass of each element in 100 g of the compound by the mass of a mole of that element to give the number of moles in the 100 g sample, and (3) obtaining the relative number of atoms contained in a molecule or formula unit by dividing the number of moles of each element by the smallest of these numbers.*

QUESTIONS AND PROBLEMS

Definitions

1. Define each of the following terms.
(a) empirical formula **(b)** molecular formula
(c) molecular weight **(d)** formula weight
(e) mole **(f)** Avogadro's number
(g) percent composition

2. Clearly distinguish between each of the following.
(a) empirical formula and molecular formula
(b) molecular weight and formula weight

3. Can the empirical formula and the molecular formula for a compound be identical? Explain.

Empirical formula and molecular formula

4. Determine the empirical formula for each of the following molecular formulas.
(a) $C_{10}H_{14}N_2$, nicotine **(b)** $C_7H_5N_3O_6$, TNT
(c) $C_{12}H_{22}O_{11}$, sucrose **(d)** $C_{10}H_8$, naphthalene
(e) $C_6H_{12}N_2O_4S_2$, cystine **(f)** $C_6H_8O_7$, citric acid
(g) $C_4H_8N_2O_2$, uracil

Formula and molecular weights

5. Calculate the molecular weight of each of the following substances.
(a) SO_3 **(b)** P_2O_5
(c) NO_2 **(d)** N_2O_3
(e) CO_2 **(f)** N_2H_4
(g) CH_4 **(h)** $SOCl_2$

6. Calculate the molecular weight of each of the substances in Question 4.

7. Calculate the formula weight of each of the following substances.
(a) CaO **(b)** $CaCO_3$
(c) NaOH **(d)** $Ca(OH)_2$
(e) Na_2SO_4 **(f)** $Al_2(SO_4)_3$
(g) Na_3PO_4 **(h)** KCN
(i) $Na_2Cr_2O_7$ **(j)** $KMnO_4$

Moles and mass

8. Calculate the number of moles of atoms in each of the following samples.
(a) 46.0 g of sodium **(b)** 16.0 g of sulfur
(c) 2.0 g of mercury **(d)** 24.3 g of magnesium

9. Calculate the number of moles of compound in each of the following samples.
(a) 180 g of water **(b)** 0.44 g of carbon dioxide
(c) 92 g of nitrogen dioxide **(d)** 14.98 g of carbon tetrachloride

10. Calculate the number of moles of formula units in each of the following.
(a) 4.0 g of sodium hydroxide
(b) 22.32 g of lead(II) oxide
(c) 0.314 g of uranium hexafluoride
(d) 0.49 g of sulfuric acid

Moles and Avogadro's number

11. Calculate the number of atoms present in each of the following.
(a) 20.0 g of mercury **(b)** 120 g of carbon
(c) 53.2 g of palladium **(d)** 0.635 g of copper

12. Calculate the number of molecules present in each of the following.
(a) 2.8 g of carbon monoxide **(b)** 90 g of water
(c) 32 g of oxygen gas **(d)** 0.040 g of CH_4 (methane)

13. Calculate the number of formula units in each of the following.
(a) 28 g of calcium oxide
(b) 0.164 g of sodium phosphate
(c) 651 g of potassium cyanide

Empirical formulas

14. Determine the empirical formula for the following salts.
  (a) 74.4% gallium, 25.6% oxygen
  (b) 62.6% lead, 8.4% nitrogen, 29.0% oxygen
  (c) 48.0% zinc, 52.0% chlorine
  (d) 25.9% iron, 74.1% bromine

15. Determine the empirical formula for each of the following molecules.
  (a) 15.8% carbon, 84.2% sulfur
  (b) 82.3% nitrogen, 17.7% hydrogen
  (c) 37.5% carbon, 12.6% hydrogen, 49.9% oxygen
  (d) 50.0% sulfur, 50.0% oxygen
  (e) 32.0% carbon, 6.7% hydrogen, 18.7% nitrogen, 42.6% oxygen
  (f) 37.8% carbon, 6.4% hydrogen, 55.8% chlorine

Percent composition

16. Calculate the percent composition of each of the following.
  (a) CH_4 **(b)** CH_2F_2
  (c) H_2O_2 **(d)** $C_6H_{12}O_6$ (dextrose)
  (e) P_2O_5 **(f)** $CaC1_2$
  (g) $AgNO_3$ **(h)** $A1_2(SO_4)_3$

17. The molecular formula of aspirin is $C_9H_8O_4$. What is its percent composition?

18. The molecular formula of DDT is $C_{14}H_9Cl_5$. What is its percent composition?

19. What is the percent composition for compounds formed by combination of the elements in the indicated quantities?
  (a) 2.50 g of a metal and 2.0 g of oxygen
  (b) 1.40 g of an element and 1.60 g of oxygen
  (c) 35.0 g of a metal and 20.0 g of sulfur

Molecular formula

20. The insecticide lindane is 24.7% carbon, 2.1% hydrogen, and 73.2% chlorine. What is its empirical formula? The molecular weight is 291 amu. What is the molecular formula of lindane?

21. Freon 12 is a gas used as a refrigerant in household refrigerators. Its molecular weight is 121 amu and it consists of 9.9% carbon, 58.7% chlorine, and 31.4% fluorine. What is the molecular formula of this refrigerant?

22. A hydrocarbon consists of 92.3% carbon and 7.7% hydrogen. The molcular weight of the hydrocarbon is 78 amu. What is its molecular formula?

23. Vitamin A contains 83.86% carbon, 10.56% hydrogen, and 5.58% oxygen. The molecule contains one atom of oxygen. What is the molecular formula?

II

CHEMICAL EQUATIONS

LEARNING OBJECTIVES

When you finish this chapter you should be able to:

(a) recognize the common symbols used in chemical equations.

(b) interpret chemical equations on both a molecular level and a mole level.

(c) balance equations given the reactants and products of a reaction.

(d) classify chemical reactions into types: combination, decomposition, substitution, metathesis, and neutralization.

(e) match the following terms with their corresponding definitions.

chemical equation
combination reaction
decomposition reaction
metathesis reaction
neutralization reaction
product
reactant
substitution reaction

In earlier chapters the elements were compared to the letters in the alphabet. Then compounds were suggested as analogs of the words of a language. Thus from the limited number of fundamental units, a large number of combined units are possible. However, just as a knowledge of words alone is insufficient to communicate in a language, the knowledge of the names of compounds alone does not really enable one to communicate effectively in chemistry. It is necessary to put words together to make sentences of our English language. Similarly, compounds are put together to make equations which are the sentences of chemistry. A chemical equation is a collection of chemical words to form a chemical sentence. These sentences are very descriptive and their number is virtually limitless in much the same manner that the number of ordinary written sentences consisting of the words of the English language is limitless.

In any language there are rules of grammar which govern the proper construction of sentences. You have learned these rules as a child and have come to accept them as natural or intuitive. You now must learn how to operate within the rules governing chemical equations. With practice these rules will also become second nature and you will use them without thinking about them.

In this chapter you will learn how to write correct chemical equations which represent actual chemical changes. When properly written, a chemical equation will help you to understand what is going on at the atomic-molecular level as well as on the macroscopic level that you can see.

11.1 THE BALANCED CHEMICAL EQUATION

In Section 3.4, word equations were used to express the chemical reaction of magnesium and oxygen in a flashbulb to yield magnesium oxide, and also to describe the conversion of mercuric oxide into mercury and oxygen. Now we can use our knowledge of chemical formulas and the quantitative calculations derivable from such formulas to make these equations more precise. By substituting the correct chemical formula for each reactant and product into the equation, we will have a better idea of what is going on at the molecular level. Of course, all equations must correspond to the experimental facts. Thus we have the following rules for writing equations:

reactants → products

1. **Use correct formulas to represent the substances involved.**
2. **Write the reactants on the left and the products on the right of the equation.**
3. **Whenever an atom of any element, either in its elemental form or in a compound, appears on one side of the equation, it must also appear on the other side of the equation and in the same amount.**

4 The sum of any charges appearing on the left of the equation must be equal to the sum of the charges on the right of the equation.

When all of the rules are followed a balanced equation results. The balanced equation for the flashbulb reaction is:

$$2Mg + O_2 \longrightarrow 2MgO$$

This can be read in a number of ways. These are:

1 Magnesium metal reacts with oxygen gas to yield magnesium oxide.
2 Two atoms of magnesium react with one molecule of oxygen to yield two formula units of magnesium oxide.
3 Two moles of magnesium atoms react with one mole of oxygen molecules to yield two moles of formula units of magnesium oxide.

The equations discussed in this chapter will deal only with electrically neutral elements, molecular compounds, and formula units. In Chapter 19 we shall consider ionic equations in which only the ions undergoing change are depicted.

11.2 SYMBOLS IN EQUATIONS

The chemicals

Just as chemical formulas are a shorthand representation of pure substances, chemical equations are shorthand expressions used to record chemical reactions. The substances on the left of the equation are called reactants regardless of whether they are elements or compounds. The substances on the right of the equation are called products and may be either elements or compounds.

$$\text{reactants} \longrightarrow \text{products}$$

Connecting symbols

$\rightleftharpoons$

On each side of the equation a plus sign is used to separate each reactant and each product. It is read as plus or and. Several symbols are used to separate the reactants and the products. The single arrow $\rightarrow$ or an equal sign $=$ is read as yields or produces. A double arrow $\rightleftharpoons$ indicates that the reaction may proceed from left to right as written or in the reverse direction. The significance of this double reaction will be discussed in Chapter 17. A listing of these symbols and others used in chemical equations is given in Table 11.1.

The coefficients

A balanced chemical equation contains numbers placed to the left of chemical formulas. These numbers, called **coefficients, indicate the number of units of**

TABLE 11.1 symbols used in chemical equations

Symbol	Meaning
$\longrightarrow$	yields or produces
$\rightleftharpoons$	reaction may proceed in both directions
$\uparrow$	a gas: written immediately after the gaseous substance
$\downarrow$	a solid substance: deposits from a solution: written immediately after the substance
(s)	solid: written immediately after the substance
(l)	liquid
(g)	gas
(aq)	substance dissolved in water
Δ	heat
+	plus

that substance relative to all other substances which are reactants or products of the reaction. As indicated in Section 11.1, the coefficients in the reaction of magnesium with oxygen can be read at either the molecular or mole level.

Symbols above the arrow

$\xrightarrow{\Delta}$

Above the arrow the symbol Δ represents the heat energy which may be required to make the reaction proceed. Occasionally a symbol for an element or a compound also appears above the arrow. This symbol represents a catalyst which is used to speed up the rate of the reaction. At the end of the reaction, the catalyst can be recovered unchanged. Further discussion of catalysts will be given in Chapter 17.

Symbols for chemical states

The physical states of the reactants and products under the reaction conditions may be denoted by writing (s), (l), or (g) to the right of the symbol. Water, for example, might be produced as a liquid in one reaction but as a gas in another reaction and the use of (l) or (g) gives the reader this information. Many reactions occur between substances dissolved in water. The symbol (aq) meaning aqueous solution is placed to the right of the chemical symbol for such solutes. This type of information is invaluable to the reader of the equation as there is a significant difference in NaCl(aq) as compared to NaCl(s) or NaCl (l).

11.3 RULES FOR BALANCING EQUATIONS

A chemical equation must be balanced in order to properly represent a chemical reaction. It is easy to balance an equation if you consider the equation as representing an occurrence on the molecular level. Each chemical formula, then, represents a reactant or product molecule.

Although there is no hard and fast set of steps which can be used to balance chemical equations, there are some general rules which are an aid. With practice and the general guidelines provided, you can balance the equations given in this text. A list of these rules is given below:

1 **Write the correct formulas for the reactants and products on the appropriate sides of the equation.** Once the correct formula is written, do not alter it or any of its subscripts during the balancing process.
2 **Disregarding hydrogen, oxygen, and polyatomic ions, find the molecule containing the largest number of atoms of a single element. Balance the number of this element by placing the proper coefficient in front of the molecule containing this element on the other side of the equation.** Remember that a coefficient placed in front of a molecular formula multiplies every atom contained in the molecule.
3 **Proceed to balance the atoms of other elements by the same process.** Check to see if in balancing one element, others have become unbalanced. Readjust the coefficients on both sides of the equation to achieve the necessary balance.
4 **Balance polyatomic ions on each side of the equation as a single unit.**
5 **Balance hydrogen and oxygen not previously considered in poly-atomic ions.**
6 **Check all coefficients to insure that they are the lowest possible whole numbers.** If the coefficients are fractions, multiply by a number to make the fraction a whole number. All other coefficients must be multiplied by the same quantity to maintain a balanced equation. If all the coefficients are divisible by a common factor, do so to achieve the lowest possible whole number.
7 **Check over the entire equation to insure that all atoms are balanced.**

11.4 BALANCING EQUATIONS

Example 1

In the conversion of magnesium and oxygen into magnesium oxide, it is possible to start to write an equation by properly representing the chemical constitution of the reactants and the product. Oxygen exists as a diatomic molecule and must be so represented in a chemical equation. Magnesium oxide consists of Mg^{2+} and O^{2-} ions in equal numbers and is represented as MgO (Figure 11.1). We then write:

$$Mg + O_2 \longrightarrow MgO$$

To balance this equation we note that there are two oxygen atoms on the left but only one on the right. The equation cannot be balanced by changing the formula of magnesium oxide to MgO_2 as such a formula is inconsistent with

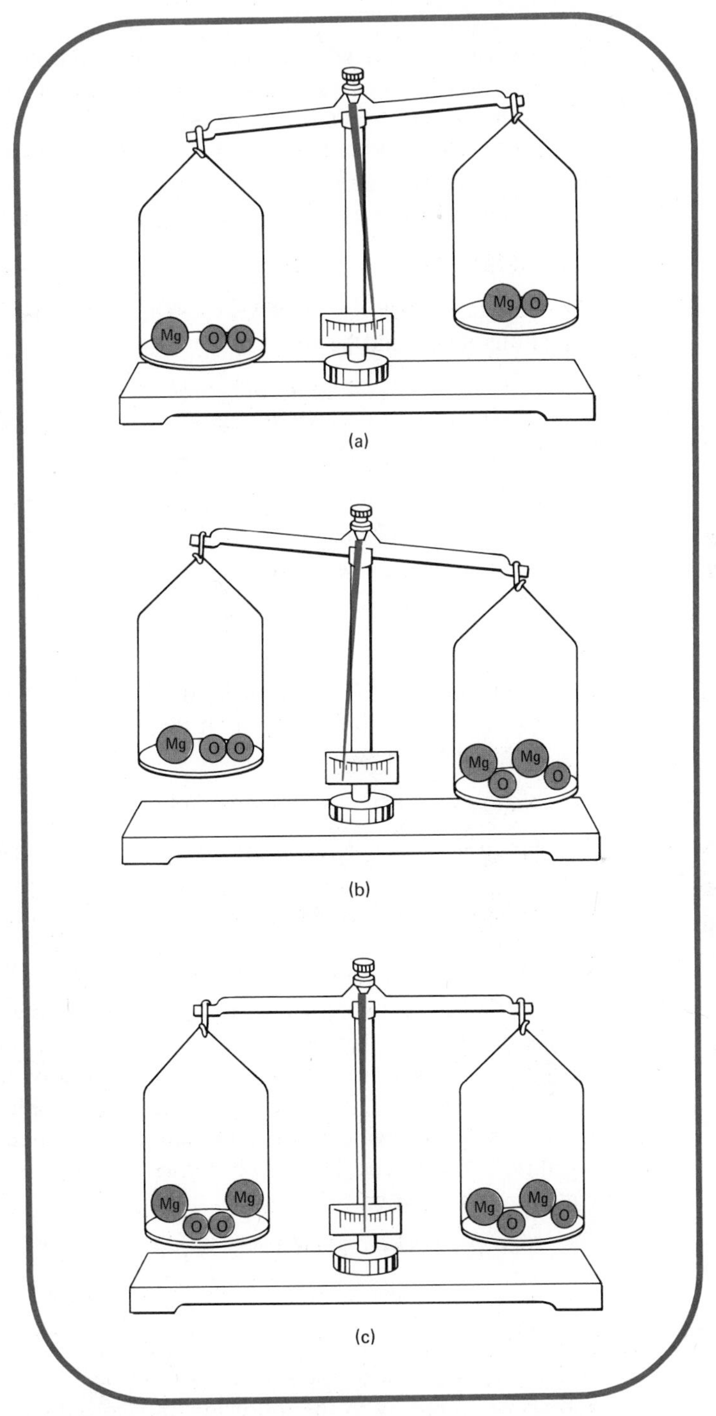

Figure 11.1 Balancing an equation. The numbers of atoms in the reactant and products must be equal.

the experimental facts. However, a 2 can be placed in front of MgO to indicate that two units of MgO are formed (Figure 11.1). When a coefficient 2 is placed in front of MgO, it means that two magnesium and two oxygen atoms are produced. Now we have two magnesium atoms on the right but only one on the left. This situation is righted by placing a 2 as the coefficient in front of magnesium (Figure 11.1).

$$2Mg + O_2 \longrightarrow 2MgO$$

The equation is now balanced. Note that there are other possible balanced equations that could be written:

$$Mg + \tfrac{1}{2}O_2 \longrightarrow MgO$$
$$4Mg + 2O_2 \longrightarrow 4MgO$$
$$6Mg + 3O_2 \longrightarrow 6MgO$$

These equations, however, are merely multiples of the equation in which the whole number coefficients are in the smallest ratio possible. Only the balanced equation with the smallest whole number coefficients is accepted.

Example 2

The solid element sodium reacts with the gaseous diatomic element chlorine to form the solid sodium chloride. If the proper symbols are used, the following equation can be written:

$$Na(s) + Cl_2(g) \longrightarrow NaCl(s)$$

The equation does not represent an equality because there are two chlorine atoms on the left and only one on the right of the arrow. The subscript of Cl_2 cannot be changed because this would constitute a misrepresentation of the known facts: chlorine exists as diatomic molecules. Similarly, sodium chloride cannot be represented as $NaCl_2$ in order to achieve an equality because the substance $NaCl_2$ is unknown and sodium chloride is known to consist of equal numbers of constituent sodium and chloride ions. The only change that is both mathematically and chemically acceptable is to place a coefficient 2 in front of NaCl. The right hand side of the equation now represents two chlorine and two sodium atoms. Finally, in order to achieve equality, a coefficient 2 must be placed in front of Na, and the resultant equation is said to be balanced:

$$2Na + Cl_2 \longrightarrow 2NaCl$$

Example 3

Aluminum hydroxide reacts with sulfuric acid to yield aluminum sulfate and water according to the following unbalanced equation:

$$Al(OH)_3 + H_2SO_4 \longrightarrow Al_2(SO_4)_3 + H_2O$$

Since there are two atoms of aluminum in the aluminum sulfate, it is necessary to place a coefficient 2 in front of the aluminum hydroxide. Furthermore, the three sulfate units in the aluminum sulfate are balanced by placing a coefficient 3 in front of the sulfuric acid.

$$2Al(OH)_3 + 3H_2SO_4 \longrightarrow Al_2(SO_4)_3 + H_2O$$

Only the oxygen not incorporated in the sulfate and the hydrogen remain unbalanced. There are six oxygen atoms contained in the two units of aluminum hydroxide which may be balanced by a coefficient of 6 for H_2O.

$$2Al(OH)_3 + 3H_2SO_4 \longrightarrow Al_2(SO_4)_3 + 6H_2O$$

This coefficient also serves to balance the hydrogen atoms on both sides of the equation as well. There are six hydrogen atoms in the two units of aluminum hydroxide and six hydrogen atoms in the three units of sulfuric acid which balance the twelve hydrogen atoms in the six water molecules.

Example 4

The hydrocarbon butane, C_4H_{10}, burns in oxygen to yield carbon dioxide and water.

$$C_4H_{10} + O_2 \longrightarrow CO_2 + H_2O$$

Butane contains the largest number of atoms of a single element. It has four carbon atoms and ten hydrogen atoms. Since hydrogen is ignored relative to other atoms, we start our balancing process with carbon. A coefficient 4 is placed in front of carbon dioxide.

$$C_4H_{10} + O_2 \longrightarrow 4CO_2 + H_2O$$

Next the hydrogen is balanced as there are no other elements besides oxygen which need to be balanced. Hydrogen is chosen over oxygen as it appears in only one compound on each side of the equation. A coefficient 5 placed in front of H_2O will balance the ten hydrogens of butane.

$$C_4H_{10} + O_2 \longrightarrow 4CO_2 + 5H_2O$$

Next the oxygen atoms may be balanced. The oxygens in the products now total thirteen, eight in the $4CO_2$ and five in the $5H_2O$. In order to balance the diatomic oxygen molecule, a coefficient of 13/2 is necessary.

$$C_4H_{10} + \tfrac{13}{2}O_2 \longrightarrow 4CO_2 + 5H_2O$$

Finally, the fractional coefficient is eliminated by multiplying all coefficients by 2.

$$2C_4H_{10} + 13O_2 \longrightarrow 8CO_2 + 10H_2O$$

Problem 11.1

Balance the equation for the combustion of propane, C_3H_8.

$$C_3H_8 + O_2 \longrightarrow CO_2 + H_2O$$

Solution

There are three atoms of carbon in one molecule of propane. Therefore, three molecules of CO_2 must be produced. A coefficient 3 must be placed in front of the formula CO_2. There are eight atoms of hydrogen contained in one molecule of C_3H_8. Therefore, four molecules of H_2O must be produced. Placing a coefficient 4 in front of the formula H_2O yields.

$$C_3H_8 + O_2 \longrightarrow 3CO_2 + 4H_2O$$

The equation is not balanced as there are ten atoms of oxygen represented on the right of the arrow and only two on the left. In order to balance the equation, a coefficient 5 must be placed in front of O_2.

$$C_3H_8 + 5O_2 \longrightarrow 3CO_2 + 4H_2O$$

Problem 11.2

Balance the reaction of magnesium hydroxide and phosphoric acid.

$$Mg(OH)_2 + H_3PO_4 \longrightarrow Mg_3(PO_4)_2 + H_2O$$

Solution

Disregarding the OH^- and the complex ion PO_4^{3-}, the compound with the greatest number of atoms is $Mg_3(PO_4)_2$ which contains 3 magnesium atoms. Balancing the $Mg(OH)_2$ with a coefficient 3 produces:

$$3Mg(OH)_2 + H_3PO_4 \longrightarrow Mg_3(PO_4)_2 + H_2O$$

Since only the phosphate ion, hydrogen, and oxygen remain, the phosphate ion is balanced next. There are two phosphates in $Mg_3(PO_4)_2$ and the coefficient has been established as 1 by the balancing of magnesium. A coefficient 2 is placed before H_3PO_4 to achieve phosphate balance.

$$3Mg(OH)_2 + 2H_3PO_4 \longrightarrow Mg_3(PO_4)_2 + H_2O$$

Finally, the hydrogen and oxygen may be balanced. Disregarding the oxygen in phosphate which has already been balanced, there are six oxygens in $3Mg(OH)_2$, and twelve hydrogens in $3Mg(OH)_2$ and $2H_3PO_4$ combined. A coefficient 6 in front of H_2O achieves the final balance.

$$3Mg(OH)_2 + 2H_3PO_4 \longrightarrow Mg_3(PO_4)_2 + 6H_2O$$

11.5 TYPES OF CHEMICAL REACTIONS

The number of known and potential reactions of the millions of known compounds is astronomically large. In order to develop a systematic understanding of chemical reactions, it is useful to classify them according to the type of process which occurs between the reacting substances.

At this point in your study of chemistry, we shall consider only the most common types of reactions. These are:

1 combination reactions,
2 decomposition reactions,
3 substitution reactions,
4 metathesis reactions,
5 neutralization reactions.

If you should continue in the study of chemistry, it would become apparent that there are other ways to classify reactions. However, this classification scheme is sufficient for the kinds of reactions we will be studying.

11.6 COMBINATION REACTIONS

The direct union of carbon and oxygen to yield carbon dioxide is an example of a combination reaction.

A combination reaction involves the direct union of two or more substances to produce one new substance. The two or more combining materials may be either elements or compounds. Of course, the combined material must be a compound. A general reaction which summarizes the combination reactions is:

$$X + Y \longrightarrow XY$$

where X and Y may be elements or compounds.

One common example of a combination reaction involves the reaction of metals with oxygen to form ionic oxides. For example, lithium (a reactive metal) reacts to yield lithium oxide.

$$4Li(s) + O_2(g) \xrightarrow{\Delta} 2Li_2O(s)$$

Magnesium contained in a flashbulb as a fine filament reacts with oxygen to give magnesium oxide.

$$2Mg(s) + O_2(g) \longrightarrow 2MgO(s)$$

Aluminum is very reactive toward oxygen and, as a consequence, all aluminum utensils have a thin oxide coating which is extremely hard. When scratched, the aluminum exposed will react with oxygen to reform the oxide.

Have you really ever seen aluminum metal?

$$4Al(s) + 3O_2(g) \longrightarrow 2Al_2O_3(s)$$

In these reactions, the oxidation numbers of the metals are changed. For lithium the change is from 0 to +1, for magnesium, from 0 to +2, and for aluminum, from 0 to +3. Further discussion of such changes in oxidation numbers will be given in Chapter 19.

Nonmetals can combine with the very electronegative element oxygen and produce covalent compounds. One such example is the reaction with sulfur to yield sulfur dioxide.

$$S(s) + O_2(g) \xrightarrow{\Delta} SO_2(g)$$

Sulfur dioxide is one of the seriously detrimental pollutants in air. It results from the burning of any sulfur-containing fuels. With proper experimental conditions, sulfur can also combine with oxygen to yield sulfur trioxide.

$$2S(s) + 3O_2(g) \longrightarrow 2SO_3(g)$$

The nonmetal carbon can also combine with oxygen to yield two possible compounds. In a limited amount of oxygen, carbon monoxide is formed. If sufficient oxygen is present, the product is carbon dioxide.

$$2C(s) + O_2(g) \longrightarrow 2CO(g)$$
$$C(s) + O_2(g) \longrightarrow CO_2(g)$$

Oxygen does not react with nitrogen at ordinary temperatures. However, at temperatures above 2000°C, such as found in the internal combustion engine, these two elements do combine to yield nitrogen oxide (nitric oxide).

$$N_2 + O_2 \longrightarrow 2NO$$

Hydrogen may combine directly with elements. One reaction of major industrial importance is the Haber process, which results in ammonia from the direct combination of hydrogen and nitrogen.

$$N_2 + 3H_2 \longrightarrow 2NH_3$$

Ammonia is used to produce fertilizer and to form a variety of nitrogen compounds.

Compounds may also undergo combination reactions. Thus metal oxides may react with water to produce metal hydroxides, which are called bases.

$$H_2O(l) + Na_2O \longrightarrow 2NaOH(aq)$$
$$H_2O(l) + MgO(s) \longrightarrow Mg(OH)_2(aq)$$
$$3H_2O(l) + Al_2O_3 \longrightarrow 2Al(OH)_3(aq)$$

Since a base is produced, the oxide from which it is derived is called a basic oxide. Oxides of nonmetals are called acid oxides or acid anhydrides as they may combine with water to yield acids.

$$N_2O_3(l) + H_2O(l) \longrightarrow 2HNO_2(aq)$$
$$N_2O_5(s) + H_2O(l) \longrightarrow 2HNO_3(aq)$$
$$P_4O_6(s) + 6H_2O(l) \longrightarrow 4H_3PO_3(aq)$$
$$P_4O_{10}(s) + 6H_2O(l) \longrightarrow 4H_3PO_4(aq)$$
$$SO_3(s) + H_2O(l) \longrightarrow H_2SO_4(aq)$$
$$SO_2(g) + H_2O(l) \longrightarrow H_2SO_3(aq)$$

Calcium oxide derived from limestone can combine with sulfur dioxide to form calcium sulfite. This process is used in some industries to remove the sulfur dioxide that would otherwise go up the stack.

$$CaO(s) + SO_2(g) \longrightarrow CaSO_3(s)$$

11.7 DECOMPOSITION REACTIONS

In a decomposition reaction, a single substance undergoes a breakdown into two or more simple substances. The reactant is a compound, but the products may be either elements or compounds. Usually heat is required to cause a decomposition reaction of a compound. A general representation of the decomposition reaction is given below.

Water can be decomposed into hydrogen and oxygen by a process called electrolysis.

$$XY \longrightarrow X + Y$$

The decomposition of mercury(II) oxide to yield mercury and oxygen, discussed in Chapter 3, exemplifies this type of reaction. This reaction was studied by early chemists. Their quantitative findings on this reaction helped establish the law of conservation of mass.

$$2HgO(s) \xrightarrow{\Delta} 2Hg(l) + O_2(g)$$

A common process used in the undergraduate laboratory to produce oxygen involves the decomposition of potassium chlorate. In this reaction manganese dioxide is used as a catalyst.

$$2KClO_3(s) \xrightarrow[\Delta]{MnO_2} 2KCl(s) + 3O_2(g)$$

Some carbonates will decompose to yield oxides and carbon dioxide when heated. For example, calcium carbonate decomposes in the following way.

$$CaCO_3(s) \xrightarrow{\Delta} CaO(s) + CO_2(g)$$

When the calcium carbonate is decomposed simultaneously in a combustion process which generates sulfur dioxide, the calcium oxide reacts with the sulfur dioxide to form calcium sulfite. Calcium carbonate is the material which deposits in pipes and in equipment from hard water. It is the major constituent of coral skeletons and is also produced by reaction of atmospheric carbon dioxide with calcium ions contained in the oceans.

Some bicarbonates decompose to yield water, carbon dioxide, and a carbonate salt.

$$2NaHCO_3(s) \xrightarrow{\Delta} Na_2CO_3(s) + H_2O(g) + CO_2(g)$$

Sodium bicarbonate may be used to put out fires because the carbon dioxide formed by its decomposition prevents oxygen from feeding the fire.

Molten nitrates may decompose in a variety of ways, as exemplified by sodium nitrate and ammonium nitrate.

$$2NaNO_3(l) \xrightarrow{\Delta} 2NaNO_2(l) + O_2(g)$$

$$NH_4NO_3(l) \xrightarrow{\Delta} N_2O(g) + 2H_2O(g)$$

The decomposition of nitrates may occur with explosive violence.

Another compound which may decompose at a very rapid rate resulting in an explosion is TNT.

$$\underset{\text{TNT}}{2C_7H_5N_3O_6(s)} \longrightarrow 3N_2(g) + 5H_2O(g) + 7CO(g) + 7C(s)$$

The explosive effect of TNT is caused by the release of a large volume of gas in a short time.

11.8 SUBSTITUTION REACTIONS

In a substitution reaction, one element substitutes for or replaces another element in a compound. Thus an element and a compound produce another element and a new compound. The elements involved may be metals or nonmetals. For example, the metal A may replace the metal ion B or the hydrogen ion B.

$$A + BY \longrightarrow B + AY$$

When the elements are nonmetals the replacement is:

$$X + BY \longrightarrow BX + Y$$

The direction of substitution of one metal by another is given by the electromotive series (Table 11.2). Further discussion of the electromotive series will be given in Chapter 19. The principle of the electromotive series is that **any metal will substitute for a metal ion below it in the series.** Thus zinc will replace the copper in copper(II) sulfate because zinc is above copper in the electromotive series.

$$Zn(s) + CuSO_4(aq) \longrightarrow ZnSO_4(aq) + Cu(s)$$

A strip of zinc immersed in a solution of copper(II) sulfate will disappear and free copper metal will come out of the solution.

Hydrogen is included in the electromotive series although it is not a metal. Any metal above hydrogen will react with an acid to form a metal salt and hydrogen gas.

$$Zn(s) + H_2SO_4(aq) \longrightarrow ZnSO_4(aq) + H_2\uparrow$$
$$Sn(s) + 2HCl(aq) \longrightarrow SnCl_2(aq) + H_2$$

TABLE 11.2 **electromotive series**

Symbol	Name	Activity
Li	lithium	↑ increasing activity
K	potassium	
Ba	barium	
Ca	calcium	
Na	sodium	
Mg	magnesium	
Al	aluminum	
Zn	zinc	
Fe	iron	
Cd	cadmium	
Ni	nickel	
Sn	tin	
H	hydrogen	
Cu	copper	
Hg	mercury	
Ag	silver	
Au	gold	

Can zinc metal replace Sn^{2+} in solution?

If the metal is below an ion of a metal in the electromotive series, then no reaction will occur. Thus silver will not react with Na_2SO_4.

$$Ag + Na_2SO_4 \longrightarrow \text{no reaction}$$

Similarly, a metal below hydrogen in the electromotive series will not react with an acid.

$$Au + HCl \longrightarrow \text{no reaction}$$

There are numerous industrial processes in which metals are produced by reacting a compound with an active metal. Usually these reactions occur in the molten or solid state at high temperatures.

$$Cr_2O_3(s) + 2Al(s) \longrightarrow 2Cr(l) + Al_2O_3(l)$$
$$3MnO_2(s) + 4Al(s) \longrightarrow 3Mn(l) + 2Al_2O_3(l)$$
$$Sb_2S_3(s) + 3Fe(s) \longrightarrow 2Sb(l) + 3FeS(l)$$

A reactivity series had also been formulated for substitution reactions of a nonmetal for a nonmetallic ion. For the halogens, the activity order is F > Cl > Br > I. Fluorine will replace chloride, bromide, or iodide ions; chlorine will replace bromide or iodide; bromine will replace only iodide. For example, chlorine gas can be used to obtain bromine from the bromide ion contained in seawater.

$$Cl_2(g) + 2NaBr(aq) \longrightarrow 2NaCl(aq) + Br_2(aq)$$

Many metals are produced by substitution reactions in which the nonmetal combined with the metal is removed by reaction with another nonmetal. Thus germanium, molybdenum, and tungsten are obtained from their oxides by reacting with hydrogen.

$$GeO_2(s) + 2H_2(g) \xrightarrow{\Delta} Ge(s) + 2H_2O(g)$$

$$MoO_3(s) + 3H_2(g) \xrightarrow{\Delta} Mo(s) + 3H_2O(s)$$

$$WO_3(s) + 3H_2(g) \xrightarrow{\Delta} W(s) + 3H_2O(g)$$

A few sulfides of metals react with oxygen to yield the free metal and sulfur dioxide.

$$HgS(s) + O_2(g) \xrightarrow{\Delta} Hg(l) + SO_2(g)$$

$$Cu_2S(l) + O_2(g) \xrightarrow{\Delta} 2Cu(l) + SO_2(g)$$

Carbon can be used to convert some metal oxides into the free metals and carbon monoxide.

$$SnO_2(s) + 2C(s) \xrightarrow{\Delta} Sn(l) + 2CO(g)$$

$$ZnO(s) + C(s) \xrightarrow{\Delta} Zn(g) + CO(g)$$

$$As_4O_6(s) + 6C(s) \xrightarrow{\Delta} As_4(g) + 6CO(g)$$

11.9 METATHESIS REACTIONS

In metathesis reactions, two compounds react to exchange atoms or groups of atoms (such as complex ions) producing two different compounds. Thus metathesis reactions are also called double substitution reactions or exchange reactions. A general representation of a metathesis reaction is:

$$AX + BY \longrightarrow AY + BX$$

where any of the letters may be single atoms or complex ions containing several atoms.

One of the reasons for a metathesis reaction to occur is the formation of an insoluble material which precipitates from solution. The exchange of partners occurs so that the least soluble compound is produced.

In order to predict when a metathesis reaction will occur, it is necessary to learn the solubilities of ionic compounds. A list of generalizations of solubilities in water at room temperature is given in Table 11.3.

Consider the reaction of silver nitrate and sodium chloride in aqueous solution. Since silver chloride is insoluble a metathesis reaction will occur to yield a precipitate. Sodium nitrate remains in solution.

$$AgNO_3(aq) + NaCl(aq) \longrightarrow AgCl\downarrow + NaNO_3(aq)$$

No apparent reaction will occur with calcium nitrate and sodium chloride

TABLE 11.3
general solubility rules

1 All nitrate and acetate ($C_2H_3O_2^-$) salts are soluble in water.
2 With the exception of AgCl, Hg_2Cl_2, and $PbCl_2$, all other chlorides are soluble. Lead(II) chloride is soluble in hot water.
3 All oxides and hydroxides are insoluble with the exception of alkali metals and Ca, Sr, Ba of the alkaline earth metals. Calcium hydroxide is moderately soluble.
4 With the exception of the alkali metals, alkaline earth metals, and the ammonium ion, all sulfides are insoluble.
5 With the exception of the alkali metals and the ammonium ions, all phosphates and carbonates are insoluble.
6 With the exception of $BaSO_4$, $SrSO_4$, and $PbSO_4$, all sulfates are soluble. Calcium and silver sulfate are slightly soluble.
7 Most alkali metal and ammonium salts are soluble.

because both of the potential products, calcium chloride and sodium nitrate, are soluble in water.

$$Ca(NO_3)_2 + NaCl \longrightarrow \text{no reaction}$$

Acids and bases may provide the ions for metathesis reactions. A salt and an acid may form a precipitate, as in the case of barium chloride and sulfuric acid.

$$BaCl_2 + H_2SO_4 \longrightarrow BaSO_4\downarrow + 2HCl$$

A salt and a base may form a precipitate as in the case of nickel(II) nitrate and potassium hydroxide.

$$Ni(NO_3)_2(aq) + 2KOH \longrightarrow 2KNO_3 + Ni(OH)_2\downarrow$$

When a metal carbonate or bicarbonate reacts with acid, the metathesis reaction should yield carbonic acid, H_2CO_3. However, carbonic acid is unstable and decomposes to yield carbon dioxide and water. Since a gas is released, the reaction occurs.

$$MgCO_3(s) + H_2SO_4(aq) \longrightarrow MgSO_4(aq) + H_2O + CO_2\uparrow$$

Magnesium carbonate is one of the ingredients in several antacids. Antacids react with the HCl in the stomach to produce carbon dioxide.

11.10 NEUTRALIZATION REACTIONS

Neutralization reactions occur whenever an acid or an acid oxide react with a base or a basic oxide. The general equation is identical to that of a metathesis reaction.

$$AX + BY \longrightarrow AY + BX$$

In addition to the fact that AX must be an acid or acid oxide, and BY must be a base or base oxide, the neutralization reaction is one in which one of the products is usually water. The formation of water is often the reason why the neutralization reaction occurs.

The various combinations of reactants possible for a neutralization reaction are:

1 acid plus base,
2 acid oxide plus base,
3 acid plus basic oxide,
4 acid oxide plus basic oxide.

Examples of each of these types are given below.

1 $HCl(aq) + KOH(aq) \longrightarrow KCl(aq) + H_2O$
acid + base ⟶ salt + water

2 $CO_2 + 2NaOH(aq) \longrightarrow Na_2CO_3(aq) + H_2O$
acid oxide + base ⟶ salt + water

3 $ZnO + 2HNO_3 \longrightarrow Zn(NO_3)_2 + H_2O$
basic oxide + acid ⟶ salt + water

4 $BaO(s) + SO_3(g) \longrightarrow BaSO_4(s)$
basic oxide + acid oxide ⟶ salt

SUMMARY

*A **balanced** chemical equation properly expresses a chemical reaction. The equations employ symbols to indicate the experimental conditions and the states of the reactants and products.*

A list of rules for balancing equations is given in Section 11.3 *with which most reactions may be readily balanced.*

*Chemical reactions may be classified into **combination**, **decomposition**, **substitution**, **metathesis**, and **neutralization** reactions.*

QUESTIONS AND PROBLEMS

Definitions

1. Define each of the following terms.
(a) word equation **(b)** chemical equation
(c) chemical symbol **(d)** reactant
(e) product **(f)** combination reaction
(g) decomposition reaction **(h)** substitution reaction
(i) metathesis reaction **(j)** neutralization reaction

2. Can an element such as oxygen be a reactant in one reaction but a product in another reaction? Explain using examples.

3. Are combination and decomposition reactions related in any way? Explain.

4. Explain the meaning of each of the following symbols used in equations.

(a) $\rightarrow$ **(b)** $=$
(c) $\rightleftharpoons$ **(d)** (g)
(e) (l) **(f)** (s)
(g) Δ **(h)** (aq)
(i) $+$ **(j)** $\uparrow$
(k) $\downarrow$

Classification of reactions

5. Classify each of the following reactions by type.

(a) $H_2 + Br_2 \rightarrow 2HBr$
(b) $2Al + 3CuSO_4 \rightarrow Al_2(SO_4)_3 + 3Cu$
(c) $2PbO_2 \rightarrow 2PbO + O_2$
(d) $4Fe + 3O_2 \rightarrow 2Fe_2O_3$
(e) $CaCO_3 + 2HCl \rightarrow CaCl_2 + H_2O + CO_2$
(f) $2KNO_3 \rightarrow 2KNO_2 + O_2$
(g) $Br_2 + 2KI \rightarrow 2KBr + I_2$
(h) $3H_2 + N_2 \rightarrow 2NH_3$
(i) $Pb(NO_3)_2 + H_2S \rightarrow PbS + 2HNO_3$
(j) $Cd + H_2SO_4 \rightarrow CdSO_4 + H_2$
(k) $AgNO_3 + HCl \rightarrow AgCl + HNO_3$
(l) $MgSO_4 \cdot 7H_2O \rightarrow MgSO_4 + 7H_2O$

Combination equations

6. Complete and balance the equations for each of the following combination reactions.

(a) $P + I_2 \rightarrow$ **(b)** $SO_2 + H_2O \rightarrow$
(c) $Na_2O + H_2O \rightarrow$ **(d)** $H_2 + Cl_2 \rightarrow$
(e) $P_2O_5 + H_2O \rightarrow$ **(f)** $Al + Cl_2 \rightarrow$
(g) $CaO + CO_2 \rightarrow$ **(h)** $BaO + SO_3 \rightarrow$
(i) $Si + O_2 \rightarrow$ **(j)** $Si + F_2 \rightarrow$

Decomposition equations

7. Complete and balance the equations for each of the following decomposition reactions.

(a) $CdCO_3 \xrightarrow{\Delta}$

(b) $CaSO_4 \cdot 2H_2O \xrightarrow{\Delta}$

(c) $NaHCO_3 \xrightarrow{\Delta}$

(d) $HgO \xrightarrow{\Delta}$

(e) $KClO_3 \xrightarrow{\Delta}$

(f) $H_2O \xrightarrow{\text{electrolysis}}$

Replacement equations

8. Complete and balance the equations for each of the following replacement reactions.

(a) $Mg + CO \rightarrow$ **(b)** $Cl_2 + KI \rightarrow$
(c) $Na + H_2O \rightarrow$ **(d)** $Zn + NiCl_2 \rightarrow$
(e) $Cd + CuSO_4 \rightarrow$ **(f)** $Al + H_2SO_4 \rightarrow$
(g) $Cd + HCl \rightarrow$

9. Using the electromotive series as a guide, determine which of the following reactions will not occur.

(a) $2Au + NiSO_4 \rightarrow Au_2SO_4 + Ni$
(b) $2Ag + ZnSO_4 \rightarrow Ag_2SO_4 + Zn$
(c) $Hg + 2HCl \rightarrow HgCl_2 + H_2$
(d) $Cu + 2NaCl \rightarrow CuCl_2 + 2Na$

Metathesis equations

10. Complete and balance the equations for each of the following metathesis reactions.

(a) $TiCl_4 + H_2O \rightarrow$ **(b)** $Sb_2S_3 + HCl \rightarrow$
(c) $Pb(NO_3)_2 + HCl \rightarrow$ **(d)** $AgNO_3 + H_2S \rightarrow$
(e) $AgNO_3 + HCl \rightarrow$ **(f)** $FeCl_3 + NaOH \rightarrow$
(g) $BaCl_2 + Na_2CO_3 \rightarrow$

Neutralization equations

11. Complete and balance each of the following neutralization equations.

(a) $NaOH + HCl \rightarrow$ **(b)** $Ca(OH)_2 + HCl \rightarrow$
(c) $BaO + HCl \rightarrow$ **(d)** $NaOH + H_2SO_4 \rightarrow$
(e) $Al(OH)_3 + H_2SO_4 \rightarrow$ **(f)** $Zn(OH)_2 + HNO_3 \rightarrow$
(g) $Fe(OH)_3 + HNO_3 \rightarrow$ **(h)** $KOH + H_3PO_4 \rightarrow$
(i) $Fe_2O_3 + H_3PO_4 \rightarrow$

Word equations

12. Using the following word equations, write balanced chemical equations.

(a) mercury(II) oxide yields mercury and oxygen
(b) sodium hydroxide and sulfuric acid yield sodium sulfate and water
(c) potassium nitrate yields potassium nitrite and oxygen
(d) phosphorus and oxygen yield diphosphorus pentoxide
(e) ammonia plus oxygen yields nitrogen oxide and water
(f) zinc oxide and sulfuric acid yield zinc sulfate and water
(g) bismuth(III) sulfide and oxygen yield bismuth(III) oxide and sulfur dioxide

Balancing equations

13. Balance the following equations.

(a) $C_4H_{10} + O_2 \rightarrow CO_2 + H_2O$

(b) $H_3PO_4 + KOH \rightarrow K_3PO_4 + H_2O$

(c) $KClO_4 \rightarrow KCl + O_2$

(d) $CaO + C \rightarrow CaC_2 + CO$

(e) $C_5H_{12} + O_2 \rightarrow CO_2 + H_2O$

(f) $NaHSO_3 + H_2SO_4 \rightarrow Na_2SO_4 + SO_2 + H_2O$

(g) $Ca(CN)_2 + H_2SO_4 \rightarrow CaSO_4 + HCN$

12 STOICHIOMETRY

LEARNING OBJECTIVES

When you finish this chapter you should be able to:

(a) inspect an equation and determine whether any reactants given are limiting reagents.

(b) distinguish between theoretical yield and actual-yield.

(c) determine the percent yield of a reaction.

(d) develop a systematic approach to solving stoichiometry problems.

(e) set up mole ratios for the substances in a balanced equation.

(f) calculate the number of moles of one substance which may be produced from a number of moles of another substance.

(g) relate the moles of a substance in a reaction to its equivalent in mass or volume.

(h) match the following terms with their corresponding definitions.

actual yield
limiting reagent
mole ratio
percent yield
stoichiometry
STP
theoretical yield

The chemist in the laboratory must know how to calculate the amounts of materials required to produce a certain quantity of a desired substance. Similarly, a chemical company must calculate how much raw material it needs to produce a certain amount of a product. In either situation, regardless of whether grams or tons, liters or tank cars are involved, the method of calculation is the same.

In Chapter 10 we learned about the relationship between the formula of a compound and the mass of a mole of the substance, and in Chapter 11 we learned how to balance equations. We will combine these concepts and techniques to show how balanced equations provide information about the quantities of materials which react and are produced in a chemical reaction. ***The mathematical calculation of the quantities of reactants and the products of a reaction given by the chemical equation is called stoichiometry.*** This word is derived from the Greek *stoichion* (element) and *metry* (to measure).

In this chapter we will approach stoichiometry problems systematically and determine how to solve these problems based on moles, mass, and volume.

12.1 THE LIMITING REAGENT

Given a certain amount of one material and a balanced equation for its reaction with other substances, it is possible to calculate the exact quantities of each substance required for the reaction and the amount of products produced. However, consider the possibility that the exact stoichiometric amounts are not used, but instead one or more of the reagents is used in excess of the amount required. How does this affect the amount of product formed? When such a situation exists, **the reagent that is not present in excess is called the limiting reagent.** The amount of product derived from the reaction can be no more than a quantity calculated from the limiting reagent. For any mixture of reactants, it is possible to determine whether one reagent is present in limiting amounts by examining the balanced equation for the reaction. The limiting reagent is that material which has the smallest mole-to-coefficient ratio. This ratio is obtained by dividing the moles of the reactant by the coefficient of that reactant in the balanced equation.

The concept of a limiting reagent may be better understood by considering an analogy such as the effectiveness of an air force in terms of pilots and airplanes. A plane cannot be of any use unless a pilot is available to fly it. Similarly, a group of trained pilots is of little use unless they have planes to fly. Consider the following equation.

$$1 \text{ pilot} + 1 \text{ plane} \longrightarrow 1 \text{ manned aircraft}$$

If a country has 100 planes and 300 pilots, we know intuitively that its air force is limited by the number of planes. We can verify our intuition by dividing these quantities by the appropriate coefficients of the equation to give quantity to coefficient ratios.

$$\frac{100 \text{ planes}}{1 \text{ plane}} = 100 \qquad \frac{300 \text{ pilots}}{1 \text{ pilot}} = 300$$

How many basketball teams may be formed from two centers, six forwards, and six guards?

The lowest ratio is for the planes, proving them to be the limiting quantity.

For the reaction of nitrogen and hydrogen to give ammonia, the balanced equation is:

$$N_2 + 3H_2 \longrightarrow 2NH_3$$

A chemist who has 2 moles of nitrogen and 3 moles of hydrogen available for the reaction can determine which reagent is limiting by looking at the ratio of the moles available to the coefficients of the equation. The ratio is lowest for hydrogen and it is the limiting reagent.

$$\frac{2 \text{ moles } N_2}{1 \text{ mole } N_2} = 2 \qquad \frac{3 \text{ moles } H_2}{3 \text{ moles } H_2} = 1$$

12.2 THEORETICAL AND PERCENT YIELDS

In the following sections, you will learn how to calculate the theoretical yield for a variety of reactions. **The theoretical yield is that amount of product that would be obtained from the complete reaction of the limiting reagent to give the indicated products.** The theoretical yield is the maximum amount which may be formed from the amounts of reagents given. It may never be realized, and seldom is, under actual reaction conditions. In real situations, some of the reactant is left over and some of the products are lost in attempts to isolate and purify them. Furthermore, other minor reactions may occur to give other products. For example, when carbon and oxygen react to form carbon dioxide, another reaction frequently occurs which forms carbon monoxide.

$$C + O_2 \longrightarrow CO_2$$
$$2C + O_2 \longrightarrow 2CO$$

If complete combustion to give carbon dioxide is the desired reaction, then the formation of carbon monoxide is the side reaction. To the extent that some of the carbon is converted to undesired material, the theoretical yield of carbon dioxide will not be obtained.

The amount of the desired product actually obtained from a chemical process is called the actual yield. Actual yields can be lowered by the competition of side reactions, by the presence of unreacted reactant when the reaction is terminated, or by mechanical losses when the desired material is separated from other products or reactants which were present in excess.

The percent yield in a chemical reaction is the actual yield divided by the theoretical yield multiplied by 100.

$$\% \text{ yield} = \frac{\text{actual yield}}{\text{theoretical yield}} \times 100$$

Theoretical yield
Actual yield
Percent yield

The percent yield is a critical consideration of all industrial chemical processes. The cost of a process and the percent yield are directly interconnected. The price of the product depends on the cost of the starting reactants, the efficiency of the process, or the percent yield. For a given investment which must be recovered in sales of product, a process which gives a 40% yield leads to a product which must be sold at twice the price of the same product which is obtained by an alternate process which gives an 80% yield.

12.3 STOICHIOMETRY AND THE BALANCED EQUATION

Every balanced equation can be used to provide a great deal more information than is directly evident from the elemental symbols and coefficients. The quantities of reactants and products involved in the reaction can be calculated on a mole, mass, or volume basis. A review of the information derivable from an equation can be summarized as follows:

$$2H_2(g) + O_2(g) \xrightarrow{\Delta} 2H_2O(g)$$

1 The reactants are hydrogen and oxygen in the gaseous state. The product is gaseous water.
2 Two molecules of hydrogen and one molecule of oxygen react to produce two molecules of water.
3 Two moles of hydrogen molecules and one mole of oxygen molecules combine to produce two moles of water. Remember that a mole contains a specific number of molecules, 6.02×10^{23}. One mole of any substance contains the same number of atoms, molecules, or formula units which represent the simplest unit of the substance.
4 In terms of grams, 4 grams of hydrogen and 32 grams of oxygen combine to yield 36 grams of water. These quantities are the gram equivalents of the moles involved.

1 mole
6.02×10^{23} molecules

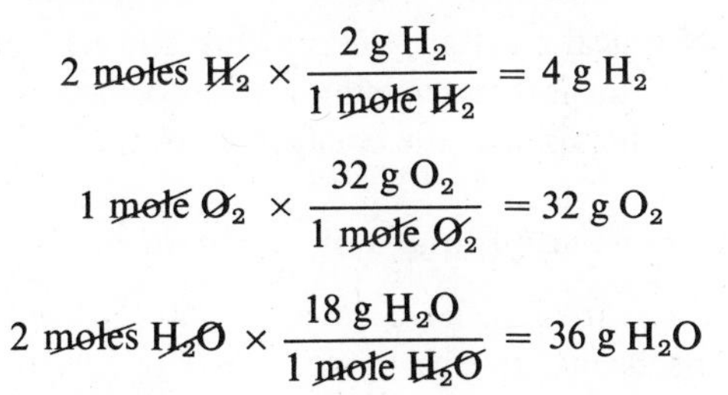

$$2 \text{ moles } H_2 \times \frac{2 \text{ g } H_2}{1 \text{ mole } H_2} = 4 \text{ g } H_2$$

$$1 \text{ mole } O_2 \times \frac{32 \text{ g } O_2}{1 \text{ mole } O_2} = 32 \text{ g } O_2$$

$$2 \text{ moles } H_2O \times \frac{18 \text{ g } H_2O}{1 \text{ mole } H_2O} = 36 \text{ g } H_2O$$

Volumes of gases and equations

There is yet another way of reading chemical equations which involve gases. Since gases are so light, it is much easier to measure them by volume than by mass. Although we have yet to study gases in detail, you have been made aware of the dependence of gas volume on pressure and temperature. Thus in order to conveniently discuss volumes of gaseous reactants and products, a standard reference pressure and temperature must be chosen. **The standard pressure is one atmosphere and the standard temperature is 0°C.** These quantities are called STP, for standard temperature and pressure.

Equal volumes of different gases at the same temperature and pressure contain the same number of molecules. Thus one liter of hydrogen at STP contains the same number of hydrogen molecules as the number of oxygen molecules in one liter of oxygen at STP. Therefore, we may directly compare the number of moles contained in various gas samples by comparing the number of liters of the gases at STP.

STP
mole
22.4 liters

At STP, 1 mole of a gas occupies 22.4 liters. Again it must be emphasized that 1 mole of any gas occupies this standard molar volume. A 2.0 g sample of hydrogen molecules is 1 mole and occupies 22.4 liters at STP; a 32.0-g sample of oxygen molecules is also 1 mole and occupies 22.4 liters at STP.

As you will learn in Chapter 14, there are definite relationships governing the effect of pressure and temperature on the volume of a gas. All gases behave the same. Identical changes in temperature and/or the pressure on equal volumes of different gases will cause identical changes in volume. Therefore, providing the experimental conditions are the same, equal volumes of gases always contain the same number of molecules or moles.

For the reaction of hydrogen and nitrogen to yield ammonia, the following equivalent statements may be made.

$$N_2 + 3H_2 \longrightarrow 2NH_3$$

$$1 \text{ mole } N_2 + 3 \text{ moles } H_2 \longrightarrow 2 \text{ moles } NH_3$$

$$22.4 \text{ liters } N_2 \text{ at STP} + 3(22.4 \text{ liters}) \ H_2 \text{ at STP} \longrightarrow 2(22.4 \text{ liters}) \ NH_3 \text{ at STP}$$

$$1 \text{ volume } N_2 \text{ at STP} + 3 \text{ volumes } H_2 \text{ at STP} \longrightarrow 2 \text{ volumes } NH_3 \text{ at STP}$$

$$1 \text{ volume } N_2 + 3 \text{ volumes } H_2 \longrightarrow 2 \text{ volumes } NH_3$$

For a chemical reaction, the pressure and temperature can be kept constant. Therefore, the last statement above is correct. Note that the relative volumes, then, are the same as the coefficients of the equation.

Methods of solving stoichiometric problems

While there are a variety of methods which could be used to solve stoichiometry problems, the best established one is the mole method. The mole method is rooted in the factor unit method which was described in Chapter 2. The number of steps involved in working problems by the mole method depends on the units in which the known quantities are measured and the units required of the unknown. The first step is to convert the known quantities into their molar quantities. Then the stoichiometric relationship between the known substances and the desired unknown is established from the balanced equation. From this relationship, the moles of the unknown are determined. Finally, the moles of the unknown are converted into the desired units. The final answer must be expressed to the proper number of significant figures. The three steps of the mole method are outlined in Table 12.1. The equivalent representation of the mole method is illustrated in Figure 12.1.

TABLE 12.1
a summary of the mole method

1 From the mass or volume (if gaseous) of the reactant or product given, convert to the mole equivalent.
2 Examine the balanced equation and calculate the moles of the desired unknown which can be obtained from the moles of the substance calculated in step 1.
3 From the moles of the desired substance calculated in step 2, determine the equivalent quantity in terms of mass or volume (if gaseous).

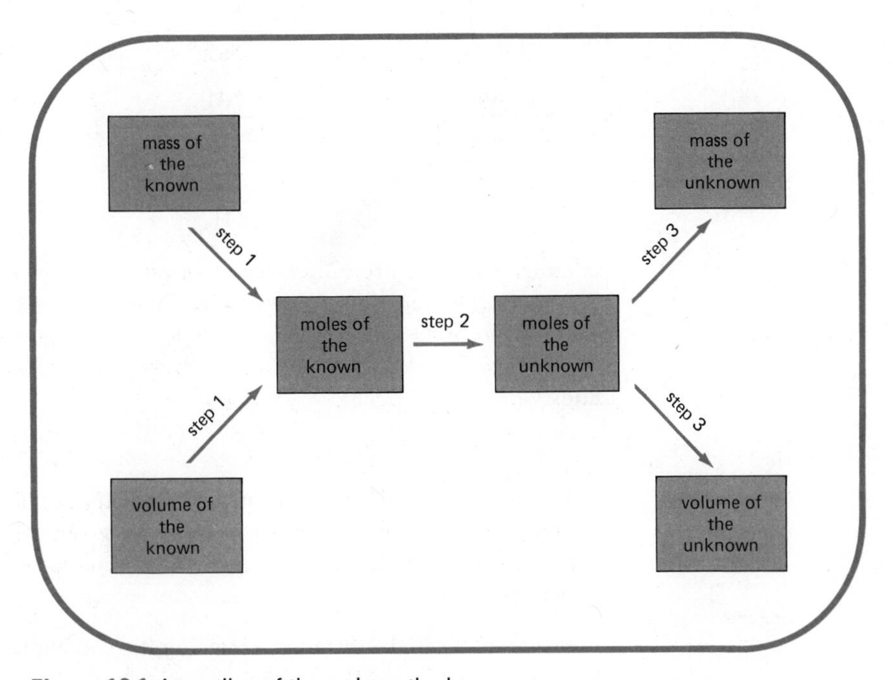

Figure 12.1 An outline of the mole method.

The number of types of stoichiometry problems which are possible are:

1 mole-mole,
2 volume-volume,
3 mole-volume or volume-mole,
4 mole-mass or mass-mole,
5 volume-mass or mass-volume,
6 mass-mass.

We shall systematically examine examples of each of these types of stoichiometry problems.

The mole ratio

The central step, step 2, of the mole method depends on relating the number of moles of one substance to the number of moles of another substance. In order to do this, a mole ratio is used. **The mole ratio is a ratio between the number of moles of two substances involved in a chemical reaction.** Thus for the reaction of hydrogen and nitrogen to give ammonia:

$$N_2 + 3H_2 \longrightarrow 2NH_3$$

There are six mole ratios which may be written. All of these ratios contain the coefficients of the corresponding substances in the equation.

$$\frac{1 \text{ mole } N_2}{3 \text{ moles } H_2} \quad \text{or} \quad \frac{3 \text{ moles } H_2}{1 \text{ mole } N_2}$$

$$\frac{1 \text{ mole } N_2}{2 \text{ moles } NH_3} \quad \text{or} \quad \frac{2 \text{ moles } NH_3}{1 \text{ mole } N_2}$$

$$\frac{3 \text{ moles } H_2}{2 \text{ moles } NH_3} \quad \text{or} \quad \frac{2 \text{ moles } NH_3}{3 \text{ moles } H_2}$$

As indicated, there are only three mole ratios which relate different pairs of substances. The remaining three are reciprocal ratios of the same quantities. Each of the mole ratios applies only to the reaction under consideration. Thus for a reaction of nitrogen and hydrogen to produce hydrazine, the mole ratios would be different.

$$N_2 + 2H_2 \longrightarrow N_2H_4$$

In using the mole method for stoichiometry problems, it is necessary to convert from the known number of moles of given starting substance to the number of moles of desired substance. These quantities can be related by the proper mole ratio.

moles of desired substance = (mole ratio) × (moles of given starting substance)

The mole ratio then must be:

$$\text{mole ratio} = \frac{\text{moles desired substance in balanced equation}}{\text{moles starting substance in balanced equation}}$$

Thus by choosing the proper mole ratio, the units will be correct as will be your answer. The mole ratio, then, is the necessary factor of the factor unit method. Note that **the mole ratio is an exact number and carries any number of significant figures desired.** Thus the number of significant figures in an answer depends only on the number of significant figures in the starting substance.

12.4 MOLE-MOLE STOICHIOMETRY PROBLEMS

mole of known —step 2→ mole of unknown

In this first type of stoichiometric calculation, you will find that the steps and math involved are easy. However, the principles established are essential to solution of more complex problems to be described in subsequent sections. Therefore, **make sure that you fully understand how to do the problems given in this section prior to trying the problems in other sections.**

The problem is to find how to calculate the number of moles of one substance which will react with or be produced from a second substance. The necessary information is given in a balanced equation which establishes the mole relationship between all reactants and products. Recall from Chapter 11 that the number in front of each formula in an equation is called the coefficient. **The coefficient gives the relative number of moles of that substance in the balanced equation with respect to the relative number of moles of every other reactant or product in the equation.** Thus in the reaction of nitrogen and hydrogen to produce ammonia, we may write:

$$3H_2 + N_2 \longrightarrow 2\,NH_3$$
$$3 \text{ moles of } H_2 + 1 \text{ mole of } N_2 \longrightarrow 2 \text{ moles of } NH_3$$

Problem 12.1 Ethane (a hydrocarbon) gas will burn in oxygen to produce carbon dioxide and water according to the following equation. How many moles of carbon dioxide can be produced from two moles of ethane?

$$2C_2H_6 + 7O_2 \longrightarrow 4CO_2 + 6H_2O$$

Solution We may bypass both steps 1 and 3 of the mole method because both the quantity given and the quantity sought are in moles. Only step 2 is needed. The coefficients of the equation indicate that 2 moles of ethane will yield 4 moles of carbon dioxide. Thus the answer to this problem is immediately obtained. Four moles of CO_2 are produced.

Problem 12.2

How many moles of ammonia (NH_3) are produced from 12 moles of hydrogen?

$$3H_2 + N_2 \longrightarrow 2NH_3$$

Solution

First establish the facts that the coefficients give you; 3 moles of hydrogen will yield 2 moles of ammonia. The mole ratio necessary for the solution of the problem must cancel the units of moles of hydrogen and leave moles of ammonia.

$$12 \text{ moles } H_2 \times (\text{mole ratio}) = ?\text{ mole } NH_3$$

$$12 \cancel{\text{moles}} \cancel{H_2} \times \left(\frac{2 \text{ moles } NH_3}{3 \cancel{\text{moles}} \cancel{H_2}}\right) = ?\text{ mole } NH_3$$

ANSWER

8 moles NH_3

Problem 12.3

Using the same equation given in Problem 12.1, determine the number of moles of oxygen required to burn completely 4 moles of ethane.

$$2C_2H_6 + 7O_2 \longrightarrow 4CO_2 + 6H_2O$$

Solution

The equation provides the information that 7 moles of oxygen are required to burn 2 moles of ethane. The ratio of the moles of the two substances in question is first established.

$$\text{mole ratio} = \frac{7 \text{ moles of } O_2}{2 \text{ moles of } C_2H_6}$$

This ratio, when multiplied by the quantity 4 moles of ethane, will give the answer in moles of oxygen.

$$4 \cancel{\text{moles}} \cancel{C_2H_6} \times \frac{7 \text{ moles } O_2}{2 \cancel{\text{moles}} \cancel{C_2H_6}} = ?\text{ mole } O_2$$

ANSWER

14 moles O_2

You could have quickly solved this problem by inspection of the equation and avoided the factor unit method. However, the factor unit method employing mole ratios is very helpful in solving the more difficult problems described in other sections.

12.5 VOLUME-VOLUME STOICHIOMETRY PROBLEMS

When two or more of the reactants and/or products are gases under the experimental conditions, problems dealing only with volumes may be done. Recall from Section 12.3 that equal volumes of gases under the same conditions of pressure and temperature contain the same number of molecules and occupy the same volume. Since a chemical reaction is carried out under conditions of constant pressure and temperature, the volumes of the gases reacting or produced are proportional to the number of moles of the gases in the balanced equation.

$$N_2 + 3H_2 \longrightarrow 2NH_3$$

$$1 \text{ mole } N_2 + 3 \text{ moles } H_2 \longrightarrow 2 \text{ moles } NH_3$$

$$22.4 \text{ liters } N_2 \text{ at STP} + 3(22.4 \text{ liters}) H_2 \text{ at STP} \longrightarrow 2(22.4 \text{ liters}) NH_3 \text{ at STP}$$

$$1 \text{ volume } N_2 \text{ at STP} + 3 \text{ volumes } H_2 \text{ at STP} \longrightarrow 2 \text{ volumes } NH_3 \text{ at STP}$$

$$1 \text{ volume } N_2 + 3 \text{ volumes } H_2 \longrightarrow 2 \text{ volumes } NH_3$$

$$X \text{ volume } N_2 + 3X \text{ volume } H_2 \longrightarrow 2X \text{ volume } NH_3$$

Note that since a mole of any gas at STP occupies 22.4 liters, the coefficients of the equation when multiplied by 22.4 gives the volumes of the moles of the gases. However, the reaction need not be done at STP. Since the temperature and pressure are constant during the reaction, the effect of these conditions on the volumes of gases applies equally to all volumes, and the ratio of the volumes involved would remain the same.

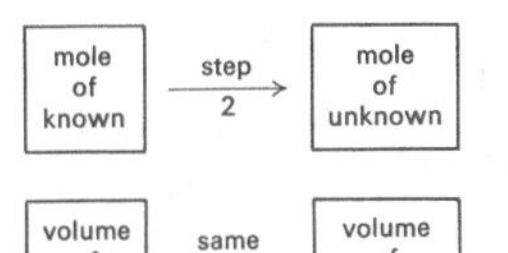

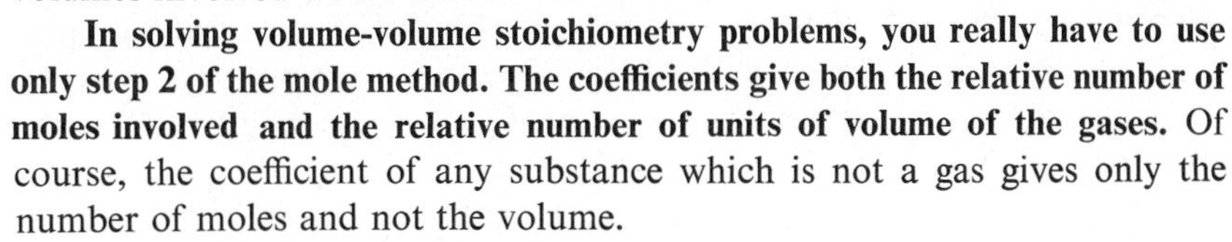

In solving volume-volume stoichiometry problems, you really have to use only step 2 of the mole method. The coefficients give both the relative number of moles involved and the relative number of units of volume of the gases. Of course, the coefficient of any substance which is not a gas gives only the number of moles and not the volume.

Problem 12.4

Methyl alcohol (wood alcohol) is prepared industrially from carbon monoxide and hydrogen at 375°C and 3000 pounds per square inch. A catalyst of chromium oxide and zinc oxide is used. Assuming that sufficient carbon monoxide is available, how many liters of gaseous methyl alcohol can be produced from 2000 liters of hydrogen gas?

$$CO(g) + 2H_2(g) \longrightarrow CH_3OH(g)$$

Solution

Since all of the reactants and products are gaseous under the reaction conditions, the coefficients of the equation give the relative number of liters of each substance required or produced. In order to solve this problem, a volume ratio which will cancel the units of liters of hydrogen and yield liters of methyl alcohol is required.

$$\overline{2000} \text{ liters } H_2 \times \text{volume ratio} = ? \text{ liter } CH_3OH$$

$$\overline{2000} \text{ liters } H_2 \times \frac{1 \text{ liter } CH_3OH}{2 \text{ liters } H_2} = ? \text{ liter } CH_3OH$$

ANSWER

$\overline{1000}$ liters CH_3OH

Problem 12.5 Ammonia is produced industrially from hydrogen and nitrogen at both high pressure and high temperature. How many liters of gaseous ammonia should be produced from $\overline{1000}$ liters of N_2?

$$N_2(g) + 3H_2(g) \longrightarrow 2NH_3(g)$$

Solution The coefficients of the equation indicate that one volume of nitrogen will produce two volumes of ammonia. Therefore, the solution of the problem is given by:

$$\overline{1000} \text{ liters } N_2 \times \frac{2 \text{ liters } NH_3}{1 \text{ liter } N_2} = ? \text{ liter } NH_3$$

ANSWER

$\overline{2000}$ liters NH_3

Problem 12.6 Water gas, a mixture of hydrogen and carbon monoxide, is produced from methane (CH_4) and steam by industrial methods. How many cubic feet of water gas can be produced by reacting $\overline{10{,}000}$ cubic feet of CH_4 with sufficient steam?

$$CH_4(g) + H_2O(g) \longrightarrow CO(g) + 3H_2(g)$$

Solution The equation gives the information that one volume of methane will yield one volume of carbon monoxide and three volumes of hydrogen. The total volume of the water gas is the sum of the volumes of the products. Thus one volume of methane will yield four volumes of water gas.

The problem may be solved using cubic feet as volume units as well as for any other volume units chosen. The coefficients represent relative volume units and not necessarily liters.

$$\overline{10{,}000} \text{ ft}^3 \, CH_4 \times \frac{4 \text{ ft}^3 \text{ water gas}}{1 \text{ ft}^3 \, CH_4} = ? \text{ ft}^3 \text{ water gas}$$

ANSWER

$\overline{40{,}000}$ ft^3 CH_4

Problem 12.7 How many liters of HCl can be produced by reacting 5 liters of hydrogen with 12 liters of chlorine?

$$H_2(g) + Cl_2(g) \longrightarrow 2HCl(g)$$

Solution The coefficients of the equation indicate that 1 volume of hydrogen will react with 1 volume of chlorine to yield 2 volumes of HCl. The problem stated involves the two reactant gases in a ratio which is not 1:1. Thus one of the reactants is in excess and cannot react completely. The 5 liters of hydrogen will react with only 5 liters of chlorine and as a consequence 7 liters of chlorine will remain.

$$5 \text{ liters } H_2 \times \frac{1 \text{ liter } Cl_2}{1 \text{ liter } H_2} = 5 \text{ liters } Cl_2 \text{ required for reaction}$$

$$12 \text{ liters } Cl_2 - 5 \text{ liters } Cl_2 = 7 \text{ liters } Cl_2 \text{ remaining}$$

For every volume of H_2, exactly 2 volumes of HCl result. Thus the volume of HCl produced from the 5 liters of H_2 is given by:

$$5 \text{ liters } H_2 \times \frac{2 \text{ liters HCl}}{1 \text{ liter } H_2} = 10 \text{ liters HCl}$$

In this section you may have found that the problems could be done quickly without using the logical procedure outlined. Furthermore, you may have ignored the use of units. **However, you will find that the sections which follow will be easier if you use all the data and the units in a logical and orderly manner. If you have done the problems casually or incompletely, return and review before proceeding further. If the problems in subsequent sections seem too difficult, return to this and the preceding section for review.**

12.6 MOLE-VOLUME STOICHIOMETRY PROBLEMS

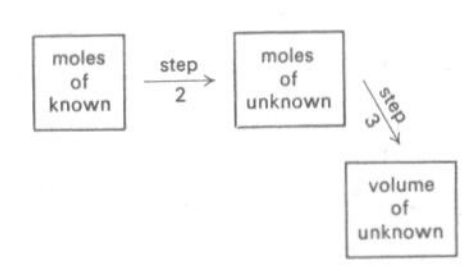

In this type of stoichiometry problem, either the known or unknown must be a gas. **If the known is in mole units, the unknown gas quantity may be calculated in volume units. If the known is in volume units, the unknown is calculated in mole units.** In either case, you need to utilize the fact that one mole of any gas at STP (Section 12.3) occupies 22.4 liters.

Problem 12.8 How many liters of oxygen at STP will be produced by the decomposition of 1.00 mole of potassium nitrate?

$$2KNO_3 \longrightarrow 2KNO_2 + O_2$$

Solution

Since the known is given in moles, step 1 of the mole method is unnecessary. First we calculate the relationship between the moles of known and unknown as required by step 2.

$$1.00\ \cancel{\text{mole}}\ \cancel{\text{KNO}_3} \times \frac{1\ \text{mole O}_2}{2\ \cancel{\text{moles}}\ \cancel{\text{KNO}_3}} = 0.500\ \text{mole O}_2$$

Note that the answer is expressed to three significant figures even though only one significant figure is illustrated in the mole ratio. The mole ratio is exact and carries any number of significant figures desired. Now the 0.5 mole of O_2 is converted into the desired volume units as required in step 3.

$$0.500\ \cancel{\text{mole}}\ \cancel{\text{O}_2} \times \frac{22.4\ \text{liters O}_2}{1\ \cancel{\text{mole}}\ \cancel{\text{O}_2}} = 11.2\ \text{liters O}_2\ \text{at STP}$$

Problem 12.9

How many moles of potassium chlorate are required to produce 33.6 liters of oxygen at STP?

$$2\text{KClO}_3 \longrightarrow 2\text{KCl} + 3\text{O}_2$$

Solution

In this problem, only steps 1 and 2 are required. First the volume of the oxygen is converted into moles, and then the moles of the oxygen are converted into moles of potassium perchlorate by using the coefficients of the balanced equation. Both steps are shown combined in one mathematical expression.

$$\underbrace{33.6\ \cancel{\text{liters}}\ \cancel{\text{O}_2} \times \frac{1\ \cancel{\text{mole}}\ \cancel{\text{O}_2}}{22.4\ \cancel{\text{liters}}\ \cancel{\text{O}_2}}}_{\text{step 1}} \times \underbrace{\frac{2\ \text{moles KClO}_3}{3\ \cancel{\text{moles}}\ \cancel{\text{O}_2}}}_{\text{step 2}} = 1.00\ \text{mole KClO}_3$$

Note that again the answer is expressed to three significant figures consistent with the three significant figures given in the known volume. The mole ratio and the moles of the molar volume are exact quantities.

12.7 MOLE-MASS STOICHIOMETRY PROBLEMS

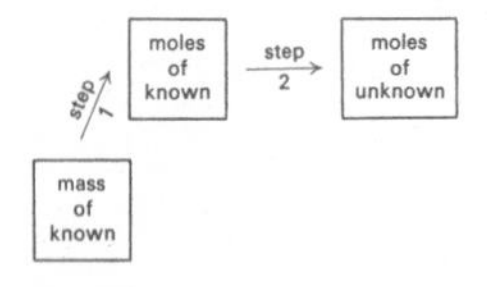

In this type of stoichiometry problem, either the known is given in mass units or the unknown is desired in mass units. The remaining quantity is in mole units. If the known is given in mole units, only steps 2 and 3 of the mole method are required. If the known is given in mass units, then steps 1 and 2 are required.

Problem 12.10 How many moles of oxygen are required to metabolically convert $9\bar{0}$ g of glucose into carbon dioxide and water?

$$C_6H_{12}O_6 + 6O_2 \longrightarrow + 6CO_2 + 6H_2O$$

Solution The molecular weight of $C_6H_{12}O_6$ is 180 amu. Thus 1 mole of glucose has a mass of 174 g. First the number of moles of glucose is calculated as required by step 1.

$$90\ \cancel{g}\ \cancel{C_6H_{12}O_6} \times \frac{1\ \text{mole}\ C_6H_{12}O_6}{180\ \cancel{g}\ \cancel{C_6H_{12}O_6}} = 0.50\ \text{mole}\ C_6H_{12}O_6$$

Now the moles of oxygen required are calculated from the mole ratio obtained from the coefficients of the balanced equation. This is step 2.

$$0.50\ \cancel{\text{mole}}\ \cancel{C_6H_{12}O_6} \times \frac{6\ \text{moles}\ O_2}{1\ \cancel{\text{mole}}\ \cancel{C_6H_{12}O_6}} = 3.0\ \text{moles}\ O_2$$

Problem 12.11 How many grams of ethyl alcohol, C_2H_5OH, can be produced from the fermentation of 10.0 moles of cane sugar, $C_{12}H_{22}O_{11}$?

$$C_{12}H_{22}O_{11} + H_2O \xrightarrow{\text{yeast}} 4C_2H_5OH + 4CO_2$$

Solution

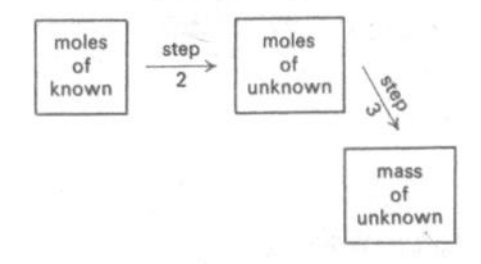

Since the number of moles of the known are given, only steps 2 and 3 are required. First the moles of the unknown C_2H_5OH are calculated by using the mole ratio. Then the moles of C_2H_5OH are converted into the desired quantity, grams of C_2H_5OH. Both steps are combined in one mathematical operation.

$$\underbrace{10.0\ \cancel{\text{moles}}\ \cancel{C_{12}H_{22}O_{11}} \times \frac{4\ \cancel{\text{moles}}\ \cancel{C_2H_5OH}}{1\ \cancel{\text{mole}}\ \cancel{C_{12}H_{22}O_{11}}}}_{\text{step 2}} \times \underbrace{\frac{46.0\ \text{g}\ C_2H_5OH}{1\ \cancel{\text{mole}}\ \cancel{C_2H_5OH}}}_{\text{step 3}}$$

$$= 1840\ \text{g}\ C_2H_5OH$$

12.8 VOLUME-MASS STOICHIOMETRY PROBLEMS

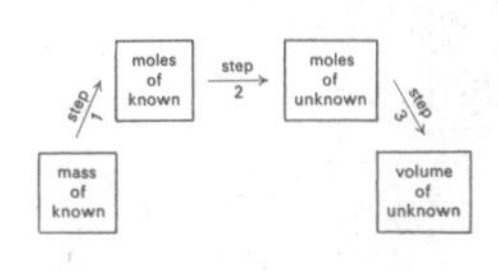

In each of the preceding four types of stoichiometry problems, only two of the three steps of the mole method were required. Now you should be thoroughly familiar with the mole method and the addition of yet another step should present no difficulties if you have carefully studied each of the earlier problems. **Now you will be given the known in mass units and asked to calculate the unknown in volume units, or given the known in volume units and asked to calculate the unknown in mass units.** In either case, you will need to know the molar volume of a gas, 22.4 liters/mole at STP.

Problem 12.12

How many liters of hydrogen gas will be produced at STP from the reaction of 3.91 g of potassium with sufficient water?

$$2K + 2H_2O \longrightarrow 2KOH + H_2(g)$$

Solution

The first step is to calculate the number of moles of potassium in the reaction.

$$3.91 \text{ g K} \times \frac{1 \text{ mole K}}{39.1 \text{ g K}} = 0.100 \text{ mole K}$$

In step 2, the mole ratio is used to determine how many moles of hydrogen may be obtained from 0.100 mole K. The mole ratio is derived from the coefficients of the equation.

$$0.100 \text{ mole K} \times \frac{1 \text{ mole } H_2}{2 \text{ moles K}} = 0.0500 \text{ mole } H_2$$

Finally in step 3, the moles of H_2 are converted to liters of the gas.

$$0.0500 \text{ mole } H_2 \times \frac{22.4 \text{ liters } H_2}{1 \text{ mole } H_2} = 1.12 \text{ liters } H_2$$

Problem 12.13

What quantity of copper(II) oxide will react with 5.60 liters of hydrogen at STP?

$$CuO + H_2 \longrightarrow Cu + H_2O$$

Solution

There are three distinct steps in solving this problem. First the volume of hydrogen must be converted into moles of hydrogen. Then in step 2, the moles of hydrogen are related to the moles of copper(II) oxide by use of a mole ratio derived from the coefficients of the equation. Finally, the moles of copper(II) oxide are converted into grams. All three steps are written within one mathematical expression.

$$5.6 \text{ liters } H_2 \times \frac{1 \text{ mole } H_2}{22.4 \text{ liters } H_2} \times \frac{1 \text{ mole CuO}}{1 \text{ mole } H_2} \times \frac{79.5 \text{ g CuO}}{1 \text{ mole CuO}}$$

$$= 19.9 \text{ g CuO}$$

12.9 MASS-MASS STOICHIOMETRIC PROBLEMS

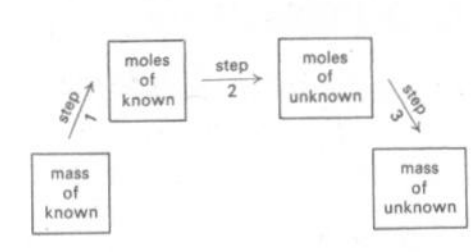

Now you should be quite adept at solving stoichiometric problems. The technique of solving mass-mass stoichiometric problems is very similar to that described for mass-volume stoichiometric problems in the preceding section. **First the mass of the known is converted to moles of the known. Then the moles of the known and unknown are interrelated by a mole ratio. Finally, the moles of the unknown are converted to mass.**

Problem 12.14 How many grams of oxygen are required to burn completely $114\bar{0}$ grams of octane?

$$2C_8H_{18} + 25O_2 \longrightarrow 16CO_2 + 18H_2O$$

Solution The number of moles of octane in the $114\bar{0}$ grams is:

$$114\bar{0}\text{ g } C_8H_{18} \times \frac{1\text{ mole } C_8H_{18}}{114.0\text{ g } C_8H_{18}} = 10.00\text{ moles } C_8H_{18}$$

The number of moles of oxygen required to burn 10.00 moles of C_8H_{18} is:

$$10.00\text{ moles } C_8H_{18} \times \frac{25\text{ moles } O_2}{2\text{ moles } C_8H_{18}} = 125.0\text{ moles } O_2$$

Finally, the mass of oxygen required is:

$$125.0\text{ moles } O_2 \times \frac{32.0\text{ g } O_2}{1\text{ mole } O_2} = 4\overline{000}\text{ g } O_2$$

Problem 12.15 How many grams of carbon dioxide are required by a plant to produce 1.80 grams of glucose by photosynthesis?

$$6CO_2 + 6H_2O \longrightarrow C_6H_{12}O_6 + 6O_2$$

Solution All three steps necessary to solve this problem may be combined in one mathematical expression.

$$1.80\text{ g } C_6H_{12}O_6 \times \underbrace{\frac{1\text{ mole } C_6H_{12}O_6}{180.0\text{ g } C_6H_{12}O_6}}_{\text{step 1}} \times \underbrace{\frac{6\text{ moles } CO_2}{1\text{ mole } C_6H_{12}O_6}}_{\text{step 2}} \times \underbrace{\frac{44.0\text{ g } CO_2}{1\text{ mole } CO_2}}_{\text{step 3}}$$

$$= 2.64\text{ g } CO_2$$

SUMMARY

Stoichiometry *is the mathematical calculation of the quantitites of reactants and the products of a reaction as indicated by a chemical equation. The reagent that is not present in excess is called the* ***limiting reagent****. It is that material which has the smallest mole-to-coefficient ratio.*

The ***theoretical yield*** *is that amount of product that would be obtained from the complete reaction of the limiting reagent to give the indicated products. The amount of the desired product actually obtained from a chemical process is called the* ***actual yield****. The* ***percent yield*** *in a chemical reaction is the actual yield divided by the theoretical yield multiplied by 100.*

Balanced chemical equations provide information about the number of molecules, number of moles, number of grams, and number of liters (if gaseous) of material involved in a reaction. The relationship between two substances of a reaction is expressed as a ***mole ratio*** *which is derived from the coefficients of the equation.*

QUESTIONS AND PROBLEMS

Definitions

1. Define each of the following terms.

(a) mole
(b) mole method
(c) mole ratio
(d) limiting reagent
(e) theoretical yield
(f) percent yield
(g) actual yield
(h) balanced equation
(i) molar volume
(j) stoichiometry
(k) STP

2. Distinguish between each of the following terms.

(a) theoretical yield and actual yield
(b) theoretical yield and percent yield
(c) mole-mole and volume-volume stoichiometry problems

Mole-mole stoichiometry

3. How many moles of oxygen are required to react completely with 10 moles of sulfur dioxide to yield sulfur trioxide?

$$2SO_2 + O_2 \longrightarrow 2SO_3$$

4. How many moles of carbon dioxide are produced in the combustion of 0.5 moles of ethane?

$$2C_2H_6 + 7O_2 \longrightarrow 4CO_2 + 6H_2O$$

5. How many moles of oxygen are required to completely convert 2 moles of iron(II) sulfide to iron(III) oxide?

$$4FeS + 7O_2 \longrightarrow 2Fe_2O_3 + 4SO_2$$

6. How many moles of nitrogen dioxide are produced along with 3 moles of oxygen in the decomposition of nitric acid by light?

$4HNO_3 \longrightarrow 4NO_2 + 2H_2O + O_2$

Volume-volume stoichiometry

7. Calculate the number of liters of nitrogen dioxide produced from the complete reaction of 56 liters of nitrogen oxide.

$2NO(g) + O_2(g) \longrightarrow 2NO_2(g)$

8. What volume of ammonia may be produced from 125 liters of nitrogen assuming that sufficient hydrogen is available?

$N_2(g) + 3H_2(g) \longrightarrow 2NH_3(g)$

9. What volume of oxygen is required to completely burn 25 liters of ethane?

$2C_2H_6(g) + 7O_2(g) \longrightarrow 4CO_2(g) + 6H_2O(g)$

10. What volume of sulfur dioxide would be produced in the reaction of iron(II) sulfide with oxygen if 1568 liters of O_2 are consumed?

$4FeS + 7O_2(g) \longrightarrow 2Fe_2O_3 + 4SO_2(g)$

11. What volume of hydrogen is required to react completely with $2\overline{000}$ liters of carbon monoxide to form methyl alcohol?

$CO(g) + 2H_2(g) \longrightarrow CH_3OH$

Mole-volume stoichiometry

12. Carefully controlled decomposition of ammonium nitrate can yield dinitrogen oxide. What volume of dinitrogen oxide at STP could be produced from the decomposition of 0.10 moles of ammonium nitrate?

$NH_4NO_3(g) \longrightarrow N_2O(g) + 2H_2O$

13. Ammonium nitrate may decompose with explosive violence to produce nitrogen gas. How many moles of ammonium nitrate are required to produce 1.12 liters of N_2 at STP?

$2NH_4NO_3(g) \longrightarrow 2N_2(g) + O_2(g) + 4H_2O$

14. Nitrogen oxide, which is produced during lightning flashes in the atmosphere, reacts rapidly with oxygen to yield nitrogen dioxide. How many liters of nitrogen dioxide at STP can result from the reaction of 8 moles of nitrogen oxide?

$2NO(g) + O_2(g) \longrightarrow 2NO_2(g)$

15. Nitrogen dioxide reacts with water to form nitric acid. Calculate the number of moles of nitric acid which can be formed from 1.68 liters of nitrogen dioxide at STP.

$3NO_2(g) + H_2O(l) \longrightarrow 2HNO_3(aq) + NO(g)$

16. Sulfur dioxide reacts with water to form sulfurous acid. What volume of sulfur dioxide at STP is required to form 0.05 moles of sulfurous acid?

$SO_2(g) + H_2O(l) \longrightarrow H_2SO_3(aq)$

Mole-mass stoichiometry

17. Carbon tetrachloride can be formed from the reaction of methane with chlorine. How many moles of carbon tetrachloride can be produced from 160 grams of methane?

$CH_4 + 4Cl_2 \longrightarrow CCl_4 + 4HCl$

18. What mass of oxygen will be produced from the decomposition of 0.025 moles of potassium chlorate?

$2KClO_3 \longrightarrow 2KCl + 3O_2$

19. How many moles of potassium chlorate can be produced from the reaction of 71.0 grams of chlorine?

$3Cl_2 + 6KOH \longrightarrow 5KCl + KClO_3 + 3H_2O$

20. How many grams of ethane can be burned by utilizing 0.70 moles of oxygen?

$2C_2H_6 + 7O_2 \longrightarrow 4CO_2 + 6H_2O$

21. How many moles of hydrogen are required to convert 159 grams of copper(II) oxide into copper?

$CuO + H_2 \longrightarrow Cu + H_2O$

Volume-mass stoichiometry

22. What volume of sulfur dioxide at STP can be produced by burning $16\bar{0}$ grams of sulfur?

$S + O_2(g) \longrightarrow SO_2(g)$

23. How many grams of zinc are required to produce 224 milliliters of hydrogen gas at STP from the following chemical reaction?

$Zn + 2HCl \longrightarrow ZnCl_2 + H_2(g)$

24. Hydrogen and oxygen are produced by the electrolysis of water. How many milliliters of hydrogen gas at STP can be produced from 1.80 grams of water?

$2H_2O \longrightarrow 2H_2(g) + O_2(g)$

25. What quantity of calcium carbonate is produced simultaneously with 4480 liters of carbon dioxide at STP from calcium bicarbonate?

$Ca(HCO_3)_2 \longrightarrow CaCO_3 + CO_2(g) + H_2O$

26. How many grams of iron can be produced in a blast furnace from iron(III) oxide from a supply of 11,200 liters of carbon monoxide at STP?

$Fe_2O_3 + 3CO(g) \longrightarrow 2Fe + 3CO_2(g)$

27. What volume of carbon dioxide at STP is produced simultaneously with 112.2 grams of calcium oxide in the thermal decomposition of limestone?

$CaCO_3 \longrightarrow CaO + CO_2(g)$

Mass-mass stoichiometry

28. How many grams of copper can be produced from the reaction of 159 grams of copper(II) oxide with hydrogen?

$CuO + H_2 \longrightarrow Cu + H_2O$

29. How many grams of carbon are required to produce 6.54 grams of zinc by the following reaction?

$ZnO + C \longrightarrow Zn + CO$

30. How many grams of calcium carbide are required to produce 5.2 grams of acetylene (C_2H_2) by the following reaction?

$CaC_2 + 2H_2O \longrightarrow Ca(OH)_2 + C_2H_2$

31. How many grams of carbon tetrachloride can be produced from 4.0 grams of methane by the following reaction?

$CH_4 + 4Cl_2 \longrightarrow CCl_4 + 4HCl$

32. How many grams of copper can be produced by roasting 1590 grams of copper(I) sulfide?

$Cu_2S + O_2 \longrightarrow 2Cu + SO_2$

Limiting reagent problems

33. What volume of hydrogen gas at STP can be produced from the reaction of 3.27 grams of zinc and 3.65 grams of hydrochloric acid?

$Zn + 2HCl \longrightarrow ZnCl_2 + H_2(g)$

34. How many grams of water will be produced from the reaction of 11.2 liters of hydrogen and 11.2 liters of oxygen at STP?

$2H_2(g) + O_2(g) \longrightarrow 2H_2O$

35. How many liters of carbon dioxide at STP can be produced from the reaction of 1.6 grams of oxygen and 1.1 grams of propane?

$C_3H_8(g) + 5O_2(g) \longrightarrow 3CO_2(g) + 4H_2O$

36. How many grams of ammonia can be produced from the reaction of 67.2 liters of hydrogen at STP and 14 grams of nitrogen?

$N_2(g) + 3H_2(g) \longrightarrow 2NH_3(g)$

37. How many grams of silver chloride can be produced from 16.99 grams of silver nitrate and 2.92 grams of sodium chloride?

$AgNO_3 + NaCl \longrightarrow AgCl + NaNO_3$

38. How many grams of cane sugar can be produced from 5.6 liters of carbon dioxide at STP and 3.6 grams of water?

$12CO_2(g) + 11H_2O \longrightarrow C_{12}H_{22}O_{11} + 12O_2(g)$

13 GASES, LIQUIDS, AND SOLIDS

LEARNING OBJECTIVES

When you finish this chapter you should be able to:

(a) compare the macroscopic features of gases, liquids, and solids.

(b) relate the macroscopic properties of the states of matter to the kinetic theory.

(c) define pressure and examine how it affects the states of matter.

(d) describe how attractive forces affect the vapor pressure, boiling point, and heat of vaporization of liquids.

(e) describe the melting point, heat of fusion, and sublimation of solids.

(f) discuss the effect of pressure on the changes of state using Le Châtelier's principle.

(g) match the following terms to their corresponding definitions.

barometer
boiling point
equilibrium
evaporation
heat of fusion
heat of vaporization
kinetic theory
Le Châtelier's principle
Maxwell-Boltzmann distribution
melting point
pressure
vapor pressure

Although all matter can exist as a gas, a liquid, or a solid, we tend to think in terms of some substances as solids, others as liquids, and still others as gases. For example, we generally think of iron as a solid, water as a liquid, and nitrogen as a gas. However, these are really just the states that these substances usually assume under the conditions of temperature and pressure of everyday life. If these conditions are altered enough, these substances can exist in different states. Steel workers, for example, are well aware that iron can be liquid. Ice skaters enjoy their sport if and only if water is solid, and doctors sometimes use liquid nitrogen to freeze skin areas.

In this chapter we will investigate the physical properties of the three states of matter and how substances can exist in various states. To do this, it will be necessary to look at the relationship of the submicroscopic particles of substances in these different states. We will see that the kinetic theory of matter can explain the existence of these three states and their properties.

13.1 THE GASEOUS STATE

While only a small fraction of substances exist in the gaseous state under normal conditions, most substances can be transformed into gases at some sufficiently high temperature. The general properties of the gaseous state are:

1. Gases have no characteristic shape or volume and can be contained in any size or shape vessel.
2. All gases expand indefinitely and uniformly to fill the available space.
3. Any gas may be compressed by the application of pressure and made to occupy a smaller volume.
4. Gases which do not react with each other will diffuse and mix completely to form mixtures of any composition desired.
5. The densities of gases under normal conditions are small and, therefore, are expressed in grams per liter.
6. At some sufficiently high pressure and low temperature characteristic for each gas, the substance may be converted to a liquid.

13.2 PRESSURE

The concept of **pressure,** which **is equal to force per unit area,** is somewhat more difficult to deal with than that of volume, temperature, and mass. It can be illustrated by some commonly encountered examples.

Using the definition of pressure can you explain why a small number of people can form an effective pressure group?

The pressure an individual exerts on the soles of his shoes depends on his mass and the area of his soles. Consider two individuals of the same mass who wear different size shoes: the individual with the larger shoes exerts less force per unit area on the soles of his shoes although the total force involved is the same in both cases. The force per unit area on the spike heels of shoes once designed for women was very high. The heels caused substantial damage to floors and carpets. At the other extreme, the effectiveness of snowshoes and water skis is due to the distribution of force over a wide area.

Measurement of pressure

In determining temperature, account is taken of the direction of heat flow from a hot body to a cold body. In a similar manner pressure is determined by the flow of mass that normally moves from regions of high pressure to low pressure. This principle can be used to define a scale of pressure just as heat flow is used to define temperature.

In measuring pressure we can make use of our observations on fluid states. The air that surrounds the earth exerts a pressure on us just as water does on a submerged swimmer. We are accustomed to this pressure and usually do not consider it, but it affects our lives in many ways. The weather patterns and air movements are controlled by the variance in pressure at different points on the earth.

The barometer, which **is the standard device for measuring the pressure of the atmosphere,** was invented by Evangelista Torricelli in the seventeenth century. It consists of a long tube closed at one end, which is filled with mercury and inverted in a vessel of mercury. If the tube is more than 76 cm long, part of the mercury will run out of the tube when it is inverted, but a column of mercury approximately 76 cm high will remain in the tube (Figure 13.1).

The reason that a column of mercury remains in the tube of the barometer is that the atmosphere exerts a pressure on the surface of the mercury. This pressure transmitted through the mercury to the base of the column supports the mercury in the tube. In the barometer above the level of the mercury there exists a near vacuum. Therefore, the pressure at the base of the mercury column is due to the mass of the mercury in the column. The mercury falls until the pressure exerted by its weight is equal to the atmospheric pressure.

Pressure should be expressed in units of force per unit area. In the barometer the downward force exerted by the mercury column is proportional to the mass of the liquid supported in the column, which in turn is proportional to the height of the column. Therefore, the pressure can be expressed in centimeters of mercury. It is understood that pressure is not the same thing as length, but **by convention the term centimeter of mercury (Hg) indicates the pressure or force per unit area exerted by the column of mercury.**

1 atmosphere
76 cm Hg
760 mm Hg
760 torr

Atmospheric pressure is a function of altitude and the local weather, and it fluctuates from day to day. **At 0°C at sea level, the average atmospheric pressure supports a column of mercury 76 cm high. This pressure is called a**

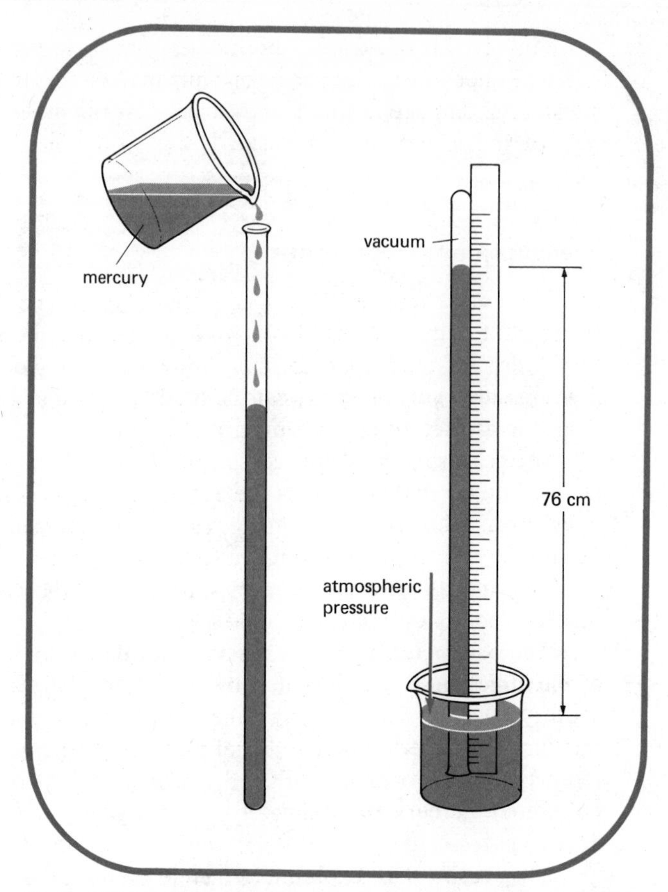

Figure 13.1 The construction of a barometer: a glass tube, filled with mercury, is closed and inserted in a beaker of mercury. When the end of the column is opened, the column drops to about 76 cm.

standard atmosphere and is referred to as 1 atm. The unit of pressure, 1 mm of Hg, is also called 1 torr. Therefore, the standard pressure (1 atm) is 760 torr.

13.3 KINETIC THEORY OF MATTER

The concept that atoms and molecules are moving is known as the kinetic theory of matter, which is a model proposed to explain the observed facts of the behavior of the gaseous state. By extension, it also applies to the liquid and solid states.

The assumptions made in the kinetic theory of matter can be summarized as follows:

1 **Gases are composed of atoms or molecules that are widely separated from one another.** The space occupied by the atoms or molecules is extremely small compared with the space accessible to them.

2 **The atoms or molecules are moving rapidly and randomly in straight lines.** Their direction is maintained until they collide with a second atom or molecule or with the walls of the container.
3 **There are no attractive forces between molecules or atoms of a gas.**
4 **Collisions of molecules or atoms are elastic:** that is, there is no net energy loss upon collision, although transfer of energy between molecules or atoms may occur in the collision.
5 **In a gas sample, individual atoms or molecules move at different speeds and possess different energies of motion—kinetic energies.** The kinetic energy of a particle is given by the expression $KE = \frac{1}{2}mv^2$, where m is the mass of the particle and v is its velocity. For a given temperature, the average kinetic energy is constant. As the temperature increases, the average kinetic energy increases and therefore the average velocity also increases. The average kinetic energy is directly proportional to the temperature on the Kelvin scale.

The ideal gas

The ideal gas is a hypothetical model gas which conforms to all of the assumptions of the kinetic theory. Thus the ideal gas particles have a very small volume compared to the total volume of the gas sample. Furthermore, the gas particles are not attracted to each other and any collisions which occur are elastic. An ideal gas, then, could never be condensed to form a liquid. However, we know that all real gases including components of our atmosphere such as oxygen and nitrogen, can be converted to a liquid. **Real gases are any gaseous substances which actually exist.**

Under conditions of high temperature and low pressure, most real gases will conform to the behavior predicted for the ideal gas. At low pressures, such as exist at high altitudes, the gas particles are far apart, as given by assumption 1 of the kinetic theory. At high temperatures the gas particles have high kinetic energies and are moving so rapidly that the attractive forces which do exist in real gases can exert very little influence.

At high pressures and low temperatures, all real gases can be condensed to a liquid. Thus they do not behave ideally. Further discussion of the behavior of real gases versus ideal gases will be presented in Chapter 14.

The Maxwell-Boltzmann distribution

Because there are a large number of atoms or molecules in a gas sample, it is possible to use statistical methods to describe the velocities and kinetic energies of the particles. Although there is a constant exchange of energies, the fraction of particles in a gas sample that have a given kinetic energy remains constant at a specified temperature. It is not necessary to specify the velocity or kinetic energy of any given particles at a given instant. The mathematical equation describing the speed distribution of atoms and

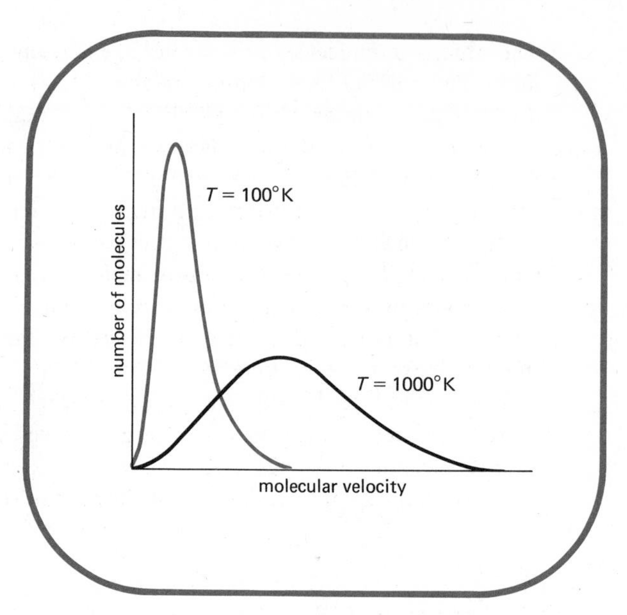

Figure 13.2 The Maxwell-Boltzmann distributions.

molecules was derived by Clark Maxwell and Ludwig Boltzmann in 1860. The actual equation and its derivation will not be given, but a graphical representation is shown in Figure 13.2, where the relative number of molecules with specified speeds is plotted on the ordinate (vertical axis) and the corresponding speeds on the abscissa (horizontal axis). The curve shows that at any temperature a wide range of molecular velocities exists but that the largest fraction have some intermediate velocity. A much smaller number of particles have very high or very low kinetic energy. Experimental determinations of the statistical distribution predicted by Maxwell and Boltzmann have verified their equation.

The effect of temperature on the distribution of speeds also is shown in Figure 13.2. At high temperatures the average speed of particles is higher than at a lower temperature. The maximum of the curve that represents the most probable speed is shifted to show a much larger number of particles at high and low velocities.

13.4 THE LIQUID STATE

At some sufficiently low temperature and high pressure, all gases will condense into liquids. The general properties of liquids are:

1 Liquids have no characteristic shape and will take on the shape of the container that they occupy.
2 Liquids maintain their volume unless the external pressure is increased. They do not expand to fill the volume of the container.

3 Liquids are only slightly compressible. At extremely high pressures the volume of liquids decreases.

4 Liquids have densities which are much higher than gases. Gas densities are usually stated in grams per liter, whereas liquid densities are stated in grams per milliliter.

5 Liquids which are soluble in each other will diffuse into one another. The diffusion rate is slower for liquids than for gases.

Kinetic theory of liquids

The kinetic theory of matter as applied to gases can be extended to the liquid state with only minor changes. All gases can be liquefied if sufficiently high pressures and low temperatures can be achieved. In the liquid state, atomic and molecular motion is more restricted than in the gaseous state as a result of attractive forces between neighboring atoms and molecules. The balance between these attractive forces and the kinetic energy of the particles produces a fluid with a semblance of cohesion but not a rigid structure.

The incompressibility of a liquid is due to the nearness of neighboring particles. Compression of a liquid would require a squeezing of the molecules or atoms, and matter tends to resist such deformations.

Although diffusion in the liquid state is slow, it does take place at a measurable rate. Since particles in the liquid state are closer together than they are in the gaseous state, an individual particle encounters more neighboring particles as it moves. Repeated collisions with neighboring particles hinder the progress of a particle in a given direction and tend to randomize its motion. In the gas phase a particle can travel relatively long distances in a straight line before collision.

Evaporation of a liquid involves transfer of matter from the liquid phase to the gaseous phase. In a liquid, the individual particles are traveling at different rates of speed. Those particles of high velocity may possess sufficient kinetic energy to break away from the attractive forces of their neighbors and become independent in the gas phase. This process is the reverse of the liquefaction of a gas.

Evaporation of liquids

Any individual who has exercised and perspired or who has stood in a breeze immediately after stepping out of the swimming pool is well acquainted with the cooling effect of evaporation. The particles which leave the liquid phase most readily are the most energetic particles. Their departure causes a decrease in the average kinetic energy of the remaining particles; this is felt as a lowering in the temperature of the liquid. The evaporation process continues at the same rate if a heat source is available to maintain the temperature of the liquid. A glass of water in a closed room without air currents will evaporate without any noticeable cooling. As the most energetic particles leave the

Why does a breeze feel refreshing after you have been working hard?

liquid phase, heat is transferred from the surroundings to the liquid, maintaining the temperature. A redistribution of particles with various kinetic energies yields the same Maxwell-Boltzmann distribution as existed before the most energetic particles left.

Evaporation of matter from the liquid phase is less likely as the temperature decreases, because the average kinetic energy of the particles and hence their ability to escape into the gaseous phase decreases. This is easily verified by experience. For example, it is more difficult to dry clothes on a clothesline on a cool day than on a hot day under identical wind conditions.

Even at the same temperature, different liquids evaporate at different rates. For example, gasoline evaporates faster than water at room temperature and water evaporates faster than lubricating oil. The ease of evaporation in the liquid phase depends on the liquid. Because the average kinetic energies of particles in two different liquids at the same temperature are identical, the escaping tendency is controlled by the attractive forces between neighbors. If attractive forces between neighboring particles are large, the escaping ability of a given particle is retarded by its neighbors.

Vapor pressure of liquids

Evaporation from an open vessel indicates that the liquid particles have an escaping tendency. This tendency is also manifested in a closed vessel. When a liquid is placed in a closed vessel, however, there is only a small decrease in the volume of the liquid with time. This should not be too unexpected because the densities of matter in the liquid and gaseous phases are drastically different. Evaporation of a small quantity of liquid results in a large volume of gas. Whenever a liquid is placed in a closed vessel, particles leave the liquid phase and enter the gaseous phase. As the number of particles in the gaseous phase increases, it becomes more likely that they will collide with the liquid surface and return to it. Eventually a balance is achieved: the rates at which the particles leave and return to the liquid phase become equal (Figure 13.3). When such a balance occurs, the system is in equilibrium. **In a system at equilibrium no net macroscopic change occurs.** Various submicroscopic processes, such as evaporation and condensation, occur continuously but in such a manner that they balance each other.

There must be some liquid present in order to measure a vapor pressure.

The particles in the gas phase exert a pressure like those of any gas. **This pressure of a gas in equilibrium with its liquid phase is called the vapor pressure of the liquid.** The vapor pressure is a measurable quantity, just as is the pressure of a gas in the absence of a second phase. It indicates the escaping tendency of the liquid, which in turn is characteristic of the individual liquid and its temperature.

One means of measuring vapor pressure involves the use of a barometer in which a vacuum exists in the space above the mercury column. If a dropper containing a liquid is placed beneath the column of mercury and the contents are ejected, the liquid will rise to the top of the column because most liquids

Figure 13.3 Equilibrium and vapor pressure.

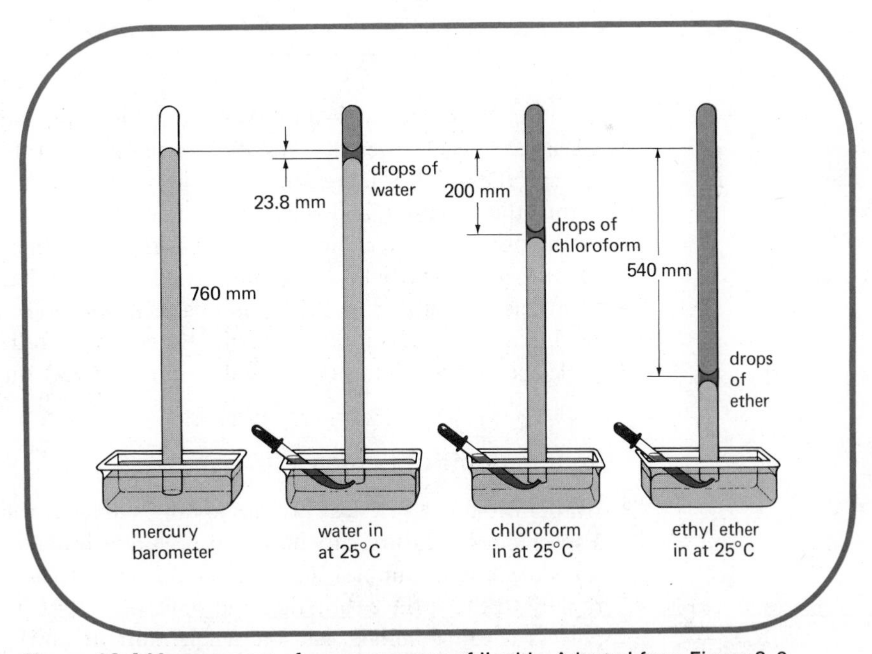

Figure 13.4 Measurement of vapor pressure of liquids. Adapted from Figure 8-9, page 157 of *General College Chemistry*, 4th edition by C. W. Keenan and J. H. Wood (New York, Harper & Row, Publishers, 1966). Reproduced with permission.

are less dense than mercury. The results of such an experiment with water, alcohol, chloroform, and ether are illustrated in Figure 13.4. As each liquid evaporates, the resultant pressure pushes the mercury column downward. The decrease in the length of the column is equal to the vapor pressure of the liquid. Again, some liquid must remain in the tube in order to measure the vapor pressure of the liquid.

The experiments illustrated in Figure 13.4 can be done at various temperatures. As would be anticipated from our observations on the phenomenon of evaporation, the vapor pressure of liquids increases with temperature. Table 13.1 lists the vapor pressures of some common liquids as a function of temperature. The molecular weights of ether, chloroform,

TABLE 13.1

vapor pressures of liquids (in cm of Hg)

°C	Ether (74 amu)	Chloroform (119.5 amu)	Alcohol (46 amu)	Water (18 amu)
0	18.5	6.2	1.2	0.5
20	44.2	14.5	4.3	1.8
40	92.0	34.7	13.2	5.5
60	173.0	72.5	34.7	14.9
80	300.0	138.0	81.4	35.5
100	486.5	246.0	178.0	76.0
120	749.5	417.5	353.5	148.9

alcohol, and water are 74 amu, 119.5 amu, 46 amu, and 18 amu, respectively. Clearly, molecular weight is not the determining feature of vapor pressures. If it were, the order of increasing vapor pressure on this basis would be chloroform, ether, alcohol, and water. Experimentally, the vapor pressure increases in the order of water, alcohol, chloroform, and ether. Therefore, attractive forces between particles are an important factor in determining the ability of a particle to escape from the liquid phase. Water molecules are very polar and tend to attract one another strongly. Ether, on the other hand, is a less polar molecule and neighboring molecules do not attract one another as strongly.

Boiling point of liquids

The normal boiling point is the boiling point at one atmosphere of pressure.

Evaporation as a physical process is rather unspectacular, but on heating to a specific temperature, any liquid eventually will undergo a very pronounced transformation. Bubbles are formed throughout the liquid, and they rise rapidly to the surface, bursting and releasing vapor in large quantities. **This process is called boiling, and the temperature at which it occurs is called the boiling point of the liquid. At the boiling point the vapor pressure of the liquid is equal to atmospheric pressure.** Owing to the large escaping tendency of the liquid at its boiling point, bubbles of vapor may be formed within the volume of the liquid. This is in contrast to the process of evaporation that occurs only at the surface.

The vapor pressure of water at 80°C is 35.5 cm of mercury. If we lived on a planet with a normal atmospheric pressure of 35.5 cm of mercury, water would boil at 80°C. At this temperature the vapor pressure of the liquid would equal that of atmospheric pressure. The boiling point of water or any other liquid can be made to occur at any temperature if the external pressure is increased or decreased appropriately. To avoid ambiguity, **the standard or normal boiling point is the boiling point at one standard atmosphere.**

The fact that liquids boil at lower temperatures under reduced pressures can be confirmed by anyone who has attempted to cook food in boiling water at high altitude. As the altitude rises, the atmospheric pressure gets lower. In Figure 13.5 the effect of altitude on the boiling point of water is illustrated.

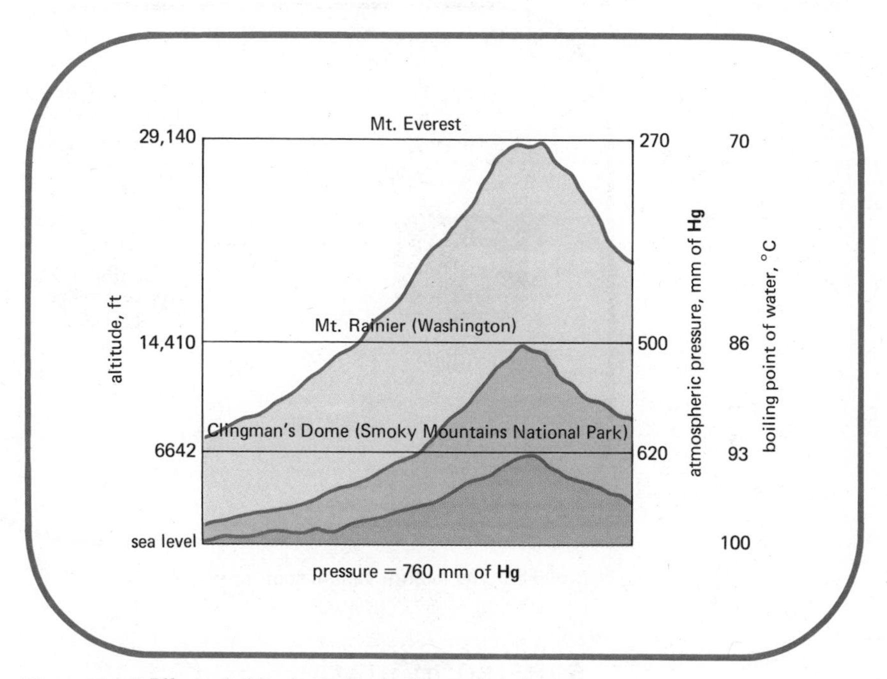

Figure 13.5 Effect of altitude on the boiling point of water.

Since water, or any liquid, at its boiling point, stays at the temperature of boiling until there is no liquid left, no amount of fuel can raise the temperature above the low boiling point. Food requires longer to cook at the lower temperature, and it is not uncommon to have to use cooking times twice as long as normal at altitudes of 7000 ft above sea level. Food can be cooked faster at the high pressures obtained in a pressure cooker. When the pressure gauge is set for 5 psi, the boiling point of the liquid in the pressure cooker is about 108°C. Under these conditions food cooks twice as fast as at 100°C.

Would you carry fresh vegetables with you while hiking in the Rocky Mountains?

Heat of vaporization of liquids

The average kinetic energies of particles in gaseous and liquid phases at the boiling point are equal. However, energy is required in order to maintain boiling with the resultant transfer of matter from the liquid to the gaseous phase. The heat added does not increase the temperature of the liquid at the boiling point but provides the energy necessary for the most energetic particles to continue to escape. **The quantity of heat energy required to transform 1 g of a substance at its boiling point from a liquid into a gas is called its heat of vaporization.** The heat of vaporization of water is 540 cal/g, a value that is rather large compared to other liquids (Figure 13.6). Its size is due to the strong attractive forces between neighboring water molecules in the liquid phase.

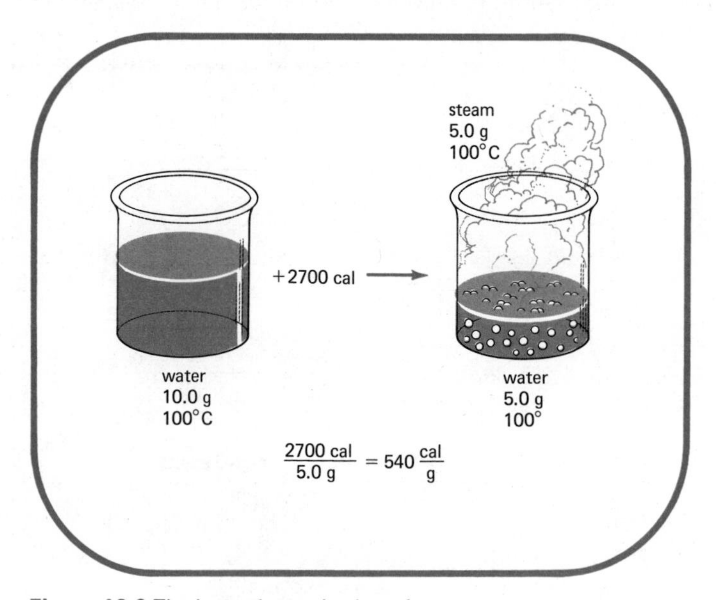

Figure 13.6 The heat of vaporization of water.

13.5 THE SOLID STATE

Most liquids when cooled form solids. The general characteristics of the solid state are:

1. Solids have a definite rigid shape and do not flow as do gases and liquids.
2. Solids maintain their volume and do not expand to fill a container.
3. Solids have high densities which are usually expressed in grams per cubic centimeter.
4. Solids are practically incompressible.
5. Solids mix or diffuse into one another at only extremely slow rates.

Kinetic theory of solids

Cooling a liquid decreases the kinetic energy of the atoms or molecules allowing the attractive forces between neighboring particles to exert a greater influence. In crystals the motion of the particles has been largely overcome so that they can vibrate only about fixed positions.

Because of the diminished motion of matter in the solid state as compared to the liquid and gaseous states, diffusion occurs at a slow rate. In addition, the rigidity and decreased motion of the solid structure decrease the probability of matter entering the liquid or gaseous states. Only the most energetic particles on the surface of the solid may leave.

Melting point of solids

When heat energy is added to a solid, the temperature increases until the solid starts to melt. **That temperature at which the added heat energy is used only to melt the solid without raising the temperature of the solid or liquid is called the melting point.** At the melting point, the solid and liquid states exist in equilibrium. Particles from the solid, which consists of ordered arrays, escape and enter the more random liquid state at the same rate particles from the liquid are deposited on the surface of the solid (Figure 13.7).

The effect of pressure on the melting point is not as dramatic as its effect on the boiling point. For most solids, the melting point of a substance increases with pressure. Water is atypical; its melting point decreases with increasing pressure. The melting point decrease is approximately 0.01°C/atm. This fact is part of the basis for the ease with which skaters can travel on the surface of the ice. Under the pressure exerted by the narrow edge of a hollow-ground skate blade, the ice melts to provide water as a lubricant.

It takes a lot of pressure to melt ice. Why are double runner skates not advisable for a beginning skater?

Figure 13.7 Equilibrium and the melting point.

Heat of fusion of solids

The amount of heat energy required to transform 1 g of a solid into a liquid at the melting point is called the heat of fusion. The heat of fusion of water is 80 cal/g (Figure 13.8). Like the heat of vaporization for water, this value is higher than that of many solids, yet another indication that water has strong attractive forces between neighboring molecules. The melting point of solids can be considered an approximate indication of the intermolecular attractive forces. For substances of similar molecular mass, those with the higher melting points have the stronger intermolecular forces. However, there are many more things that affect the melting point of a solid. Foremost among these is the packing or the geometrical arrangement of one particle with respect to its neighbors.

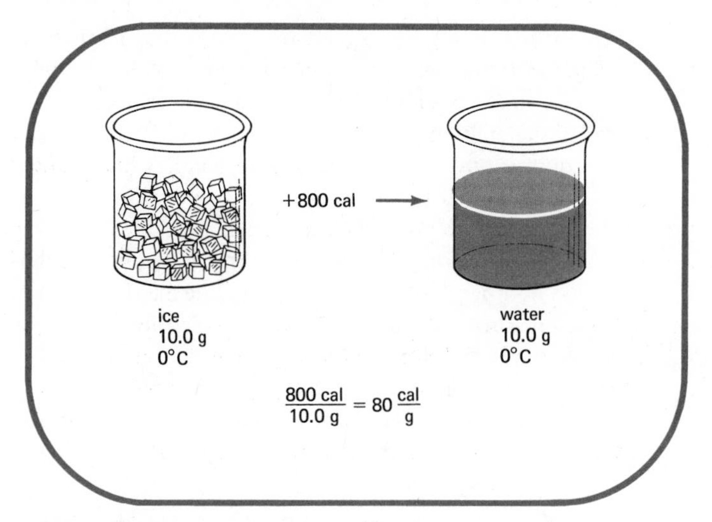

Figure 13.8 The heat of fusion of water.

Sublimation

Before the invention of the clothes dryer, clothes were dried in northern states by hanging them outdoors in the winter.

Solids, like liquids, have a vapor pressure. However, the idea that a solid can evaporate is not as common among the nonscientific community as the concept of evaporation of liquids. The evaporation of ice is the reason that even frozen clothes will dry outdoors in subzero weather. The vaporizations of solid carbon dioxide (dry ice) and mothballs are other common examples. **Direct vaporization of a solid is called sublimation.** As is the case with liquids, the vapor pressure is a measure of an equilibrium process. If a solid such as ice is placed in a closed container, the most energetic surface particles escape. However, eventually the gaseous particles return to the solid. The pressure of the vapor gradually increases to a maximum value; then the rate of escape is equal to the rate of return to the solid.

The larger the attractive force between molecules or atoms, the smaller will be the vapor pressure of the solid. With increasing temperature, the vapor pressure of the solid increases because the average kinetic energy of the particles increases. However, the vapor pressure of most solids rarely gets very large because the solid usually melts at high temperatures. For most solids, the tendency to enter the liquid phase is larger than that for entering the gaseous phase. Dry ice is an exception in that upon heating, it is converted directly to the gaseous state; the liquid state is not observed at atmospheric pressure.

13.6 HEAT ENERGY AND CHANGES IN STATE

The relationship between the heat energy added to a substance and the temperature change is an interesting one. Heating a solid, liquid, or gas produces an increase in the temperature of that particular state. The temperature increase is due to the higher average energy of the particles making up the substance. However, when a change of state occurs, such as when a solid melts or a liquid vaporizes, the heat energy is employed in breaking down the attractive forces between the particles and giving them increased freedom of movement.

Changes of state and temperature variations can be graphically illustrated. Consider the changes that occur when energy is added to a 1-g sample of a

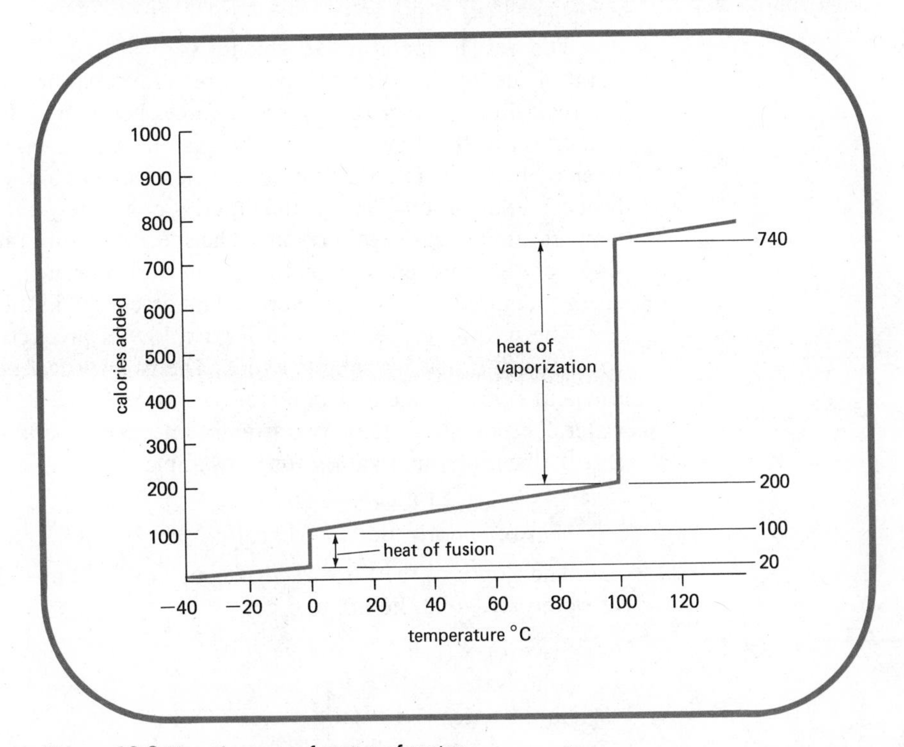

Figure 13.9 The changes of states of water.

solid. In the graph in Figure 13.9, the horizontal axis represents temperature scale. The vertical axis indicates the number of calories added. At the beginning of the experiment, the temperature of the solid is $-40°C$, which may be any temperature below the melting point. As heat is added, the temperature of the solid increases, due to increased motion of particles in the solid state. The ratio of the number of added calories to the temperature increase for the 1-g sample is given by the slope of the line, which by definition is the specific heat of the solid.

Eventually a temperature is reached at which the addition of heat energy no longer leads to a temperature increase. This temperature is the melting point, and its attainment is reflected by the first vertical line shown in Figure 13.9. The line continues as long as the solid is present in equilibrium with the liquid. When the last of the solid disappears, an increase in temperatures occurs. Therefore, the length of the vertical line represents the number of calories necessary to melt the solid and is equal to the heat of fusion. The heat of fusion indicates the energy required to break down the attractive forces in the solid and generate a more mobile liquid.

After complete conversion of the solid into liquid, the slope of the heating curve represents the number of calories necessary to cause a change in temperature of the liquid state. By definition, this ratio is the specific heat of the liquid state, which reflects the increase in the average kinetic energy of the particles in the liquid state.

The next temperature at which a vertical line is obtained is the boiling point of the liquid. At this temperature, additional heat energy again is used to break down further the attractive forces between molecules and allow for separation of the molecules into the gas phase. The number of calories represented by the length of the second vertical portion of the heating curve is the heat of vaporization. Finally, the liquid is completely converted to gas, and a temperature increase again results. The specific heat of the gas is equal to the slope of the final portion of the graph and represents the amount of heat energy required to raise the average kinetic energy in the gaseous phase.

The heating curve drawn in Figure 13.9 is precisely the one found experimentally for a 1-g sample of ice. The two vertical portions of the curve occur at 0°C, the melting point of ice, and at 100°C, the boiling point of water. The lengths of these two straight lines correspond to the heat of fusion and the heat of vaporization for the sample.

$$\text{heat of fusion} = 80 \text{ cal/g}$$
$$\text{heat of vaporization} = 540 \text{ cal/g}$$

The slopes of the three rising regions of the curve give the specific heat of water as a solid, liquid, and gas.

HEAT CAPACITY (PER °C)

solid water = 0.5 cal/g
liquid water = 1.0 cal/g
gaseous water = 0.5 cal/g

Problem 13.1 How many calories are required to convert 25.0 g of ice at 0°C to steam at 100°C?

Solution This problem must be approached by considering stepwise the physical transformations that will ultimately accomplish the desired overall conversion. These are:

1 melt the ice at 0°C to water at 0°C;
2 heat the water from 0°C to 100°C;
3 boil the water and convert to steam at 100°C.

The energy required for step 1 is related to the heat of fusion of water which is $8\bar{0}$ cal/g.

$$8\bar{0} \text{ cal/g} \times 25.0 \text{ g} = 2\bar{0}00 \text{ cal}$$

The energy required to heat water as in step 2 is related to the specific heat of water which is 1.00 cal/g per degree Celsius.

$$1.00 \frac{\text{cal}}{\text{g °C}} \times 25.0 \text{ g} \times 100\text{°C} = 25\bar{0}0 \text{ cal}$$

The energy required to boil water in step 3 is related to the heat of vaporization which is $54\bar{0}$ cal/g.

$$54\bar{0} \text{ cal/g} \times 25.0 \text{ g} = 13{,}500 \text{ cal}$$

Thus the overall energy requirement is the sum of the three calculated quantities.

$$\begin{array}{rr} & 2\bar{0}00 \text{ cal} \\ & 25\bar{0}0 \text{ cal} \\ & +\ 13{,}500 \text{ cal} \\ \hline \text{Total:} & 18{,}\bar{0}00 \text{ cal} \end{array}$$

Problem 13.2 Your skin is exposed to $1\bar{0}$ grams of steam at 100°C. The steam condenses on your skin and the water formed eventually reaches your body temperature of 37°C. During this process, the steam and water release heat energy to your skin and cause a burn. Which process contributes most to the burn, the condensation of steam to hot water, or the cooling of the hot water?

Solution The heat liberated by condensing steam into water is related to the heat of vaporization.

$$1\bar{0} \text{ g} \times 54\bar{0} \text{ cal/g} = 5400 \text{ cal}$$

The cooling of the hot water liberates heat related to the specific heat of water. The temperature change is 100 − 37 = 63 degrees.

$$1.00 \frac{\text{cal}}{\text{g °C}} \times 63\text{°C} \times 1\bar{0}\text{ g} = 630 \text{ cal}$$

Thus we see that the heat liberated by condensing steam is nine times greater than that of cooling the water.

13.7 LE CHÂTELIER'S PRINCIPLE

Le Châtelier in 1884 proposed a principle that predicts the behavior of an equilibrium when it is subjected to an external force: **if an external force is applied to a system at equilibrium, the system will readjust to reduce the stress imposed upon it if possible.** This principle explains the effect of pressure on the boiling point of a liquid and the melting point of a solid. In the first case, evaporation of a liquid always involves a tremendous increase in volume. Application of pressure will cause the equilibrium system of a liquid and its vapor at its normal boiling point to shift toward the liquid state, because the resultant decrease in volume tends to diminish the pressure on the equilibrium system. Therefore, in order to boil a liquid at high pressures, higher temperatures are necessary. In the case of the melting point of a solid, most substances decrease in volume in going from the liquid to the solid state, and application of pressure will cause an equilibrium system of liquid and solid to shift toward the solid state. This shift decreases the volume of the system and counteracts the applied pressure. Therefore, in order to melt most solids under pressure, higher temperatures are necessary. That only small increases in melting points are usually observed is due to the small change in volume between solids and liquids. The abnormal behavior of water is the result of the increase in volume of this compound in going from water to ice. Thus an increase in pressure shifts ice-water systems toward water.

SUMMARY

Pressure *is equal to force per unit area. The pressure of the atmosphere is measured by the use of a* ***barometer****. At sea level the average atmospheric pressure will support a 76-cm column of mercury in a barometer. One* ***standard atmosphere*** *is 76 cm of Hg or 760 torr. A* ***torr*** *is 1 mm of Hg.*

Matter is moving in all three states. This concept of moving matter is known as the ***kinetic theory of matter****. The distribution and number of particles having various velocities or energies is given by the* ***Maxwell-Boltzmann distribution****.*

Evaporation *of liquids involves escape of the most energetic particles from the surface of the liquid. At* ***equilibrium*** *the pressure of the gas above its liquid state is called the* ***vapor pressure****. The temperature at which the vapor pressure equals the atmospheric pressure is the* ***boiling point****. The number of calories required to vaporize one gram of a liquid at its boiling point is the* ***heat of vaporization****.*

The temperature at which added heat energy is used only to melt the solid without raising the temperature is called the ***melting point****. The amount of heat energy required to convert one gram of solid into a liquid at the melting point is called the* ***heat of fusion****.*

Direct vaporization of a solid is called ***sublimation.***

Le Châtelier's principle *states that if an external force is applied to a system at equilibrium, the system will readjust to reduce the stress imposed upon it if possible.*

QUESTIONS AND PROBLEMS

Definitions

1. Define each of the following terms.

(a) pressure **(b)** barometer
(c) vapor pressure **(d)** boiling point
(e) evaporation **(f)** melting point
(g) sublimation **(h)** heat of vaporization
(i) heat of fusion **(j)** equilibrium

2. List the physical characteristics of gases, liquids, and solids. In which ways are they similar? How do they differ?

3. Describe the kinetic molecular model of gases, liquids, and solids.

Evaporation of liquids

4. What factors control the drying of clothes at temperatures below the freezing point of water?

5. Explain the difference felt by a swimmer emerging from the water on a windy day versus a calm day assuming the temperature is the same.

6. The rate of evaporation of a beaker of water inside a larger sealed container at constant temperature decreases with time. Explain this phenomenon and contrast it with the rate of evaporation of the same beaker in a room at constant temperature.

7. A beaker of water in a closed room at constant temperature does not decrease in volume on a given day. Why does the water not evaporate?

Vapor pressure of liquids

8. The space above the column of mercury in a barometer is considered to be a vacuum. Is this idea strictly correct?

9. Why is mercury used in a barometer rather than another liquid such as water?

10. The vapor pressure of liquid A at 25°C is equal to that of liquid B at 50°C. If the molecular masses of A and B are identical, which of the two liquids has the stronger intermolecular attractive forces?

11. Is the vapor pressure of water at 25°C at an altitude of 10,000 ft in the mountains less than, equal to, or more than the vapor pressure of water at sea level at 25°C?

Boiling points of liquids

12. What is the highest temperature at which water vapor will condense to yield water under 1 atm pressure? At which temperature will water boil under pressure of 148.9 cm of mercury?

13. Will a pan of water boil faster with its lid on or off? Why?

14. Explain the operation of a pressure cooker.

15. Explain why the time required to fry meat at a high altitude is no different than at sea level, whereas the time required to boil potatoes differs significantly as a function of altitude.

Heat of vaporization

16. Steam at 100°C causes more severe burns than water at 100°C. Why?

17. Calculate the amount of heat required to convert 10 grams of water at its normal boiling point into steam.

18. How many calories are released by 10 grams of steam when it condenses to water at 100°C?

Melting point of solids

19. Why does a solid A melt at 200°C whereas a solid B melts at 400°C?

20. A highly trained figure skater skates on only a small portion of the edge of a blade and may exert 300 atmospheres pressure. What is the melting point of the ice under this pressure?

21. A speed skating blade is long but is very thin. Explain why this design is chosen.

22. Mercury thermometers cannot be used below −39°C. Explain why.

Heat of fusion

23. How many calories are required to melt $1\overline{00}0$ grams of ice?

24. How many calories must be removed from water at 0°C to freeze $1\overline{00}$ grams of water?

Sublimation

25. Some solids are stored in sealed containers. Why?

26. How do sublimation and evaporation differ?

Le Châtelier's principle

27. Explain the prediction of the effect of pressure on boiling point using Le Châtelier's principle.

28. Explain the effect of pressure on the melting point of solids using Le Châtelier's principle.

Changes in states

29. What stage in the conversion of ice to steam requires the largest input of energy?

30. How many calories are required to heat $1\bar{0}$ g of ice at −20°C to steam at 100°C?

31. How many calories are required to heat $1\bar{0}$ g of ice at 0°C to steam at 120°C?

32. How many calories are required to convert water at 20°C into steam at 100°C?

14 THE GAS LAWS

LEARNING OBJECTIVES

When you finish this chapter you should be able to:

(a) calculate changes in gas volume resulting from pressure changes and vice versa at constant temperature.

(b) calculate changes in volume resulting from temperature changes and vice versa at constant pressure.

(c) calculate changes in pressure resulting from temperature changes and vice versa at constant volume.

(d) apply the combined gas laws to changes in gas samples in which two experimental conditions change.

(e) use Dalton's law of partial pressures in problems involving mixtures of gases.

(f) determine the molecular weight of a gas from its density under stated conditions.

(g) calculate the density of a gas given its molecular formula.

(h) recognize the limitations of the gas laws in dealing with real gas.

(i) match the following terms with their corresponding definitions.

Avogadro's hypothesis
Boyle's law
Charles' law
Dalton's law of partial pressures
Gay-Lussac's law
molar volume
partial pressure
STP

We live at the bottom of an ocean of gases called the atmosphere. These gases vary in temperature and pressure and are the reason why weather changes from day to day and even hour to hour. All living things depend on components of the atmosphere, such as carbon dioxide and oxygen, for their very existence.

Since gases are odorless and colorless, we often forget that they are there. In fact, we often use gases without being aware of them. Many common occurrences reflect the behavior of gases. For example, the increase in the pressure of an automobile tire when a car is driven at high speeds is due to the way gases respond to heat. The car itself is propelled because of the volume changes of the gases formed in the combustion of gasoline. A less common transportation mode, the hot-air balloon, depends on the density changes of the air in the balloon when it is heated. In the kitchen, gases are responsible for many changes which occur during cooking. A cake rises when carbon dioxide is released and expands within the batter in a hot oven. Popcorn pops due to increased pressure created on the hull by the gases trapped within the kernel. Gases are involved in the conversion of cream to whipped cream, the insulation of down sleeping bags, the operation of refrigerators, the welding of metals, and in scuba diving, and oxygen tents and operating rooms of hospitals.

In order to better understand the gaseous world about us, this chapter is devoted to examining the quantitative relationships involved in the behavior of gases.

14.1 BOYLE'S LAW

In the course of experimenting with vacuums and vacuum pumps, Robert Boyle in 1662 made the first systematic study of the relationship between volume and pressure in gases. He measured what he termed "the spring of air," the pressure with which a gas sample pushes back when it is compressed. Boyle's observations indicated that there is a fixed relationship between gas volume and pressure. This relationship has been verified many times with a variety of gases and is known as **Boyle's law. At constant temperature, the volume of a given quantity of an ideal gas varies inversely as the pressure exerted on it.** Thus if the pressure of a given volume of a gas is doubled, the volume will be decreased by one-half. Conversely, if the pressure is decreased by one-half, the volume will double.

Boyle's law was one of the first physical laws of chemistry.

A mathematical expression of Boyle's law is:

$$V \propto \frac{1}{P}$$

where V stands for volume, P for pressure, and $\propto$ is a proportionality symbol.

The equation can be changed to an exact equality by using a proportionality constant k. The value of k depends on the temperature and the quantity of gas being studied.

$$V = k \times \left(\frac{1}{P}\right)$$

An alternative expression, obtained by multiplying both sides of the equation by P is:

$$PV = k$$

Therefore, **the product of the volume and pressure of a constant amount of gas at a constant temperature is constant.** If the quantity of the gas sample and the temperature remains the same, the product of the pressure and volume will remain unchanged under different experimental conditions. The mathematical statement given below is another way of expressing Boyle's law.

$$\boxed{P_2 \times V_2 = k = P_1 \times V_1}$$

TABLE 14.1 hypothetical Boyle's law data

P (atm)	V (liters)	PV (atm liters)
10	1.0	10
5.0	2.0	10
1.0	10	10
0.50	20	10
0.20	50	10
0.10	100	10

An example of some conveniently hypothetical data illustrating Boyle's law is given in Table 14.1.

For a change in pressure, the new volume may be calculated from the rearranged equation.

$$V_2 = V_1 \times \left(\frac{P_1}{P_2}\right)$$

$$V_2 = V_1 \times P_{\text{factor}}$$

Similarly, if a new pressure must be calculated to effect a desired volume change, then:

$$P_2 = P_1 \times \left(\frac{V_1}{V_2}\right)$$

$$P_2 = P_1 \times V_{\text{factor}}$$

In either a pressure or volume calculation, the necessary volume factor or pressure factor may be determined by recalling the inverse proportionality of Boyle's law. Thus if a pressure is increased, the volume must decrease. Therefore, only a pressure factor less than unity will give the correct answer. Memorization of the formulas given is not necessary.

Problem 14.1

A gas sample occupies 15 liters at 25°C and 3.0 atmospheres. If the pressure is then decreased to 1.0 atmosphere, what will be the volume of the gas?

Solution

Since the temperature and the sample of the gas are unchanged in the described process, we may use Boyle's law. First arrange the information in a clear form so that you can analyze how to solve the problem.

$P_1 = 3.0$ atm	$V_1 = 15$ liters
$P_2 = 1.0$ atm	$V_2 = ?$ liter
a pressure decrease	∴ the volume must increase

The pressure has decreased from 3 atm to 1 atm and the volume must increase by a pressure factor composed of these two numbers.

$$V_2 = V_1 \times P_{\text{factor}}$$

$$V_2 = (15 \text{ liters})\left(\frac{3 \text{ atm}}{1 \text{ atm}}\right)$$

$$V_2 = (15 \text{ liters})(3)$$

$$V_2 = 45 \text{ liters}$$

Problem 14.2

What pressure is required to make a 15-liter sample of a gas at 2 atmospheres occupy a volume of 5 liters? Assume a constant temperature.

Solution

First arrange the data in tabular form.

$V_1 = 15$ liters	$P_1 = 2$ atm
$V_2 = 5$ liters	$P_2 = ?$ atm
a volume decrease	∴ the pressure must increase

The volume has decreased from 15 liters to 5 liters, so the pressure must increase by a volume factor composed of these two numbers.

$$P_2 = P_1 \times V_{\text{factor}}$$

$$P_2 = (2 \text{ atm})\left(\frac{15 \text{ liters}}{5 \text{ liters}}\right)$$

$$P_2 = (2\ \text{atm})\left(\frac{15}{5}\right)$$

$$P_2 = 6\ \text{atm}$$

Any problem which involves a specific gas sample at a constant temperature, but with a volume change and a pressure change, can be solved by using Boyle's law. These problems depend only upon remembering the inverse proportionality between pressure and volume. Figure 14.1 may help you to remember this relationship.

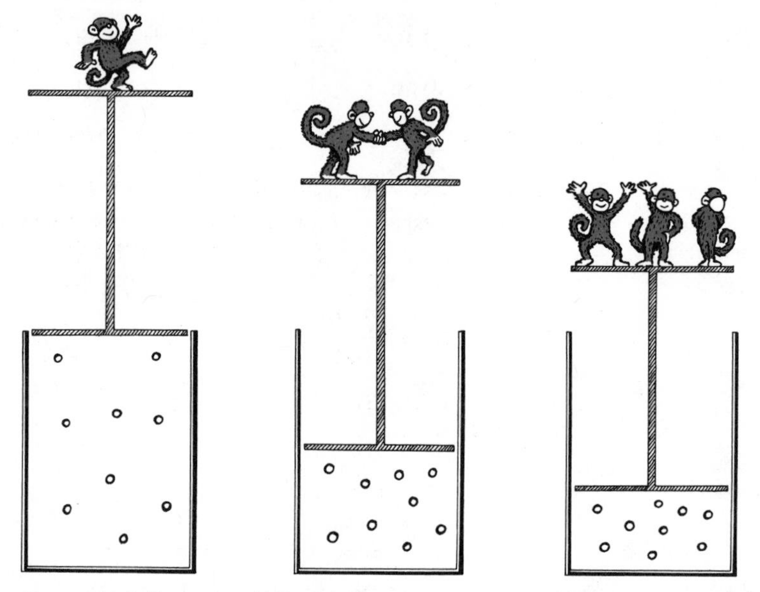

Figure 14.1 Illustration of Boyle's law.

14.2 CHARLES' LAW

Gases expand when heated under constant pressure and contract when cooled. The French physicist, J. A. C. Charles, in 1787 observed this relationship between volume and temperature at constant pressure. He and later scientists established that at constant pressure, the volume of a gas increases $\frac{1}{273}$ for every degree Celsius rise in temperature. An example of some conveniently hypothetical data is given in Table 14.2. Note that although there is a uniform relationship between volume and the Celsius temperature, it is not a direct proportion. However, the volume is directly proportional to the temperature on the Kelvin scale (Section 2.9). No real gases behave exactly according to this data because they condense to liquids at very low temperatures.

TABLE 14.2

hypothetical Charles' law data

T (°C)	V (ml)	T (°K)	V/T (ml/°K)
546	819	819	1
273	546	546	1
100	373	373	1
10	283	283	1
2	275	275	1
1	274	274	1
0	273	273	1
−1	272	272	1
−2	271	271	1
−10	263	263	1
−100	173	173	1
−273	(0)	(0)	——

From the data of Charles and other scientists, **Charles' law** states that **at constant pressure the volume of a fixed mass of a gas is directly proportional to the absolute (Kelvin) temperature.** Stated in terms of a proportionality sign, we have the mathematical expression:

$$V \propto T$$

By using a proportionality constant, k, the expression becomes:

$$V = kT$$

The proportionality constant is dependent only on the pressure and the sample of gas considered. By dividing both sides of the equation by T, the expression obtained is:

$$\frac{V}{T} = k$$

The volume of a gas at constant pressure is directly proportional to the absolute temperature.

Since the constant remains unchanged, the relationship given below must be obeyed under any experimental conditions. It is an expression giving Charles' law.

$$\boxed{\frac{V_2}{T_2} = k = \frac{V_1}{T_1}}$$

For a change in temperature, the new volume may be calculated by the use of a temperature factor.

$$V_2 = V_1 \times \left(\frac{T_2}{T_1}\right)$$

$$V_2 = V_1 \times T_{\text{factor}}$$

Alternatively, if it is necessary to calculate the temperature required to alter a volume, we use a volume factor.

$$T_2 = T_1 \times \left(\frac{V_2}{V_1}\right)$$

$$T_2 = T_1 \times V_{\text{factor}}$$

Remember in the use of either formula, **the temperature must be in Kelvin units.** The formulas need not be memorized if one recalls that the proportion is a direct one.

Problem 14.3 A sample of a gas occupies $6\bar{0}0$ ml at 27°C and 1 atm. What will be its volume at 127°C if the pressure is kept constant?

Solution Since the pressure and quantity of the gas remains constant, Charles' law applies to the solution of the problem. Arrange the data in tabular form and calculate the temperature in Kelvin units.

$T_1 = 27 + 273 = 300°K$	$V_1 = 6\bar{0}0$ ml
$T_2 = 127 + 273 = 400°K$	$V_2 = ?$ ml
the temperature has increased	the volume should increase

The volume must increase as a consequence of the increase in the temperature. Therefore, the volume is multiplied by a temperature factor to increase the volume.

$$V_2 = V_1 \times T_{\text{factor}}$$

$$V_2 = (6\bar{0}0 \text{ ml})\left(\frac{400°K}{300°K}\right)$$

$$V_2 = (6\bar{0}0 \text{ ml})\left(\frac{400}{300}\right)$$

$$V_2 = 8\bar{0}0 \text{ ml}$$

Problem 14.4 A gas sample occupies $3\bar{0}0$ ml at 27°C and 2 atm. What temperature is necessary to decrease the volume to $2\bar{0}0$ ml at 2 atm?

Solution The conditions for Charles' law apply here. The data is arranged as follows:

$V_1 = 3\bar{0}0$ ml	$T_1 = 27 + 273 = 300°K$
$V_2 = 2\bar{0}0$ ml	$T_2 = ?$ °K
the volume has decreased	the temperature must decrease

An equality using a proportionality constant which is dependent only on the volume and the amount of the gas can be written.

$$V = kT \quad \text{or} \quad \frac{V}{T} = k$$

For a variety of combinations of pressures and temperatures, an equation giving Charles' law may be written as:

$$\frac{V_2}{T_2} = k = \frac{V_1}{T_1}$$

The two equations which are useful in solving Charles' type problems are:

$$V_2 = V_1 \times \left(\frac{T_2}{T_1}\right) = V_1 \times T_{\text{factor}}$$

$$T_2 = T_1 \times \left(\frac{V_2}{V_1}\right) = T_1 \times V_{\text{factor}}$$

Again, as in the two previous laws discussed, one does not need the formulas. You need only recall that the temperature must decrease by a volume factor. Therefore, we write:

$$T_2 = T_1 \times V_{\text{factor}}$$

$$T_2 = (300°\text{K}) \left(\frac{200 \text{ ml}}{300 \text{ ml}}\right)$$

$$T_2 = (300°\text{K}) \left(\frac{200}{300}\right)$$

$$T_2 = 200°\text{K}$$

The Kelvin temperature can be converted to Celsius temperature.

$$200°\text{K} = (200–273)°\text{C} = -73°\text{C}$$

An illustration of Charles' law which may aid you in recalling the direct relationship between volume and temperature is given in Figure 14.2.

14.3 GAY-LUSSAC'S LAW

Gay-Lussac's law states that at constant volume the pressure of a fixed mass of a gas is directly proportional to the absolute (Kelvin) temperature. Mathematically stated, the direct proportionality between pressure and temperature is given by:

$$P \propto T$$

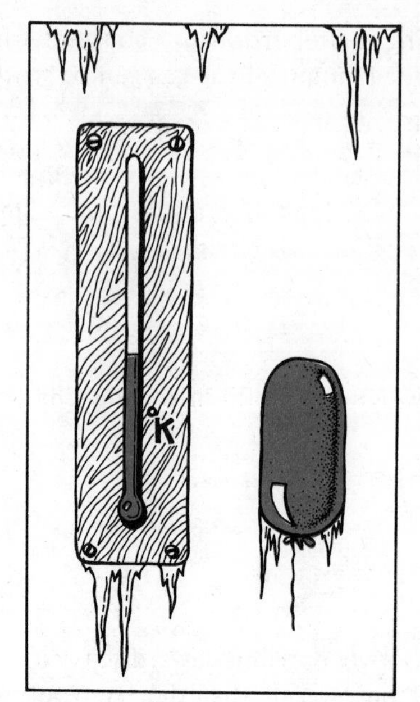

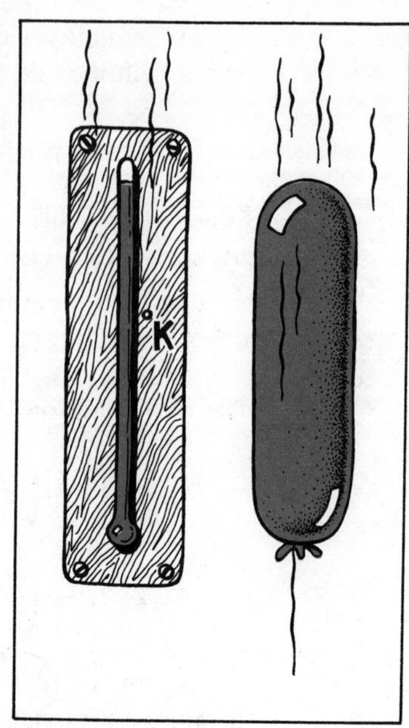

Figure 14.2 Illustration of Charles' law.

The direct proportion can be changed to an equality by using a proportionality constant k. The value of k depends on the volume of the sample and the quantity of gas.

$$P = kT$$

An alternative expression is obtained by dividing both sides of the equation by T.

$$\frac{P}{T} = k$$

Why does the pressure in a tire increase during high speed driving?

Therefore, the quotient of the pressure divided by the temperature will be a constant for a specific volume containing a given mass of gas, regardless of the pressure or temperature.

$$\boxed{\frac{P_2}{T_2} = k = \frac{P_1}{T_1}}$$

For a change in temperature, the effect on the pressure may be calculated from the following equivalent equation:

$$P_2 = P_1 \times \left(\frac{T_2}{T_1}\right) = P_1 \times T_{\text{factor}}$$

Similarly, for a change in pressure, the related necessary temperature change may be calculated from an alternative rearranged equation.

$$T_2 = T_1 \times \left(\frac{P_2}{P_1}\right) = T_1 \times P_{\text{factor}}$$

In either case, you only need remember that there is a direct proportion between the temperature in degrees Kelvin and the pressure.

Problem 14.5 What will happen to the pressure of a 5-liter sample of a gas at 5 atmospheres if it is heated from 250°K to 300°K and the volume is held constant?

Solution Gay-Lussac's law conditions are stated and the data may be arranged as follows:

$T_1 = 250°\text{K}$	$P_1 = 5$ atm
$T_2 = 300°\text{K}$	$P_2 = ?$ atm
the temperature increases	the pressure must increase

The pressure increases by a temperature factor. Therefore, we write:

$$P_2 = P_1 \times T_{\text{factor}}$$

$$P_2 = (5\text{ atm})\left(\frac{300°\text{K}}{250°\text{K}}\right)$$

$$P_2 = (5\text{ atm})\left(\frac{300}{250}\right)$$

$$P_2 = 6\text{ atm}$$

14.4 COMBINED GAS LAWS

Both Boyle's and Charles' laws may be combined into one mathematical expression which then also gives the Gay-Lussac law as well.

$$\frac{PV}{T} = k$$

Therefore, for a fixed sample of a gas, it follows that:

$$\boxed{\frac{P_2 V_2}{T_2} = k = \frac{P_1 V_1}{T_1}}$$

This equation need not be memorized because any variable can be changed by factors of the other two variables.

$$P_2 = P_1 \times V_{factor} \times T_{factor}$$
$$V_2 = V_1 \times P_{factor} \times T_{factor}$$
$$T_2 = T_1 \times P_{factor} \times V_{factor}$$

In order to solve problems in which two variables change, it is only necessary to consider separately what effect each variable will have. Thus for a volume to change as a consequence of pressure and temperature changes, you have to determine what effect the pressure change will have on the volume. If the pressure increases, the volume will decrease, and vice versa. After the proper pressure factor has been determined, the temperature factor can be considered. If the temperature decreases, the volume will decrease, and vice versa. Application of the temperature factor gives the final correct answer.

Problem 14.6 A 1000-ml sample of a gas at −73°C and 2 atm is heated to 123°C and the pressure is decreased to 0.5 atm. What will be the final volume?

Solution

$V_1 = 1000$ ml	$V_2 = ?$ ml
$T_1 = 200$°K	$T_2 = 400$°K
$P_1 = 2$ atm	$P_2 = 0.5$ atm

Since the temperature increases, the volume must increase by a factor of 400/200. The pressure decreases and the volume must increase by a factor of 2/0.5.

$$V_2 = V_1 \times T_{factor} \times P_{factor}$$

$$V_2 = 1000 \text{ ml} \times \frac{400°\text{K}}{200°\text{K}} \times \frac{2 \text{ atm}}{0.5 \text{ atm}}$$

$$V_2 = 1000 \text{ ml} \times \frac{400}{200} \times \frac{2}{0.5}$$

$$V_2 = 8000 \text{ ml}$$

14.5 DALTON'S LAW OF PARTIAL PRESSURES

John Dalton, of atomic theory fame, also studied some of the properties of gases. His **law of partial pressures states that each gas in a gas mixture exerts a**

partial pressure which is equal to the pressure which it would exert if it were the only gas present under the experimental conditions; the sum of the partial pressures of all of the gases is equal to the total pressure. Dalton's law expressed mathematically is:

$$P_{\text{total}} = P_1 + P_2 + P_3 + \cdots$$

where the subscripted P values are the partial pressures of the gases in the mixture.

One example of the law of partial pressures is the air of our atmosphere. Air consists of nitrogen and oxygen and, in lesser amounts, argon and carbon dioxide. The total pressure exerted by air is a sum of the partial pressures of the individual gases.

$$P_{\text{air}} = P_{N_2} + P_{O_2} + P_{A} + P_{CO_2}$$

A certain partial pressure of oxygen in the atmosphere is necessary for animals to survive. At one atmosphere, the partial pressures of nitrogen, oxygen, and argon are approximately 592 torr, 160 torr, and 8 torr, respectively, and that of CO_2 is very low. At higher altitudes, the total pressure of air decreases as a result of the decrease in the partial pressures of all the components of air. For example, at an altitude of one mile, the total pressure is 630 torr and the partial pressures of nitrogen, oxygen, and argon are 491 torr, 132 torr, and 7 torr, respectively. At this altitude, the human body cannot operate efficiently and becomes tired easily. This change occurs because at this partial pressure of oxygen, the red blood cells absorb a smaller amount of oxygen and the body needs cannot be fulfilled. However, if one stays at the high altitude for a week or more, the body adjusts by forming more red blood cells.

Could a human survive at a reduced pressure of 500 torr if the partial pressure of oxygen were 200 torr?

Problem 14.7

A 5-liter rigid container holds a mixture of nitrogen and oxygen. The total pressure of the mixture is 700 torr at 25°C. If a chemical means of removing oxygen were employed, and the pressure of the remaining nitrogen is 450 torr, calculate the original partial pressures of oxygen and nitrogen.

Solution

The pressure after removal of oxygen is due solely to nitrogen. The 450-torr pressure is that which nitrogen exerts when alone and is therefore the pressure that it would also exert in a mixture. Thus by definition, the partial pressure of nitrogen is 450 torr.

$$P_{N_2} = 450 \text{ torr}$$

Since the total pressure of the mixture is equal to the sum of the partial pressures of nitrogen and oxygen, the partial pressure of oxygen may be calculated.

$$P_{total} = P_{N_2} + P_{O_2}$$
$$700 \text{ torr} = 450 \text{ torr} + P_{O_2}$$
$$(700 - 450) \text{ torr} = P_{O_2}$$
$$250 \text{ torr} = P_{O_2}$$

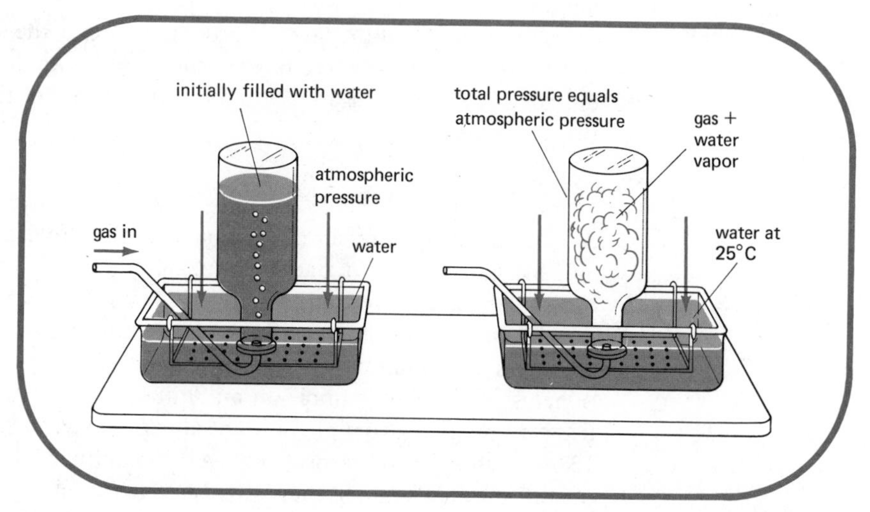

Figure 14.3 Dalton's law of partial pressures. The total pressure of the gaseous phase over the water is equal to the pressure exerted by the collected gas plus the pressure exerted by the water vapor. Using Dalton's law, the pressure of the collected gas may be determined.

Gases are commonly collected by water displacement in college chemistry laboratories (Figure 14.3). One such gas is oxygen. Potassium chlorate, when heated with manganese dioxide, decomposes to produce potassium chloride and oxygen.

$$2KClO_3 \xrightarrow{MnO_2} 2KCl + 3O_2$$

The oxygen, which is collected by downward displacement of water from an inverted jar, is not pure but contains some water vapor mixed with it. The pressure of the collected gas is equal to the external atmospheric pressure when the level of the water inside the jar is equal to the level in the large water bath. It is a sum of a partial pressures of the two gases contained in the jar, which are water vapor and oxygen.

$$P_{total} = P_{atm} = P_{O_2} + P_{H_2O}$$

In order to determine the pressure of the gas, it is necessary to subtract the partial pressure of the water at that temperature (Appendix 3).

Problem 14.8

A 200-ml sample of oxygen is collected at 26°C over water. The vapor pressure of water is 25 torr at 26°C and the atmospheric pressure is 750 torr at the time of the experiment. Calculate the partial pressure of oxygen and determine the volume that the dry oxygen would occupy at 26°C and 750 torr.

Solution

The partial pressure of oxygen is determined by use of the law of partial pressures.

$$\begin{aligned} P_{\text{total}} = P_{\text{atm}} &= P_{O_2} + P_{H_2O} \\ 750 \text{ torr} &= P_{O_2} + 25 \text{ torr} \\ (750 - 25) \text{ torr} &= P_{O_2} \\ 725 \text{ torr} &= P_{O_2} \end{aligned}$$

Now the problem may be treated as a Boyle's law type. If 200 ml of dry oxygen at 725 torr was subjected to a new pressure of 750 torr, the volume would decrease.

$$V_2 = (200 \text{ ml}) \left(\frac{725}{750}\right)$$

$$V_2 = 193 \text{ ml}$$

This means that only 193 ml of the 200 ml is actually O_2.

14.6 AVOGADRO'S HYPOTHESIS

Amadeo Avogadro in 1811 suggested that **equal volumes of gases under the same conditions of temperature and pressure contain the same number of submicroscopic particles.** For example, equal volumes of helium and argon at **1 atm pressure and 0°C (called standard temperature and pressure, or STP)** contain the same number of atoms. The atomic concept had been considered by many individuals since the time of Democritus in ancient Greece. However, the derivation of the concept was philosophical and was not based on any experimental data. Avogadro's hypothesis used the atomic concept in a quantitative manner and provided a basis for further experimental tests.

Avogadro's principle can be used to determine the relative weight of atoms and molecules. Because equal volumes of gases under the same conditions of temperature and pressure contain the same number of submicroscopic particles, the weight of the gas particles must be in the same ratio as the gas densities. The densities of the monatomic gases, helium and argon, at standard temperature and pressure are 0.179 g/liter and 1.79 g/liter, respectively, and indicate that one argon atom is ten times as massive as one helium atom. If the absolute weight of one atom of one element were known, the

TABLE 14.3 density and relative mass

Formula	Density at STP (g/ml)	Relative mass of molecule	Relative mass of atom
He	0.179	—	4
Ne	0.90	—	20
Ar	1.79	—	40
H_2	0.090	2	1
N_2	1.25	28	14
O_2	1.43	32	16
F_2	1.70	38	19

absolute weight of an atom of the other could be calculated. However, even without a knowledge of absolute atomic weights, the relative atomic weights of all gaseous elements can be established from density measurements and the knowledge of whether the element is monatomic or diatomic (Table 14.3).

Hydrogen, the lightest element known, has a density of 0.090 g/liter at STP. Comparison with the density of helium, 0.179 g/liter, indicates that the helium atom is twice as massive as the hydrogen molecule. Because hydrogen is diatomic, the weight of the hydrogen atom is one-fourth that of the helium atom. Oxygen, which is diatomic, has a density of 1.43 g/liter at STP. Therefore, the oxygen atom is 4 times as massive as helium and 16 times as massive as the hydrogen atom.

14.7 MOLAR VOLUME

STP
1 mole
22.4 liters

The concepts of the mole, Avogadro's hypothesis, and Avogadro's number allow the determination of the molecular weight of an unknown compound. The volume occupied by 1 mole (32 g) of oxygen molecules at standard temperature and pressure has been determined to be 22.4 liters. **Since equal volumes of gases contain the same number of particles, 22.4 liters must be the volume occupied by 1 mole of any gaseous substance at standard temperature and pressure** (Table 14.4).

TABLE 14.4 the molar volume

Substance	Volume at STP	Mass
O_2	22.4 liters	32 g
N_2	22.4 liters	28 g
CO_2	22.4 liters	44 g
H_2	22.4 liters	2 g
He	22.4 liters	4 g
Ar	22.4 liters	40 g

The density of methane, CH_4, is 0.71 g/liter at standard temperature and pressure; therefore, the mass of 22.4 liters is 16 g.

$$0.71 \text{ g/liter} \times 22.4 \text{ liter} = 16 \text{ g}$$

Thus one mole of methane has a mass of 16 g. The mass of a methane molecule on the atomic mass scale is 16 amu.

Problem 14.9 A 1.12-liter sample of a gas at 2 atmospheres and 0°C has a mass of 3.00 grams. What is the molecular weight of the gas?

Solution It is first necessary to change the stated experimental conditions to STP in order to make use of the molar volume quantity of 22.4 liters/mole.

$$V_2 = V_1 \times P_{\text{factor}}$$

$$V_2 = 1.12 \text{ liters} \times \frac{2 \text{ atm}}{1 \text{ atm}}$$

$$V_2 = 2.24 \text{ liters}$$

Now it is possible to calculate the number of moles present in the sample.

$$\text{number of moles} = 2.24 \text{ liters} \times \frac{1 \text{ mole}}{22.4 \text{ liters}}$$

$$\text{number of moles} = 0.100 \text{ mole}$$

Once the number of moles which are present in the sample are known, the mass of the sample can be used to calculate the mass of a mole.

$$\text{mass of one mole} = \frac{3.00 \text{ g}}{0.100 \text{ mole}}$$

$$\text{mass of one mole} = 30.0 \text{ g/mole}$$

The molecular weight is then 30.0 amu.

Problem 14.10 Calculate the density of CO_2 at STP.

Solution The molecular weight of carbon dioxide is:

$$\text{molecular weight of } CO_2 = \text{atomic weight of C} + 2(\text{atomic weight of O})$$
$$\text{molecular weight of } CO_2 = 12.0 \text{ amu} + 2(16.0 \text{ amu})$$
$$= 44.0 \text{ amu}$$

Therefore, 22.4 liters of CO_2 at STP must have a mass of 44 g. The density of CO_2 can be calculated as follows:

$$\text{density of } CO_2 = \frac{44.0 \text{ g}}{22.4 \text{ liters}}$$

$$\text{density of } CO_2 = 1.97 \text{ g/liter}$$

14.8 DEVIATIONS FROM IDEALITY

Most real gases deviate from ideal gas behavior at high pressures and low temperatures. The reasons for these deviations can be found in the assumptions of the kinetic theory. At high pressure, the volume of free space is comparable to the actual volume of the gas particles. The volume of the particles, therefore, can no longer be regarded as negligible. The available free space is such that additional pressure will involve actual compression of the gas particles. This further compression is resisted by the impenetrability of atoms and molecules. Although Boyle's law is obeyed in compressions from 1 atm to 2 atm, it is unlikely that it will be obeyed in compressions from 500 atm to 1000 atm. In the first case, a diminution in volume by a factor of 2 will result; the second case will lead to a change by a factor less than 2.

The deviations at low temperatures reflect the fact that, contrary to the assumptions of the kinetic theory, there actually are attractive forces between molecules and atoms. Indeed, if there were no attractive forces, the liquid and solid states could not exist and all matter would be in the gaseous state. The assumption of no attractive forces is approximately correct at high temperatures where such forces are small with respect to the average kinetic energy of the gas particles. As the temperature is lowered, the average kinetic energy decreases until it becomes comparable to the energy of attraction between the gas particles. The forces of attraction become more important as the velocity of the particles decreases.

SUMMARY

Boyle's law: *At constant temperature, the volume of a given quantity of an ideal gas varies inversely as the pressure exerted on it.*

Charles' law: *At constant pressure, the volume of a fixed mass of a gas is directly proportional to the absolute (Kelvin) temperature.*

Gay-Lussac's law: *At constant volume, the pressure of a fixed mass of a gas is directly proportional to the absolute (Kelvin) temperature.*

The three individual gas laws may be combined into a single gas law which interrelates the pressure, volume, and temperature of a quantity of a gas.

***Dalton's law of partial pressures**: Each gas in a mixture exerts a pressure which is equal to the pressure which it would exert if it were the only gas present under the experimental conditions. The sum of the partial pressures of all of the gases in the mixture is equal to the total pressure.*

***Avogadro** hypothesized that equal volumes of gases under the same conditions of temperature and pressure contain the same number of submicroscopic particles.*

*At the **standard temperature and pressure** of 0°C and 1 atmosphere, or STP, the volume occupied by one mole of a gas is 22.4 liters.*

Real gases deviate from ideal gas behavior at high pressures and low temperatures.

QUESTIONS AND PROBLEMS

Definitions

1. What is the difference between a real gas and an ideal gas?

2. Define the following terms.

(a) molar volume **(b)** real gas
(c) STP **(d)** Charles' law
(e) Boyle's law **(f)** Dalton's law of partial pressures
(g) Gay-Lussac's law

Boyle's law

3. A sample of a gas at 600 torr and 25°C occupies 300 ml. What volume will the gas occupy at 800 torr and 25°C?

4. A sample of a gas at 0.5 atmosphere and 100°C occupies 2 liters. What pressure is necessary to cause the gas to occupy 0.5 liters?

Charles' law

5. A sample of a gas at −91°C and 1 atmosphere occupies 2.0 liters. What volume will the gas occupy at 0°C at the same pressure?

6. A sample of a gas at 300°K and 700 torr occupies 60 ml. What temperature is necessary to increase the volume to 75 ml?

Gay-Lussac's law

7. A sample of a gas in a rigid container is heated from 273°K to 273°C. If the gas was initially at 1 atmosphere, what is the final pressure?

8. A sample of a gas in a rigid container at −23°C is at a pressure of 500 torr. What temperature will be necessary to change the pressure to 800 torr?

Combined gas laws

9. A gas occupies 250 ml at 700 torr and 300°K. What volume will the gas occupy at 350 torr and 450°K?

10. A gas occupies 800 ml at 1 atmosphere and 250°K. At what pressure will the gas occupy 400 ml at 500°K?

11. A gas occupies 2 liters at 127°C and 2 atmospheres. At what temperature will the gas occupy 6 liters at 1 atmosphere?

Dalton's law of partial pressures

12. A mixture of three gases has the following partial pressures: oxygen, 100 torr; nitrogen, 300 torr; hydrogen, 150 torr. What is the total pressure of the mixture? If sufficient carbon dioxide were added until the pressure reached 750 torr, what would be the partial pressure of the carbon dioxide?

13. A nitrogen sample collected over water at 26°C and 775 torr has a volume of 300 ml. The vapor pressure of water at 26°C is 25 torr. What volume of nitrogen at 775 torr has been collected?

14. If 300 ml of nitrogen gas at 760 torr was bubbled through water at 25°C and collected, what would be the volume of the wet gas sample?

Molar volumes

15. How many moles of a gas are present under each of the following conditions?
    (a) 1.12 liters at 2 atm and 0°C
    (b) 560 ml at 5 atm and 182°C
    (c) 22.4 liters at 273°C and 2 atm

16. How many molecules of sulfur dioxide (SO_2) are contained in a 2-mole sample of the gas?

17. A 1.12-liter sample of a compound has a mass of 2.2 g at STP. What is the mass of a molecule of this compound on the atomic mass scale?

18. A 5.6-liter sample of a gas at 182°C and 38 cm of mercury has a mass of 4.8 g. What is the mass of a mole of this gas?

19. Calculate the density of carbon monoxide, CO, at STP.

20. Calculate the density of C_2H_6 at 0°C and 2 atmospheres.

21. Calculate the density of CH_4 at 273°C and 2 atmospheres.

LEARNING OBJECTIVES

When you finish this chapter you should be able to:

(a) examine why water has very large attractive forces.

(b) describe the difference between heavy water and ordinary water.

(c) identify hydrates and distinguish between the ways in which water is contained in the substances.

(d) relate efflorescence and deliquescence to vapor pressure of water.

(e) list some of the important reactions which produce water.

(f) match the following terms to their proper definitions.

efflorescence
hard water
heavy water
hydrate
hydrogen bonds
hygroscopic

The triatomic molecule of water is among the simplest of molecules but is one of the most important for all life on this planet. Water surrounds us in our biosphere and is within us in our cells.

Approximately 71 percent of the earth is covered with water in either the liquid or solid form. The oceans, lakes, rivers, groundwater and polar ice caps dominate the surface of the earth. Of this water, about 98 per cent is present in the oceans.

The oceans are postulated to be the birthplace of life. Most likely this is why the properties of water are very important to life as we know it. As life of the sea developed, it gradually moved onto the smaller land mass. At the present time, land contains more life of all kinds than the oceans from which they evolved.

Water is an important heat regulator on earth. Large quantities of heat energy can be absorbed or released by the oceans. Thus the temperature of the land surrounded by water is regulated and does not undergo the dramatic temperature changes that would be expected without the water. Water can absorb heat energy without undergoing great temperature changes itself. As a consequence, life in the water has a comfortable thermal environment.

The oceans dilute and dissipate the chemicals of the earth which naturally dissolve in surface waters flowing to the ocean. In addition, civilization is placing chemicals in the oceans. Although at one time the vastness of the ocean caused people to be careless about dumping wastes into the ocean, more recent studies of the harmful effects on ocean life have given us cause for concern.

Living organisms are composed predominantly of water. Approximately 80 percent of the weight of cells is water. All of the biological chemistry of life as we know it requires water. The water transports ingredients for chemical reactions to reaction sites. Furthermore, water is an active participant in many reactions within cells. Water also serves as a temperature regulator in an organism in much the same way that the oceans regulate the temperature of land.

Water has a remarkable fitness for supporting life. There is no other substance among the millions known that even begins to approach the capabilities of water. Therefore, in the quest for finding life as we know it on other planets, the scientists look for conditions under which water may be found.

The many useful functions of water are a consequence of its molecular structure. This structure in turn results from the electronic characteristics of its component atoms, hydrogen and oxygen. In this chapter the structure and properties of this versatile compound will be examined.

15.1 STRUCTURE OF WATER

The water molecule consists of two hydrogen atoms and one oxygen atom. Both hydrogen atoms are bonded by a single covalent bond. Since oxygen is more electronegative than hydrogen, the oxygen atom attracts the shared electron pair more than the hydrogen atom. The two O—H bonds in water, therefore, are polar covalent.

A molecule with three atoms such as H_2O could be linear or angular.

H—O—H (linear)   H—O with H below (angular)

Because water is polar, the water molecule is actually angular and not linear. If water were linear, the polar characteristics of each bond would have the net effect of balancing each other. The arrow symbol (+→) used to indicate the polar nature of the bond has the head of the arrow toward the negative atom and the cross near the positive atom. A linear water molecule would be nonpolar in total where the angular molecule would be polar.

+→ ←+
H—O—H   H—O with H below (+→ and ↑ dipole arrows)

Since the electronic configuration of an oxygen atom is $1s^22s^22p^4$, it has two unpaired electrons and one pair of electrons in the *p* orbitals. By sharing electrons with two hydrogen atoms, it fills all its valence orbitals. Since the *p* orbitals are mutually perpendicular, it would be expected that H_2O would have a bond angle of 90° (Figure 15.1). Actually, water has a bond angle of 105°. The exact bond angle of a compound cannot be predicted accurately

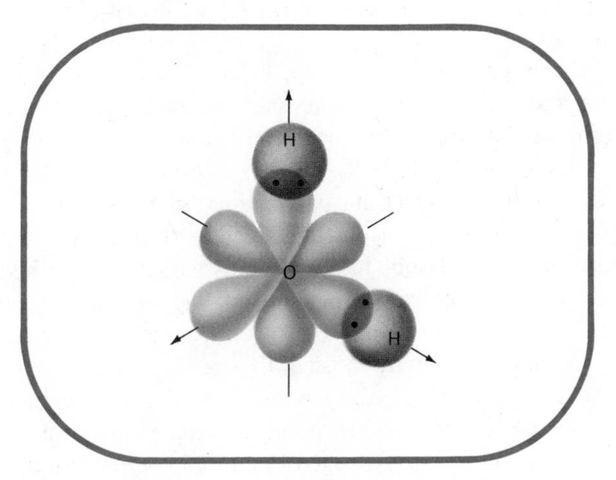

Figure 15.1 Model of covalent bonding in water.

from the geometry of the orbitals of the atoms. When atoms combine to form molecules, their orbitals do not remain exactly as they were in the free atoms. Each partner influences the other in achieving their new molecular identity.

15.2 HYDROGEN BONDING

In Chapter 8 you learned that the properties of the elements and compounds of a group resemble each other. Consider now the data for water and the other compounds of Group VIA with hydrogen given in Table 15.1. In every case, if you considered the series H_2S, H_2Se, and H_2Te and predicted a property of H_2O on the basis of a group relationship, the value for water

TABLE 15.1

properties of the hydrogen compounds of group VIA elements

Property	H_2O	H_2S	H_2Se	H_2Te
Molecular weight	18.0	34.1	81.0	129.6
Boiling point (°C)	100.0	−61	−41	−2
Melting point (°C)	0	−86	−66	−49
Molar heat of fusion (kcal/mole)	1.44	0.57	0.60	1.0
Molar heat of vaporization (kcal/mole)	9.72	4.46	4.62	5.55

would be substantially different from the observed value. Thus the boiling point of water should be in the vicinity of −81°C and not the +100°C observed. Similarly, the freezing point might be expected to be about −106°C and certainly not 0°C. In each case the properties are much higher than expected. Furthermore, both the heats of fusion and vaporization are larger than expected.

The reason for the strange behavior of water is that its molecules are not free and independent of one another. A fairly strong force called hydrogen bonding exists between the water molecules. **Hydrogen bonding involves the attraction of the partially positive hydrogen atom for the unshared electron pairs of a very electronegative atom such as oxygen** (Figure 15.2). Other strongly electronegative atoms such as fluorine, nitrogen, and chlorine also form hydrogen bonds in compounds which contain hydrogen.

Thus the many hydrogen bonds *between molecules* (indicated by dashes in Figure 15.2) hold many molecules together in clusters. As a result, the properties of water are those of these clusters and not of free water molecules. The energy of the hydrogen bond is about 6 kcal/mole. The energy of a covalent hydrogen-oxygen bond in water is 118 kcal/mole.

Hydrogen bonding occurs in DNA and RNA molecules and contributes to their structure.

The amount of hydrogen bonding in water differs in the gas, liquid, and solid states. In ice, the water molecules are extensively hydrogen bonded and a

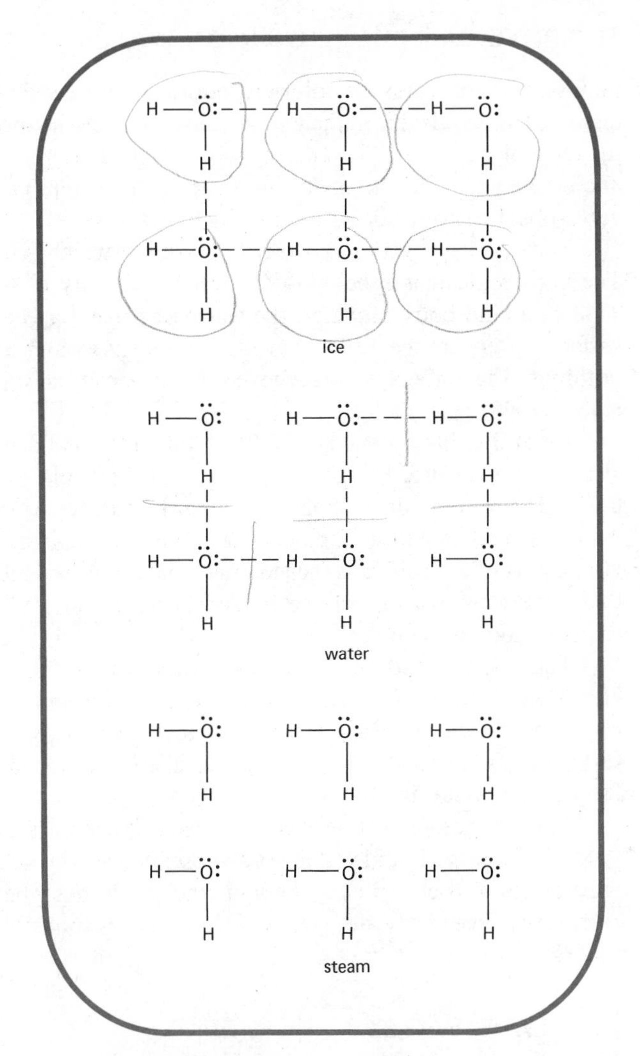

Figure 15.2 Hydrogen bonds in water.

relatively large amount of energy is required to break a sufficient number of hydrogen bonds to produce the liquid. This accounts for the high melting point and the high heat of fusion. In the course of heating water, the number of hydrogen bonds broken continues to increase. At the boiling point, more energy is required to break the hydrogen bonds remaining and this accounts for the high boiling point and heat of vaporization.

The remaining elements of Group VIA are not sufficiently electronegative to cause the H—S, H—Se, and H—Te covalent bonds to be very polar. Thus the hydrogen atoms in these bonds are not very positive and are not attracted to the Group VIA elements to form hydrogen bonds. The properties of the compounds H_2S, H_2Se, and H_2Te are normal.

15.3 PHYSICAL PROPERTIES OF WATER

Pure samples of water are colorless, odorless, and tasteless. Small amounts of dissolved minerals are sufficient to cause changes in these features. In fact, the taste of fresh water in a spring which people often regard as pure is really due to the particular impurities present in the water. Very pure water is not very appealing to drink because of its blandness.

The density of water at 4°C is 1.0000 g/ml, which is its maximum density. The density decreases below 4°C until the density of 0.99987 g/ml at 0°C. Thus in a cold body of water, the warmest place may be at the bottom. The surface water may be at 0°C but the more dense water at 4°C will be on the bottom of the body of water. Above 4°C the density of water decreases until it reaches 0.958 g/ml at 100°C.

Ice at 0°C has a density of 0.917 g/ml and is less dense than liquid water at that temperature. For that reason ice easily displaces its own weight and floats. The volume of ice floating above the surface of the water is approximately 8 percent. Since 92 percent of a mass of ice is below the surface, the visible part of an iceberg is indeed only "the tip of the iceberg." There is more than fifteen times as much ice hidden beneath the sea for every part of an iceberg which we can see.

It's a good thing that water isn't like other substances.

The lower density of ice is responsible for the maintenance of aquatic life. If ice were more dense than water, it would sink to the bottom of the lakes and escape the warming rays of the sun and the destructive action of surface motion and winds. Lakes would then freeze from the bottom up and fish could not survive the winter.

Since ice has a lower density than water, it occupies a larger volume in the solid state than as liquid water. As a consequence, any liquid water which is in the crevices of rocks will expand and crack the rocks when frozen. The forces exerted by freezing water are great and are responsible for some erosion processes.

15.4 HEAVY WATER

Heavy water is composed of deuterium, ^2H, and oxygen. It is also known as deuterium oxide and is symbolized by D_2O. Ordinary water contains one

TABLE 15.2 water and heavy water

Property	H_2O	D_2O
Molecular weight	18	20
Density (g/ml at 20°C)	0.997	1.108
Boiling point (°C)	100.00	101.41
Melting point (°C)	0.00	3.79
Heat of vaporization (kcal/mole)	9.720	9.944
Heat of fusion (kcal/mole)	1.436	1.500

molecule of D_2O for every 7000 molecules of ordinary water. The properties of pure D_2O are listed in Table 15.2 and are compared with H_2O.

Heavy water was first obtained from the electrolysis of purified water samples. In this reaction, an electric current is applied to water to break it down into its component elements, O_2 and H_2 gases. The oxygen-deuterium bonds of D_2O are broken slightly slower than the oxygen-hydrogen bonds of H_2O. As the electrolysis proceeds, the unreacted water becomes richer in D_2O and becomes denser. Other processes for preparing heavy water are now being used on a large scale. All are based on the slight difference in the chemical properties of D_2O compared to H_2O. The production of 1 pound of D_2O requires the processing of approximately 50,000 pounds of pure water.

The atomic energy program requires large amounts of heavy water. In the nuclear chain reactors, heavy water is used as a moderator which reduces the velocity of the neutrons and controls the atomic reaction.

Heavy water and compounds formed from it are used as tracers in the study of chemical and biological processes. If, for example, a compound is made which contains deuterium rather than hydrogen, then its presence can be traced in the course of a chemical reaction by a small difference in physical properties.

15.5 HYDRATES

Hydrates are crystalline compounds which contain water in definite proportions by weight. Examples of some common hydrates are listed in Table 15.3. Each hydrate is known by a common name but the systematic name is one which indicates the number of molecules of water contained per formula unit.

TABLE 15.3 composition of hydrates

Formula	Common name
$CuSO_4 \cdot 5H_2O$	blue vitriol
$Na_2SO_4 \cdot 10H_2O$	Glauber's salt
$KAl(SO_4)_2 \cdot 12H_2O$	alum
$MgSO_4 \cdot 7H_2O$	epsom salts
$Na_2CO_3 \cdot 10H_2O$	washing soda
$CaSO_4 \cdot 2H_2O$	gypsum
$Na_2B_4O_7 \cdot 10H_2O$	borax

Water in hydrates may be incorporated in three ways to yield the definite proportion by weight which is a characteristic of hydrates. These are:

1 coordinate covalent bonding to the metal cation;
2 hydrogen bonding with the anion;
3 crystal entrapment.

Examples of coordinate covalent bonding to the metal cation include all four water molecules of $BeSO_4 \cdot 4H_2O$ and four of the five water molecules of $CuSO_4 \cdot 5H_2O$. Coordination of six water molecules to the metal cation is very common in transition metal ions. Examples include $NiSO_4 \cdot 6H_2O$ and six of the seven water molecules of $CoSO_4 \cdot 7H_2O$. Thus in these two sulfate salts, the metal ions exist as $Ni(H_2O)_6{}^{2+}$ and $Co(H_2O)_6{}^{2+}$.

Hydrogen bonding to the anion generally involves a smaller number of water molecules. In both $CuSO_4 \cdot 5H_2O$ and $CoSO_4 \cdot 7H_2O$, one of the water molecules is hydrogen bonded to the sulfate ion.

Crystal entrapment involves the location of water molecules at definite positions within the crystal. In the mixed salt $KAl(SO_4)_2 \cdot 12H_2O$, six of the water molecules are coordinated to aluminum and six water molecules are spaced in the lattice in positions surrounding the potassium ion. The water molecules around potassium are not coordinate covalent bonded.

Efflorescence

The loss of water of hydration to air is called efflorescence. Most hydrates will release some or all of their water upon heating. When the water is given off, the crystalline character of the salt changes. For example, blue vitriol, a blue salt of formula $CuSO_4 \cdot 5H_2O$, produces a white salt, $CuSO_4$.

$$\underset{\text{(blue)}}{CuSO_4 \cdot 5H_2O} \xrightarrow[\Delta]{} \underset{\text{(white)}}{CuSO_4} + 5H_2O$$

The hydrates have vapor pressures due to the water present. This vapor pressure increases as the temperature increases. If a hydrate has a vapor pressure which is higher than the partial pressure of water in the air, the hydrate will lose all or part of its water to the air. Glauber's salt loses all of its water rapidly when exposed to air, which has a partial pressure of water of less than 14 mm. Blue vitriol is unstable if the partial pressure is less than 8 mm; it undergoes transformations to successively yield $CuSO_4 \cdot 3H_2O$ and eventually $CuSO_4 \cdot H_2O$. Washing soda loses nine water molecules to yield $Na_2CO_3 \cdot H_2O$.

Deliquescence

Some substances when exposed to air with a high vapor pressure of water will absorb water to form hydrates. Thus any substance which effloresces to yield a lower hydrate can absorb water to reform the higher hydrate. **Any substance which absorbs water from the air is hygroscopic.** Certain substances will remove so much water from the air that the solid disappears and becomes dissolved completely in a water solution. **A substance that absorbs water from the air to form a solution is deliquescent.** An example of a deliquescent substance is anhydrous calcium chloride, which first forms $CaCl_2 \cdot 6H_2O$, which, in turn, absorbs more water to form the solution. Sodium hydroxide is also deliquescent.

15.6 FORMATION OF WATER

Water is so abundant on earth that there is no need to produce it by chemical reactions for commercial purposes. However, it is formed in a number of reactions which occur commonly. These reactions are:

1 combustion of hydrogen in rocket engines;
2 combustion of hydrocarbons in energy production;
3 biological metabolism of food;
4 acid-base neutralization reactions.

Combustion of hydrogen

Hydrogen will combine directly with oxygen to form water and liberate energy.

$$2H_2 + O_2 \longrightarrow 2H_2O + \text{energy}$$

If the hydrogen is pure, it will burn smoothly at the point of contact with oxygen and produce a very hot colorless flame. This process can be demonstrated easily in an undergraduate chemistry lab. However, if the hydrogen and oxygen are mixed prior to ignition, the process occurs explosively.

The reaction of hydrogen and oxygen provides the power supply of both the second and third stages of the workhorse of the American space program, the Saturn V rocket. The hydrogen and oxygen are stored in the rocket as liquids. At the conclusion of the third-stage burn, the speed of an Apollo spacecraft is nearly 25,000 miles per hour.

Combustion of hydrocarbons

Organic compounds contain carbon, hydrogen, and some limited amounts of other elements, such as oxygen, sulfur, and phosphorus. When organic compounds burn in air, the oxygen combines with carbon to form carbon dioxide and with hydrogen to form water. The hydrocarbons contain only carbon and hydrogen and serve as one of the major sources of energy in our modern society. The natural gas we burn in our homes is predominantly the hydrocarbon methane, CH_4. Methane is a fuel which burns cleanly to liberate large quantities of energy.

$$CH_4 + 2O_2 \longrightarrow 2H_2O + CO_2 + 211 \text{ kcal/mole}$$

Other hydrocarbon fuels such as gasoline, kerosene, and home heating oil are mixtures. For simplicity, gasoline can be represented by an average molecular formula of C_8H_{18}. This mixture liberates 1300 kcal/mole when it burns.

$$2C_8H_{18} + 25O_2 \longrightarrow 16CO_2 + 18H_2O + \text{energy}$$

Kerosene is a mixture of heavier hydrocarbons and is used in the first stage of the Saturn V rocket. It may be represented by an average molecular formula of $C_{12}H_{26}$. This mixture liberates approximately 1950 kcal/mole.

Metabolism of food

All animals metabolize food to provide the energy necessary for life processes. Although the chemical reactions are many and beyond the scope of this text, they can be generalized as combustion type reactions which produce carbon dioxide, water, and energy. As an example, consider sucrose (table sugar):

$$C_{12}H_{22}O_{11} + 12O_2 \longrightarrow 11H_2O + 12CO_2 + \text{energy}$$

One mole of sucrose liberates approximately 350 kcal of energy.

Neutralization reactions

As indicated in Chapter 9, acids and bases react with each other in a neutralization reaction to form a salt and water. For example, sodium hydroxide and hydrochloric acid yield sodium chloride and water.

$$NaOH + HCl \longrightarrow NaCl + H_2O$$

Many of the metabolic reactions which occur in our body produce acids which must be neutralized in order to prevent acidosis. The chemicals in the blood serve to counter the buildup of acid. In Chapter 18 the chemistry of acids and bases will be presented in greater detail.

15.7 REACTIONS OF METALS WITH WATER

Some metals react with water to displace one or both hydrogen atoms by a metal ion. Because it rains everywhere, the metals used in construction must not be too reactive toward water. A list of the reactivity of metals in this type of reaction is given in Table 15.4 and is called the electromotive series. Additional details of this series will be presented in Chapter 19.

Those metals at the top of the electromotive series will replace hydrogen in water under some experimental conditions. The very active, such as potassium, will react violently with water.

$$2K + 2H_2O \longrightarrow 2KOH + H_2$$

The elements lithium through sodium in this list will all react directly with water. The metals magnesium through to cadmium react only with steam under proper experimental conditions.

$$Mg + H_2O(g) \longrightarrow MgO + H_2$$

Note that iron will react with steam. In order to prevent iron from reacting with water and oxygen, steel products often contain other metals as alloys

TABLE 15.4
electromotive series

Symbol	Name	Activity
Li	lithium	↑ increasing activity
K	potassium	
Ba	barium	
Ca	calcium	
Na	sodium	
Mg	magnesium	
Al	aluminum	
Zn	zinc	
Fe	iron	
Cd	cadmium	
Ni	nickel	
Sn	tin	
(H)	hydrogen	
Cu	copper	
Hg	mercury	
Ag	silver	
Au	gold	

or as coatings. Thus the less active nickel is used in making the alloy stainless steel, and tin is used to coat the steel of so-called tin cans.

All of the metals listed above hydrogen in the electromotive series will react with aqueous solutions of acids. Those below hydrogen will not react with acids such as HCl, H_2SO_4, H_3PO_4, and $HClO_4$. However, they will react with oxidizing acids such as HNO_3 and a mixture of HNO_3 and HCl called aqua regia.

15.8 HARD WATER

Limestone ($CaCO_3$) is an abundant mineral in the earth's crust. Most groundwater comes in contact with it and dissolves small amounts of the calcium ion. The presence of this ion in the water supply leads to many household and industrial problems. **Groundwater containing dissolved Ca^{2+} is called hard water.** It forms precipitates when heated or when soap is added.

When soap, which is the sodium salt of a complex acid called **stearic acid** ($C_{18}H_{35}O_2H$), is added to a solution containing Ca^{2+}, an insoluble salt called **calcium stearate** is formed. Its more familiar name is **scum** or **bathtub ring.**

$$Ca^{2+} + 2C_{18}H_{35}O_2^- \longrightarrow Ca(C_{18}H_{35}O_2)_2$$

If enough soap is added, eventually all the Ca^{2+} is precipitated, and the residual soluble sodium salt will be an effective cleansing agent. However, the precipitate is unsightly, and hard water is uneconomical to use if the concentration of Ca^{2+} is too high.

Hard water is a more serious problem for industries since they use more water than homes.

When heated, hard water forms a precipitate of $CaCO_3$. This creates industrial problems as $CaCO_3$ precipitates.

$$Ca^{2+} + 2HCO_3^- \rightleftharpoons CaCO_3 + CO_2 + H_2O$$

If the bicarbonate ion HCO_3^- is present, this reaction can be used to remove the calcium ion before the water is used. **Hard water containing the bicarbonate ion therefore is called temporary hard water.** However, this process is impractical for large-scale industry. Both temporary and permanently hard water can be softened by the addition of sodium carbonate (Na_2CO_3), commonly called washing soda. The carbonate ion (CO_3^{2-}) precipitates out the calcium ions.

$$Ca^2 + CO_3^{2-} \rightleftharpoons CaCO_3$$

Temporary hardness can be eliminated by the addition of a base that reacts with bicarbonate in an acid–base reaction to produce carbonate ions that then precipitate the Ca^{2+}.

$$HCO_3^- + OH^- \rightleftharpoons H_2O + CO_3^{2-}$$

The use of silicate minerals called zeolites to soften water by an ion exchange process has provided a remarkably convenient way of removing Ca^{2+} and replacing it by ions such as Na^+. The ion exchanger material consists of a large, covalently bonded, solid substance containing many negatively charged sites. The electrical neutrality of the substance is maintained by the presence of sodium ions. When water containing Ca^{2+} is passed through the ion exchanger, the Na^+ is replaced by Ca^{2+}. The hard water is softened and now contains sodium ions.

SUMMARY

Water is an angular polar molecule which forms ***hydrogen bonds.*** *The hydrogen bonds account for the high heat capacity, melting point, boiling point, heat of vaporization, and heat of fusion of water.*

Heavy water *contains deuterium instead of ordinary hydrogen. It is more dense than ordinary water, and its bonds are broken less easily.*

Hydrates *are crystalline compounds which contain water in definite proportion by weight. When hydrates* ***effloresce,*** *they lose some or all of their water. Substances which absorb water from the air are* ***hygroscopic.***

Water reacts with the metals listed above hydrogen in the electromotive series. The activity of the metals in the series decrease from top to bottom.

Water containing dissolved Ca^{2+} *is called* ***hard water.*** *Water which also contains bicarbonate ion is* ***temporary*** *hard water.*

QUESTIONS AND PROBLEMS

Definitions

1. Define each of the following terms.

(a) polar compound
(b) water
(c) hydrate
(d) deliquescence
(e) efflorescence
(f) hygroscopic
(g) hard water
(h) zeolite
(i) hydrocarbon
(j) heavy water
(k) electromotive series
(l) neutralization reaction
(m) hydrogen bonding

2. Explain the difference between each of the following.

(a) efflorescence and deliquescence
(b) hard water and soft water
(c) hydrogen bonding and polar covalent bonding
(d) deliquescent and hygroscopic

Hydrogen bonding

3. Describe physically the hydrogen bonding which occurs in water.

4. How many hydrogens can the oxygen of water become associated with by hydrogen bonding?

5. Recalling that fluorine is very electronegative, explain why the following boiling points are observed for the indicated compounds.

Compound	Boiling Point (°C)
HF	+20
HCl	−85
HBr	−67

6. What order of boiling points would you expect for NH_3, PH_3, and AsH_3, given the fact that nitrogen is sufficiently electronegative to give rise to hydrogen bonds in NH_3?

Physical properties of water

7. Describe how the density of water varies with temperature.

8. In the middle of winter the warmest place in a lake is at the bottom. Explain why.

9. Explain why ice floats.

10. A glass of water with ice is filled to the brim. The ice is above the brim. As the ice melts, will the glass overflow? Will the level change at all?

Heavy water

11. Explain how heavy water can be produced by electrolysis.

12. How is heavy water used as a tracer?

Hydrates

13. Describe the different ways in which water is found in hydrates.

14. A sample of a hydrate which is known to be efflorescent is exposed to air in a room over a period of time and remains unchanged. Explain why.

15. A sample of a hydrate slowly becomes wet and eventually becomes a liquid. Explain why.

16. A sample of a known deliquescent substance remains unchanged while exposed to the air in a room. Explain why.

Formation of water

17. List four ways in which water is commonly formed in reactions. Describe how each reaction is important to humanity.

Reaction of water with metals

18. What is the electromotive series?

19. Arrange the metals Al, Cd, and Sn in order of their reactivity with water.

20. Why is iron covered with tin to produce cans?

21. Disregarding cost, what metals would best be used to coat structural metals to prevent reaction with water?

22. Why are the metals silver and gold used to produce art objects to last for the ages?

16 SOLUTIONS AND COLLOIDS

LEARNING OBJECTIVES

When you finish this chapter you should be able to:

(a) differentiate between solutions and colloids.

(b) identify solutions involving all three states of matter.

(c) calculate the concentration of a solution using percent by mass, molality, molarity, and normality.

(d) explain the effect of pressure, temperature, and solvent polarity on solubility.

(e) interpret the effect of solute on the vapor pressure, boiling point, freezing point, and osmotic pressure of solutions using the kinetic molecular theory.

(f) identify colloids involving all three states of matter.

(g) demonstrate the effect of colloidal particles on the property of colloids.

(h) match the following terms with their corresponding definitions.

alloy
amalgam
colloid
emulsion
equivalents
foam
gel
liquid aerosol
molality
molarity
normality
osmotic pressure
percent by mass
saturated
sol
solid aerosol
solid foam
solid sol
solute
solution
solvent
unsaturated

Solutions are homogeneous mixtures. The atmosphere consists of a solution of oxygen in other gases, which must be mixed in the proper proportions to maintain the life of land animals. Fish and marine life are supported by the solution of oxygen gas dissolved in water. Water as commonly found on earth is a solution that contains dissolved minerals. These minerals, to a large extent, determine the palatability of water and are absorbed by plants from soil water. All the life processes in animals, such as digestion, metabolism, blood circulation, and waste removal, involve chemical reactions of substances dissolved in water. Most of the metals used by humanity are not pure elements but rather solutions of metals in metals.

In this chapter we will identify solutions of all three states of matter, express their composition in several units of concentration, and study how the properties of solutions depend on their composition.

Colloids are an intermediate state of aggregation of matter between heterogeneous mixtures and solutions. Colloids are suspended heterogeneous substances which require a long time to settle out. Among the commonly encountered colloids are whipped cream, milk, jello, foam rubber, and ruby. In this chapter colloids will be identified involving all three states of matter. The properties of colloids will be presented.

16.1 SOLUTIONS

A solution is a homogeneous mixture of substances which can vary in composition. The homogeneity of a mixture can be checked by the uniformity of smaller and smaller microscopic portions of the mixture. If the particles of different substances present in the mixture are of molecular size and are randomly distributed, then the mixture is a solution. The properties, then, of a solution are uniform down to the molecular level. As will be seen in Section 16.9, which deals with colloids, there are mixtures in which aggregates of molecules are dispersed among the molecules of another substance. These mixtures are fairly uniform but not down to the molecular level. The presence of such clusters of matter identifies mixtures as colloids.

Solution
Solvent
Solute

The substance present in the largest quantity in a solution is referred to as the solvent; the substance dispersed in the solvent is called the solute. There are substances that form mixtures with widely varying compositions. A drop of water will dissolve in a glass of alcohol, as will a drop of alcohol in a glass of water. Both systems are solutions of water and alcohol. In the former case the solvent is alcohol, whereas in the latter case the solvent is water.

Gaseous solutions

All gases that do not react with each other to form new substances will mix with each other in all proportions. In such solutions, the individual molecules

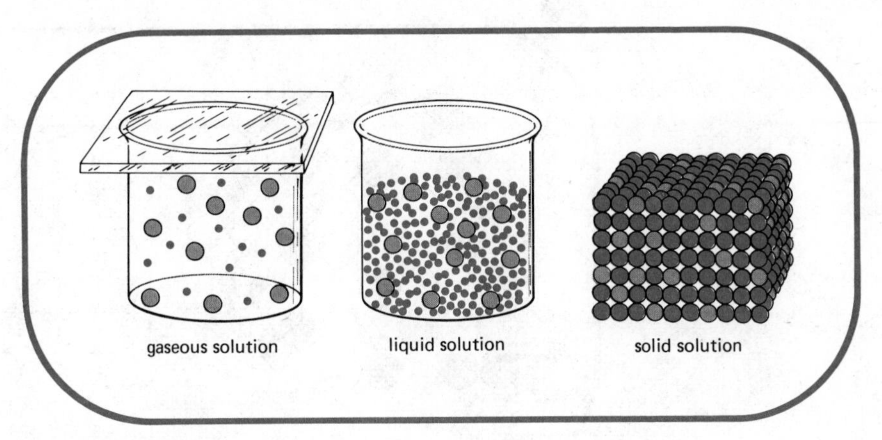

Figure 16.1 Models of solutions.

are far apart from each other (Figure 16.1) and move independently. Air is the most common example of a solution consisting only of gases. Air contains approximately 78 percent nitrogen molecules, 21 percent oxygen molecules, and 1 percent argon atoms.

There are no true solutions of a liquid or a solid in a gas. Such systems are colloids and are discussed in Section 16.9.

Liquid solutions

Gases, liquids, and solids may dissolve in liquid solvents. Because the molecules of liquids are close together, they are less independent of one another (Figure 16.1). Air dissolved in water, which maintains aquatic life, is an example of a solution of a gas in a liquid. Another example is carbon dioxide in carbonated beverages. Two common examples of liquids that dissolve in water are ethyl alcohol and acetic acid. Ethyl alcohol is a component of all alcoholic beverages; acetic acid is present in vinegar. Both salt and sugar are well known examples of solids which will dissolve in water. Although all gas mixtures which do not react chemically form a solution, not all gases, liquids, and solids dissolve in all liquids. This limitation is called the solubility of the solute and will be discussed in Section 16.7.

Do you think that you have ever seen a liquid that does not contain a solute?

Solid solutions

Although solid solutions might appear to be less common, there are, in fact, numerous examples of this type of solution. Many **metals dissolve in one another to form solid solutions called alloys. These alloys are usually substitutional solutions in which atoms of the solute metal are located in positions in the crystal where atoms of the solvent metal might otherwise have been** (Figure 16.2). Examples of such alloys are brass (zinc and copper) and sterling silver (silver and copper). Other alloys are listed in Table 16.1.

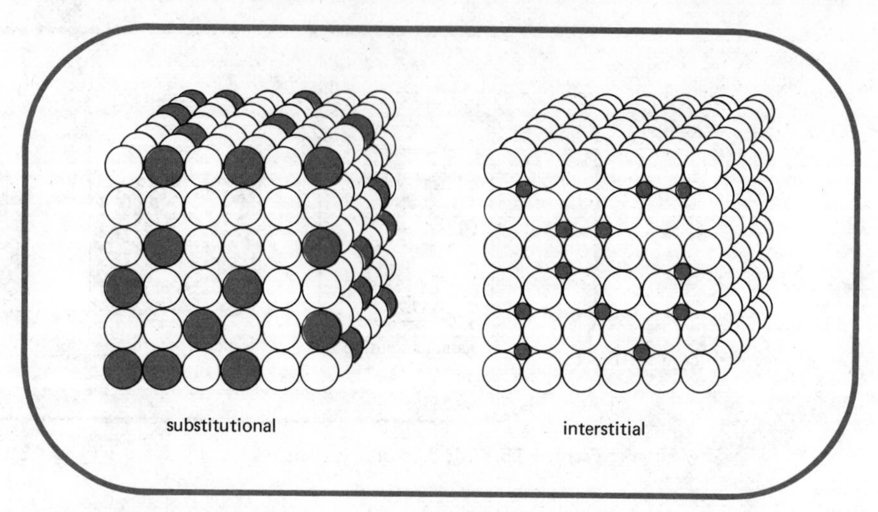

Figure 16.2 Substitutional and interstitial solutions.

The majority of metals are not pure but alloys.

Not all alloys are solid solutions. Some are heterogeneous mixtures that contain microscopic crystals of each of the components. Such heterogeneous mixtures result when the size difference between the atoms exceeds 14 percent. If the size difference exceeds this number, substitutional solutions cannot be formed. For example, the radii of zinc and copper are 1.25 Å and 1.17 Å, respectively, and the two elements form brass alloys which are substitutional solids. However, a copper-lead alloy is not a substitutional solid solution as

TABLE 16.1 composition and uses of alloys

Name	Composition by weight	Use
coinage silver	90% Ag, 10% Cu	coins
sterling silver	92.5% Ag, 7.5% Cu	tableware
pewter	85% Sn, 6.8% Cu, 6% Bi, 1.7% Sb	bowls, pots
plumber's solder	67% Pb, 33% Sn	solder
type metal	82% Pb, 15% Sb, 3% Cu	typesetting
naval brass	60% Cu, 39% Zn, 1% Sn	ship fittings
Babbitt metal	90% Sn, 7% Sb, 3% C	bearings
Wood's metal	50% Bi, 25% Pb, 12.5% Sn, 12.5% Cd	plugs in automatic sprinklers
Duralumin	95.5% Al, 4% Cu, 0.5% Mg	structural
Nichrome	80% Ni, 20% Cr	heating elements
stainless steel	80.6% Fe, 0.4% C, 18% Cr, 1% Ni	utensils
steel	99% Fe, 1% C	structural
vanadium steel	98.9% Fe, 1% C, 0.1% V	automobiles
molybdenum steel	98% Fe, 2% Mo	auto axles
spring steel	98.6% Fe, 1% Cr, 0.4% C	saw blades

the atomic radii are 1.17 Å and 1.54 Å, respectively, and differ by more than 14 percent.

There are alloys in which the size difference between the atoms is so large that the smaller atoms can be located in the spaces between the larger atoms. Such alloys are called interstitial solutions (Figure 16.2). Examples of interstitial solutions are carbon steel and boron steel in which the small atoms of the carbon or boron exist in the spaces between the larger iron atoms. As little as 0.5 percent carbon is required to increase markedly the hardness of steel.

A limited number of liquids will dissolve in solid metals to form solutions called amalgams. Mercury is a liquid metal that forms amalgams with solid metals; these solutions may be either solid or liquid depending on the amount of mercury used. Mercury readily dissolves in gold. At one time, mercury was used by prospectors to amalgamate gold and separate it from sand in the panning process. A person wearing gold jewelery should be careful in handling mercury. The mercury will rapidly dissolve in the gold and change the gold's appearance to a dull silver.

Even some gases will dissolve in certain metals. The best known examples are the dissolution of hydrogen in metals such as platinum, palladium, or nickel. Such solutions are of the interstitial type. Because there are many spaces in the metal, large quantities of hydrogen can be dissolved. Such solutions are important in catalyzed reactions of hydrogen.

Concentration of solutions

Solutions may be qualitatively described as concentrated or dilute. These adjectives indicate only that **a concentrated solution contains more solute than a dilute solution of the same substance.** More quantitative terms are used to describe concentrations in industry, hospitals, pharmacies, and chemical laboratories.

A variety of methods are used to express the concentration of solutions. Each one has been chosen for convenience of use under some particular set of circumstances. The types of concentrations described in the next four sections are:

1. percent by mass (weight),
2. molality,
3. molarity,
4. normality.

16.2 PERCENT BY MASS

The percent by mass (or weight) of a solute in a solution is equal to the mass of the solute divided by the mass of the total solution with the quotient multiplied by 100.

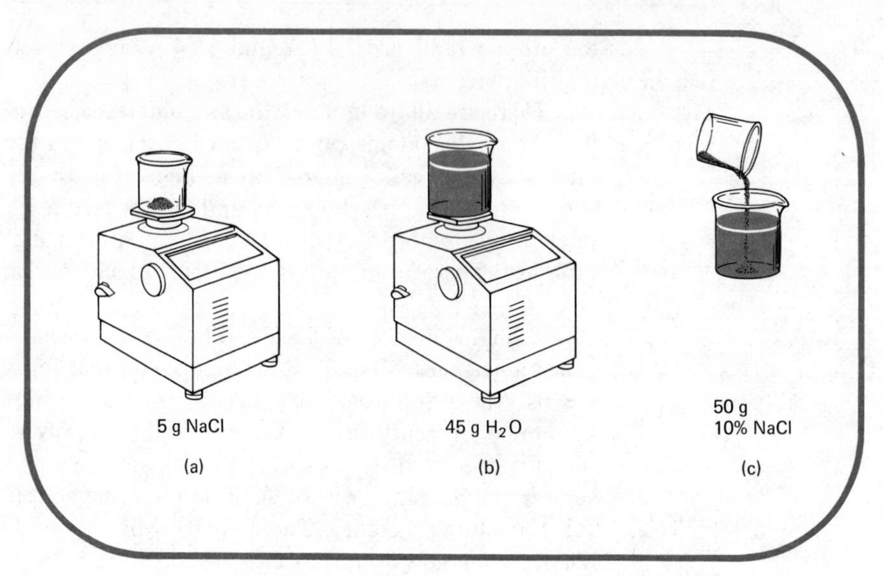

Figure 16.3 Making a 10% mass solution of sodium chloride: (a) weigh 5 g of sodium chloride; (b) weigh 45 g of water; (c) add the sodium chloride to the water and stir until it dissolves completely.

$$\% \text{ by mass} = \frac{\text{mass of solute}}{\text{mass of solution}} \times 100$$

The mass of the solution is equal, of course, to the mass of the solute plus the mass of the solvent.

In preparing or determining the percent by mass of a solution, only mass is involved. Thus no measurements of volumes are needed. If 5 g of sodium chloride is dissolved in 45 g of water, the resultant solution has a mass of 50 g. This is a 10% sodium chloride solution (Figure 16.3).

You do not have to be a chemist to use percent concentration units.

$$\frac{5 \text{ g NaCl}}{50 \text{ g solution}} \times 100 = 10\% \text{ NaCl solution}$$

Problem 16.1 A glucose solution for intravenous feeding is prepared from $1\bar{0}$ g of glucose and $19\bar{0}$ g of water. What is the mass percent concentration of glucose?

Solution The total mass of the solution is $2\bar{0}\bar{0}$ g.

$$19\bar{0} \text{ g water} + 1\bar{0} \text{ g glucose} = 2\bar{0}\bar{0} \text{ g glucose solution}$$

The mass percent is calculated as follows:

$$\frac{1\bar{0} \text{ g glucose}}{2\bar{0}\bar{0} \text{ g solution}} \times 100 = 5.0\% \text{ glucose solution}$$

Problem 16.2 How many grams of sodium chloride are present in 250 g of a salt solution which is 0.90% sodium chloride?

Solution A 0.90% solution contains 0.90 g of sodium chloride for every 100 g solution. Therefore, the amount of sodium chloride in any quantity of sodium chloride solution may be calculated by using a factor.

$$250 \text{ g solution} \times \frac{0.90 \text{ g NaCl}}{100 \text{ g solution}} = 2.2 \text{ g NaCl}$$

A closely related percent concentration is the percent by volume, which can be used when both a liquid solute and solvent are involved. The most common usage is in alcoholic beverages. An 11 percent by volume wine contains 11 ml of alcohol per 100 ml of wine solution. **The term proof is equal to twice the percent by volume.** A liquor which is 40 percent alcohol is 80 proof.

16.3 MOLALITY

Molality is equal to the number of moles of solute per kilogram of solvent. This concentration is abbreviated as *m*.

$$m = \text{molality} = \frac{\text{moles of solute}}{\text{kilogram of solvent}}$$

Note that this concentration does not involve any volume measurement. In fact, even the final volume of the solution is unknown. Only the mass of the final solution may be obtained by summing the mass of the moles of solute used and the mass of the solvent.

A one molal solution of sodium chloride is prepared by dissolving one mole (58.5 g) of sodium chloride in 1.00 kilogram of water (Figure 16.4).

$$\frac{1 \text{ mole NaCl}}{1 \text{ kilogram } H_2O} = 1m \text{ NaCl in } H_2O$$

In dealing with molality, factors involving mass of solute to mass of solution and mass of solute to mass of solvent are both necessary.

Problem 16.3 Calculate the molality of a glucose ($C_6H_{12}O_6$) solution obtained by dissolving 18 g of glucose in 500 g of water.

Solution The molecular weight of glucose is 180 g. Therefore, the number of moles is 0.10.

$$18 \cancel{g} \times \frac{1 \text{ mole}}{180 \cancel{g}} = 0.10 \text{ mole}$$

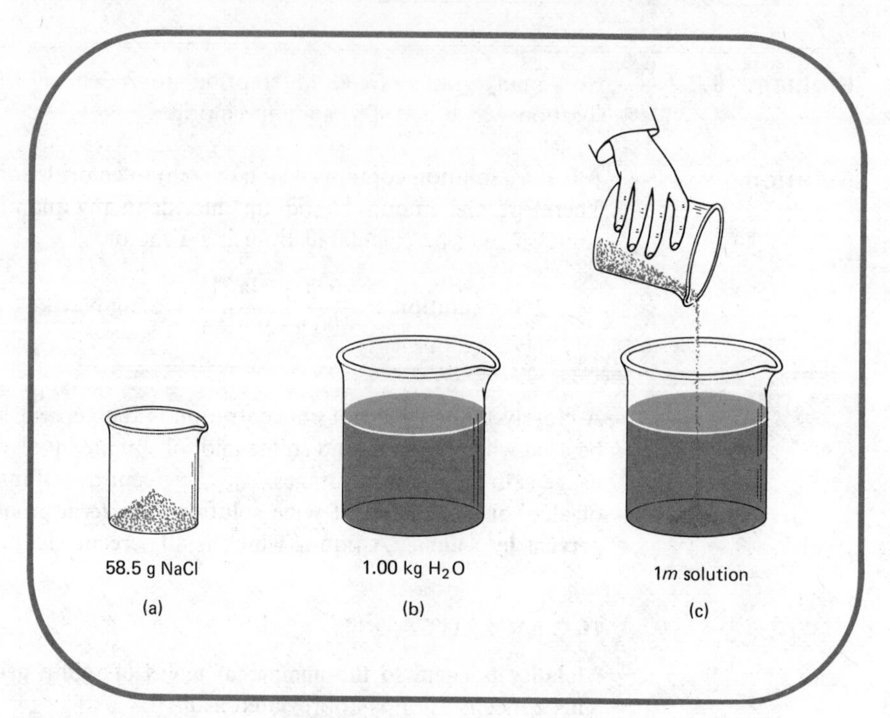

Figure 16.4 Making a one molal solution of sodium chloride: (a) weigh 58.5 g (1 mole) of sodium chloride; (b) weigh 1.00 kg of water; (c) add the sodium chloride to the water and stir until it dissolves.

The number of kilograms of water is 0.500.

$$5\bar{0}0\ \text{g} \times \frac{1\ \text{kg}}{1000\ \text{g}} = 0.500\ \text{kg}$$

Now the molality may be calculated.

$$\frac{0.10\ \text{mole glucose}}{0.500\ \text{kg H}_2\text{O}} = \frac{0.20\ \text{mole glucose}}{\text{kg H}_2\text{O}} = 0.20m$$

Problem 16.4 How many grams of water are required to dissolve 2.3 g of ethyl alcohol (C_2H_6O) and produce a 0.50 molal solution?

Solution In this problem the molality and the quantity of solute involved are given. Since the mass of the solvent is desired, it is necessary to use the inverse of the molality as a factor.

$$\frac{0.50 \text{ mole alcohol}}{1 \text{ kg } H_2O} = m$$

$$\frac{1 \text{ kg } H_2O}{0.50 \text{ mole alcohol}} = \frac{1}{m} = \frac{2.0 \text{ kg } H_2O}{\text{mole alcohol}}$$

The molecular weight of ethyl alcohol is 46 g. The number of moles of ethyl alcohol in a 2.3-g sample is 0.050.

$$2.3 \not{g} \times \frac{1 \text{ mole}}{46 \not{g}} = 0.050 \text{ mole}$$

Now the number of kilograms and grams of water needed may be calculated.

$$0.050 \text{ m}\not{\text{o}}\text{le alc}\not{\text{o}}\text{hol} \times \frac{2.0 \text{ kg } H_2O}{\text{m}\not{\text{o}}\text{le alc}\not{\text{o}}\text{hol}} = 0.10 \text{ kg } H_2O$$

$$0.10 \text{ kg } H_2O \times \frac{1000 \text{ g } H_2O}{1 \text{ kg } H_2O} = 1\bar{0}0 \text{ g } H_2O$$

Problem 16.5 How many grams of sulfuric acid are required to prepare $\overline{200}$ g of a 1.00 molal solution in water?

Solution In this problem it is necessary to obtain a factor giving the mass of sulfuric acid to the total mass of solution. A 1.00*m* solution of sulfuric acid would contain 1 mole, or 98.0 g, dissolved in 1000 g of water to produce 1098 g of solution. Thus the factor relating the mass of solute to the mass of solution is:

$$\frac{98.0 \text{ g } H_2SO_4}{1098 \text{ g solution}}$$

Therefore, in order to prepare 200 g of solution, the number of moles of sulfuric acid must be:

$$\overline{200} \not{g} \text{ sol}\not{u}\text{tion} \times \frac{98.0 \text{ g } H_2SO_4}{1098 \not{g} \text{ sol}\not{u}\text{tion}} = 17.8 \text{ g } H_2SO_4$$

16.4 MOLARITY

Molarity is defined as the number of moles of solute per liter of solution. The abbreviation for molarity is *M*.

$$\text{molarity} = \frac{\text{moles of solute}}{\text{liter of solution}} = M$$

In order to prepare a one molar solution, it is necessary to add sufficient

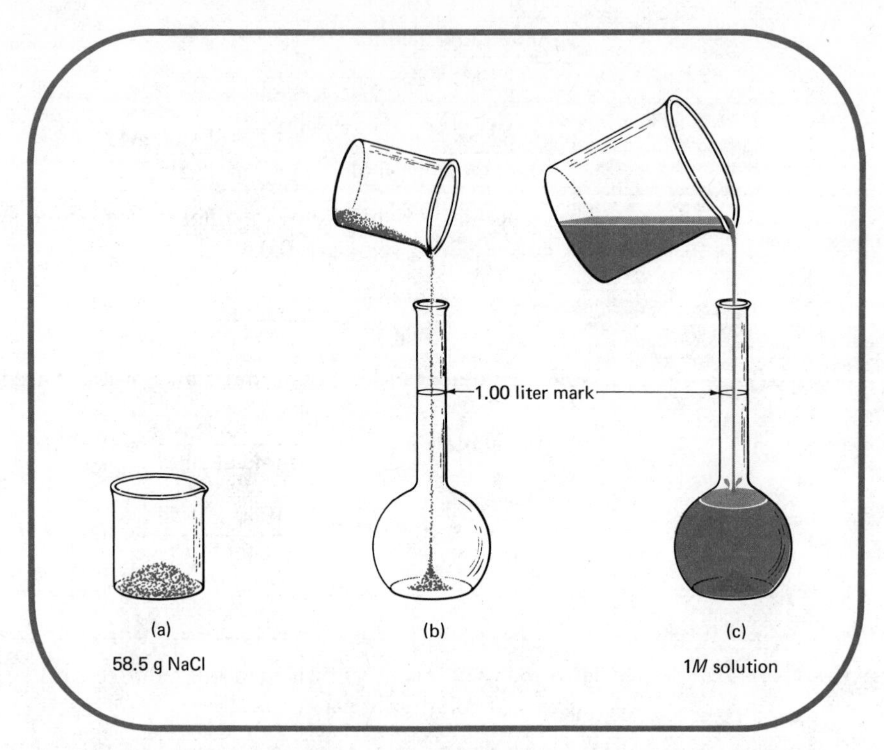

Figure 16.5 Making a one molar solution of sodium chloride: (a) weigh 58.5 g (1 mole) of sodium chloride; (b) add to a one-liter volumetric flask; (c) add water and mix thoroughly until one liter of solution results.

solvent to one mole of the solute to make a total volume of exactly one liter. Thus there is no information given in the molarity symbolism about the volume of solvent used to make the final volume of solution one liter.

Molarity and molality for dilute solutions in water are almost identical. Why?

Molarity is a very convenient concentration unit because it gives information about the volume of the solution that must be used to obtain the desired number of moles of solute. Volumes are easily measured with simple laboratory glassware (Figure 16.5).

Problem 16.6 What is the molarity of a solution prepared from 9.8 g of sulfuric acid and sufficient water to yield 200 ml of solution?

Solution The number of moles of sulfuric acid is 0.10.

$$9.8\,\cancel{g}\,H_2SO_4 \times \frac{1.0\text{ mole }H_2SO_4}{98\,\cancel{g}\,H_2SO_4} = 0.10\text{ mole }H_2SO_4$$

The volume of solution in liters is:

$$\bar{2}00\,\cancel{ml} \times \frac{1\text{ liter}}{1000\,\cancel{ml}} = 0.200\text{ liter}$$

The molarity is calculated from the number of moles of solute and the volume of solution.

$$\frac{0.10 \text{ mole } H_2SO_4}{0.200 \text{ liter solution}} = 0.50M \ H_2SO_4$$

Problem 16.7 How many grams of sodium hydroxide are required to produce 25$\bar{0}$ ml of a 0.20*M* solution?

Solution First the number of moles of NaOH required for this volume and molarity must be determined, and then the number of grams may be calculated. The number of moles is:

$$25\bar{0} \text{ ml} \times \frac{1 \text{ liter}}{1000 \text{ ml}} \times \frac{0.20 \text{ moles}}{\text{liter}} = 0.050 \text{ moles}$$

The number of grams is calculated from the formula weight.

$$0.050 \text{ moles} \times \frac{40.0 \text{ g}}{\text{mole}} = 2.0 \text{ g}$$

Problem 16.8 Calculate the number of liters of a 0.10*M* glucose solution required to provide 9$\bar{0}$ g of glucose ($C_6H_{12}O_6$).

Solution First the number of moles contained in 9$\bar{0}$ g must be calculated.

$$9\bar{0} \text{ g} \times \frac{1 \text{ mole}}{180 \text{ g}} = 0.50 \text{ mole}$$

Now the volume which contains 0.50 moles of glucose may be calculated by using the inverse factor of the molarity.

$$0.50 \text{ mole} \times \frac{1 \text{ liter}}{0.10 \text{ mole}} = 5.0 \text{ liters}$$

16.5 NORMALITY

Normality is equal to the number of equivalents of solute per liter of solution and is abbreviated as *N*.

$$\text{normality} = \frac{\text{equivalents solute}}{\text{liter of solution}} = N$$

The definition of an equivalent is based on the type of solute and the use of the solution. Equivalents are used in discussing acid-base reactions and oxidation-reduction reactions. In this chapter, the definition of an equivalent is limited to acid-base solutes.

One equivalent of an acid will neutralize one equivalent of a base.

An equivalent is equal to that mass of an acid which contains one mole of hydrogen atoms. An equivalent of a base is that mass which will react with one mole of hydrogen atoms supplied by an acid. The equivalent mass (or weight) of an acid is calculated by dividing the molecular mass (or weight) of the acid by the number of moles of hydrogen atoms contained in one mole of the acid. Similarly the equivalent mass (or weight) of a base is calculated by dividing the molecular mass (or weight) of the base by the number of moles of hydrogen atoms which will react with one mole of the base. A listing of the molecular weights and equivalent weights of several acids and bases is given in Table 16.2.

TABLE 16.2

molecular and equivalent weights of acids and bases

Compound	Molecular weight (g)	Equivalent weight (g)
HCl	36.5	$\frac{36.5}{1} = 36.5$
H_2SO_4	98.0	$\frac{98.0}{2} = 49.0$
H_3PO_4	98.0	$\frac{98.0}{3} = 32.7$
NaOH	40.0	$\frac{40.0}{1} = 40.0$
$Ca(OH)_2$	74.0	$\frac{74.0}{2} = 37.0$
$Al(OH)_3$	78.0	$\frac{78.0}{3} = 26.0$

The advantage of the normality unit is that one equivalent of an acid will react with one equivalent of a base. If molarity is used, a factor must always be applied before determining whether equal numbers of equivalents of acid or base are present for neutralization. For example, while **one liter of a 1*M* solution of HCl will neutralize one liter of a 1*M* solution of NaOH, a 1*M* solution of H_2SO_4 contains more hydrogen and will require two liters of a 1*M* NaOH solution. However, one liter of a 1*N* solution of any acid will neutralize one liter of a 1*N* solution of any base** (Figure 16.6).

Problem 16.9 What is the normality of a solution of H_3PO_4 obtained by dissolving 49.0 g of the acid in sufficient water to produce 3.00 liters of solution?

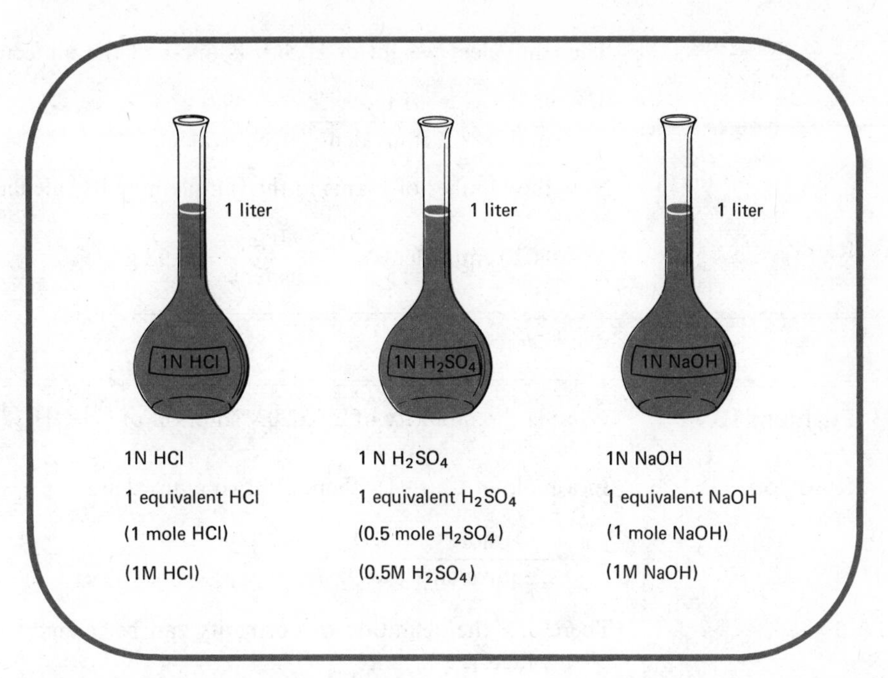

Figure 16.6 Normality of acids and bases.

Solution First the equivalent weight of H_3PO_4 must be used to determine how many equivalents are present in 49.0 g.

$$\text{molecular weight } H_3PO_4 = 98.0 \text{ g}$$

$$\text{equivalent weight } H_3PO_4 = 98.0 \text{ g}/3 = 32.7 \text{ g}$$

$$49.0 \cancel{g} \times \frac{1 \text{ equivalent}}{32.7 \cancel{g}} = 1.50 \text{ equivalents}$$

Now the normality may be calculated.

$$\frac{1.50 \text{ equivalents}}{3.00 \text{ liters}} = 0.500N$$

Problem 16.10 Calculate the number of grams of H_2SO_4 present in 100 ml of a $0.20N$ H_2SO_4 solution.

Solution First the number of equivalents of H_2SO_4 in the solution must be calculated.

$$\frac{0.20 \text{ equivalents}}{\cancel{\text{liter}}} \times 0.100 \cancel{\text{liter}} = 0.020 \text{ equivalents}$$

The equivalent weight of H_2SO_4 is one-half the molecular weight.

$$\frac{98.0\text{ g}}{\text{mole}} \times \frac{1\text{ mole}}{2\text{ equivalents}} = \frac{49.0\text{ g}}{\text{equivalent}}$$

Now the number of grams in the sample may be calculated.

$$0.020\text{ equivalent} \times \frac{49.0\text{ g}}{\text{equivalent}} = 0.98\text{ g}$$

Problem 16.11 What is the molarity of a $0.020N$ solution of $Ca(OH)_2$?

Solution In a mole of $Ca(OH)_2$ there are two equivalents.

$$\frac{1\text{ mole }Ca(OH)_2}{2\text{ equivalents }Ca(OH)_2}$$

Therefore, the definition of normality can be related to molarity by the use of this factor.

$$\frac{0.02\text{ equivalent }Ca(OH)_2}{1\text{ liter solution}} \times \frac{1\text{ mole }Ca(OH)_2}{2\text{ equivalents }Ca(OH)_2} = \frac{0.01\text{ mole }Ca(OH)_2}{1\text{ liter solution}} = 0.01M$$

16.6 KINETIC MOLECULAR THEORY OF SOLUTIONS

The physical process that occurs when a solute is dissolved in a solvent can be described by the kinetic molecular theory of matter. If a solid, such as potassium chloride, is placed in water, it may be seen to dissolve. The potassium and chloride ions in the solid are attracted to each other because they bear opposite charges. In the process of dissolving, these ions must be separated from one another. This occurs when the very polar water molecules approach and eventually surround the ions (Figure 16.7). The electronegative oxygen of water is oriented toward the positive potassium ion while the electropositive hydrogens of water are oriented toward the negative chloride ion. Thus the forces maintaining the ions together in the solid are exceeded by a somewhat stronger interaction between the ions and the solvent molecules, causing the solute to become dispersed throughout the solvent. The ions are removed from the site of the potassium chloride crystal by the water molecules.

The amount of potassium chloride which will dissolve in water depends

Figure 16.7 The solution of potassium chloride in water.

on the temperature. If, for example, 30 g of potassium chloride is placed in 100 g of water at 25°C, the solid would dissolve completely. If an additional 6 g of potassium chloride were added, it, too, would dissolve although at a somewhat slower rate. Now if any additional potassium chloride is added, no further apparent dissolution occurs and the solid remains at the bottom of the container. At this point, the solution is said to be saturated. **A solution is saturated when the concentration of dissolved solute is such that it can exist in contact with excess undissolved solute.** Thus 36 g of potassium chloride in 100 g of water at 25°C is a saturated solution.

How does a saturated solution differ from an unsaturated solution?

An unsaturated solution is one in which the concentration of solute is less than that of a saturated solution under the same conditions. Thus $2\bar{0}$ g of potassium chloride in $1\bar{0}\bar{0}$ g of water at 25°C is an unsaturated solution. Furthermore, 35 g of potassium chloride in the same amount of solute at 25°C is also an unsaturated solution.

Although no further change is apparent in a saturated solution, two dynamic processes are continuing. Some of the solid is still dissolving in the solvent, while some of the dissolved solute is returning to the solid lattice. The system is in a state of dynamic equilibrium, as can be demonstrated by placing an irregularly shaped crystal in the saturated solution and watching it slowly change in shape to a more regularly shaped crystal. It retains its mass despite its obvious change. Some of the solid is dissolving while a compensating crystallization of the solute is occurring (Figure 16.8). Thus a saturated

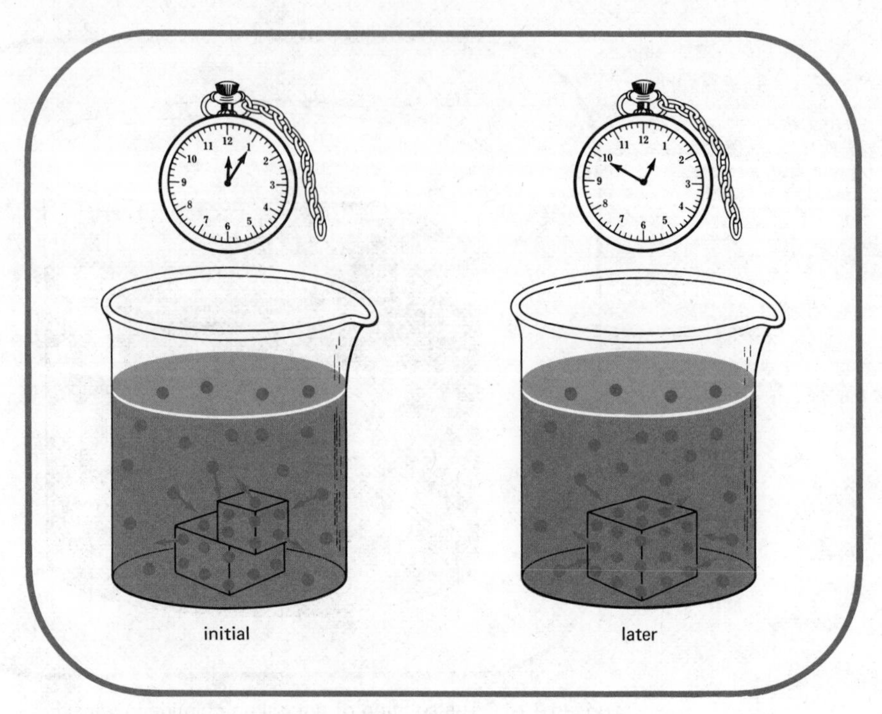

Figure 16.8 Equilibrium in a saturated solution.

solution is one in which the rate at which dissolution of the solute occurs is equal to the rate of crystallization of the dissolved solute.

$$\text{undissolved solute} \underset{\text{crystallization}}{\overset{\text{dissolution}}{\rightleftharpoons}} \text{dissolved solute}$$

In an unsaturated solution, the addition of solute leads to further dissolving because the rate of dissolution is greater than the rate of crystallization. However, as the solute continues to dissolve, the amount of dissolved solute increases and the rate at which it crystallizes increases. Eventually at some point, the rates become equal and the solution is then saturated.

16.7 SOLUBILITY

Effect of pressure on solubility

As was pointed out in discussing the properties of the states of matter, the most dramatic physical changes that occur with pressure changes are encountered in the gaseous state. Similarly, the solubility of gaseous solutes in liquid and solid solvents show the largest dependence on pressure. The solubility of a gas in a liquid can be expressed in terms of mass or volume of gas per unit volume of solvent or in any other units. Whatever units are chosen, **the solubility of a gas is directly proportional to the pressure of that gas above the surface of the solution. This relationship is known as Henry's law.** It is an example of another phenomenon that can be interpreted in terms of Le Châtelier's principle. If the pressure above the surface of a liquid is increased, the strain imposed on the system can be relieved by diminishing the amount of gas (Figure 16.9). Dissolution of the gas in the liquid does this.

Figure 16.9 Effect of pressure on the solubility of a gas.

Carbonated beverages are practical examples of the operation of Henry's law. All carbonated beverages are bottled under pressure, and when the bottle is opened, the pressure above the solution diminishes. As a result, the solubility of the gas decreases, the solution effervesces, and the dissolved carbon dioxide bubbles off. At 0°C, a liter of water will dissolve 1.7 liters of carbon dioxide under standard conditions. If the pressure is increased to 5 atm, the solubility of carbon dioxide increases fivefold. In Table 16.3, the solubilities of some common gases in water are listed.

Carbonated beverages are bottled under pressure. Why?

The condition called the "bends" that deep-sea divers may experience if they do not readjust gradually to the lower pressure when returning from

TABLE 16.3

solubility of gases in 1 liter of water at 0°C and 1 atmosphere

Gas	(g)
hydrogen	0.0019
nitrogen	0.029
oxygen	0.070
carbon dioxide	3.4
ammonia	1001

the ocean's depth, is due to the effect of pressure on the solubility of a gas. The air that divers breathe must be compressed. As a consequence, large quantities of nitrogen dissolve in the blood and other body fluids. If the pressure on the diver suddenly decreases by a rapid return to the surface, the nitrogen comes out of the solution rapidly and forms bubbles within the body, causing great pain and sometimes death. Artificial mixtures for breathing consisting of helium and oxygen are used for lengthy stays in the ocean by aquanauts. Helium is only one-fifth as soluble as nitrogen in water. When the pressure on the diver is lessened, there is less dissolved gas to form bubbles.

Effect of temperature on solubility

GASEOUS SOLUTES

The solubility of many gases in water decreases with increasing temperature. As water is warmed, the dissolved air can be seen to form bubbles and escape from the surface of the liquid. Although the solubility in other solvents is often the same, there are cases known where the opposite effect is observed.

Thermal pollution, a concern of ecologists, causes a decrease in the solubility of oxygen. Waste heat in many industrial plants is disposed of by transferral to water that is passed into lakes and streams. As these waterways warm up, oxygen is less soluble and aquatic animals cannot survive. To make matters worse, the animals' metabolic rates increase in warm waters. This causes them to need more rather than less oxygen.

Why does a warm soft drink taste flat?

LIQUID AND SOLID SOLUTES

There is no general rule for the solubility changes of liquids and solids with temperature. Quite often solubility increases with increasing temperature. For example, the solubility of potassium chloride (KCl) increases from 28 g/100 g of water at 0°C to 57 g/100 g at 100°C. Similarly, the solubility of sodium chloride (NaCl) increases from 35 g/100 g to 40 g/100 g of water over the same temperature range. The solubilities of salts such as cerium sulfate [$Ce_2(SO_4)_3$] decrease with increasing temperature (Figure 16.10).

Effect of solvent polarity on solubility

A maxim of the chemistry laboratory is that "likes dissolve likes." This generalization is reasonable since molecules of solute that are similar to molecules of solvent are expected to be better able to coexist in the same phase. Water is classified as a polar solvent because it consists of a collection of polar molecules. It is a good solvent for polar solutes, ionic compounds, and substances that can produce ions in water. Carbon tetrachloride (CCl_4), which can be used to remove grease spots from clothes, is a poor solvent for sodium chloride. However, fats and waxes readily dissolve in this nonpolar solvent because they are relatively nonpolar substances.

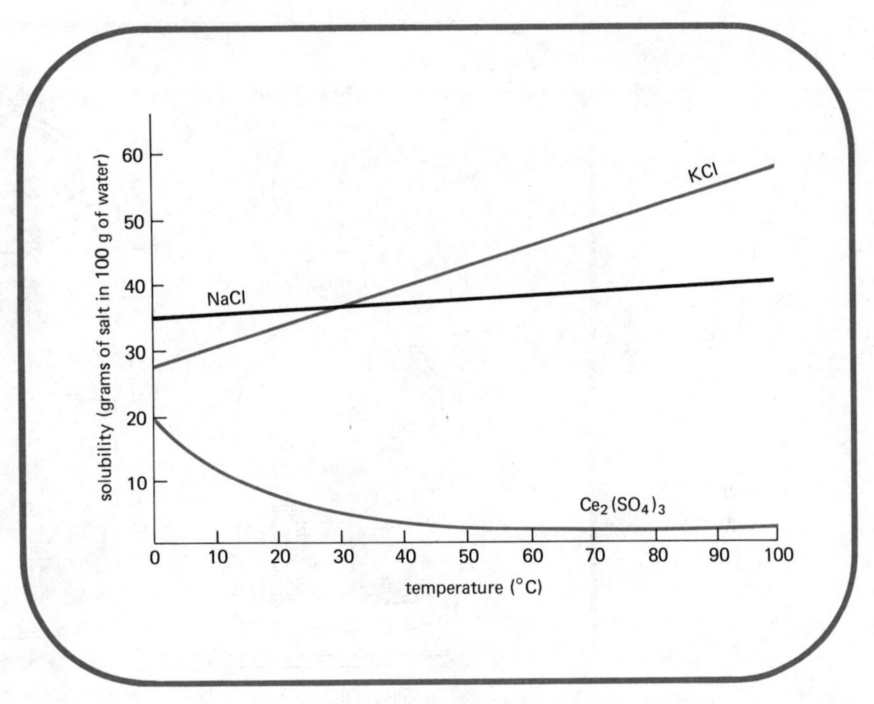

Figure 16.10 Solubility of solids in water as a function of temperature.

Problem 16.12 The structure of ethyl alcohol is given below. Explain why ethyl alcohol is very soluble in water.

```
   H  H
   |  |    __
H—C—C—O
   |  |    ‾‾ \
   H  H        H
```

Solution The structural feature —Ō—H resembles water. Ethyl alcohol should be polar, as is water. In addition, the nonbonded electron pairs on the oxygen in ethyl alcohol will form hydrogen bonds with water, thus making it soluble.

16.8 PROPERTIES OF SOLUTIONS

Vapor pressure of solutions

The addition of a solid solute such as salt or sugar to water decreases the vapor pressure of water. This fact can be easily demonstrated by enclosing a beaker

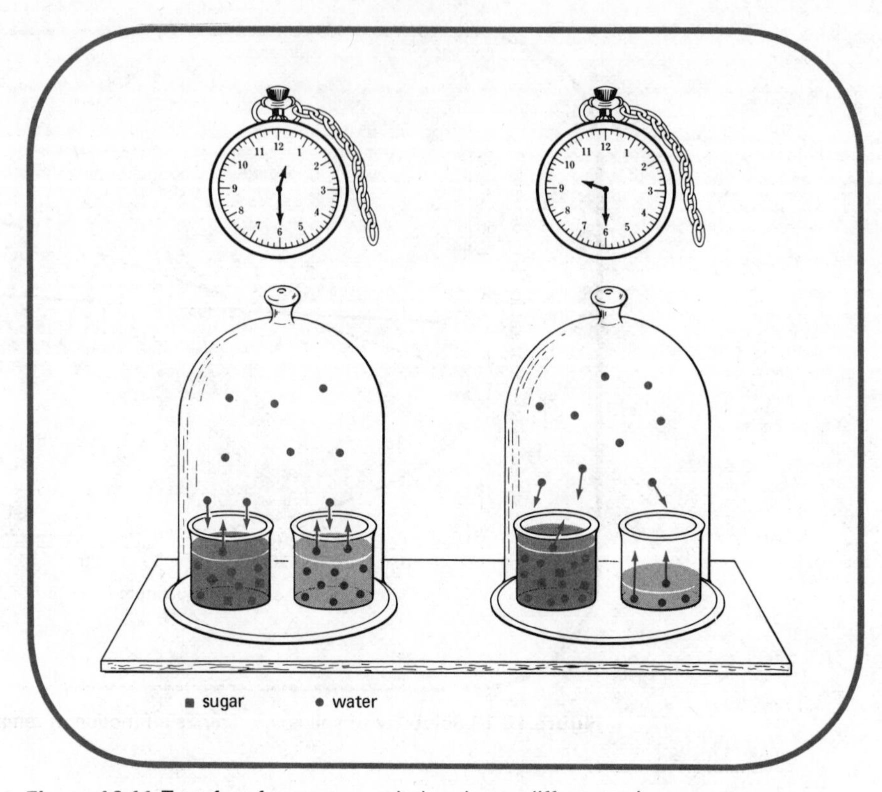

Figure 16.11 Transfer of water to a solution due to differences in vapor pressure.

of water together with a beaker containing an equal volume of, say, a sugar solution, as shown in Figure 16.11. In time, the volume of the sugar solution increases while that of the pure water decreases. Eventually, complete transfer of the liquid to the beaker containing the sugar occurs. The only possible explanation for the observed results is that the escaping tendency of water, its vapor pressure, has been reduced by the solute contained in one beaker. Since both beakers are in contact with the water molecules in the gaseous phase, both capture these water molecules at equal rates. Of course, water molecules are escaping from the liquid phase of each beaker. However, more are escaping from the beaker of pure water. The net transfer results from the slower rate of escape from the solution as compared to the pure liquid.

For solutions containing nonvolatile solutes such as sugar, the lowering of the vapor pressure of the solvent is directly proportional to the concentration of the solute in the solution. This relationship is known as Raoult's law.

Boiling point of solutions

The decreased vapor pressure of a solution of a nonvolatile solute means that the boiling point of the solvent must be elevated (Figure 16.12). It will

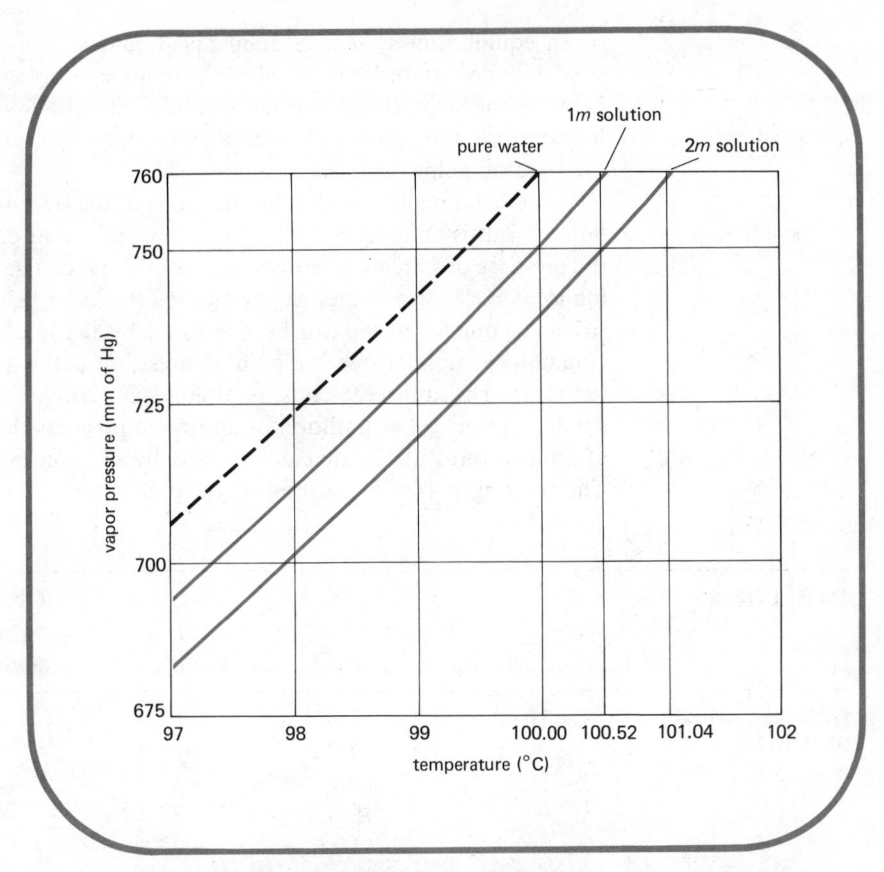

Figure 16.12 Effect of dissolved solute on the vapor pressure and boiling point of water.

require a higher temperature to raise the vapor pressure to atmospheric pressure. There is a direct relationship between the escaping tendency of the solvent molecules (vapor pressure) and the number of solute particles. Therefore, the increase in the boiling point of the solvent also is directly proportional to the concentration of solute.

Freezing point of solutions

The escaping tendency of a solvent decreases upon addition of a solute both for the liquid-vapor and liquid-solid transformations. However, the observed physical change in the case of freezing points is a decrease rather than the increase noted for boiling points. **The escaping tendency of solvent molecules from the solid state to the liquid state is unaffected by solute contained in the liquid phase. However, the tendency of the solvent liquid to enter the solid phase is decreased because of the presence of the dissolved solute.** If solute is added to an equilibrium system of solid and liquid at constant temperature, the solid will melt. In order to equalize the relative escaping tendencies of solvent between the two phases, the temperature must be lowered. At some lower temperature, the liquid and solid phases can coexist, and this temperature is the freezing point of the solution.

Salt used in ice control on roads also increases the rate of corrosion of the body of a car.

There are many practical applications of the freezing point depression of liquids. Salt is commonly spread on snow and ice in order to melt them. In the presence of salt, snow cannot exist at 0°C. Of course, this method of melting ice is ineffective if the temperature of the ice is below that at which the freezing point of water can be depressed by the addition of salt. Another common example of freezing point depressions is the use of antifreeze in car radiators. The antifreeze consists of ethylene glycol ($C_2H_6O_2$), which is very soluble in water. The addition of antifreeze prevents the water from freezing at temperatures above that determined by the concentration of antifreeze. The freezing points of various ethylene glycol–water solutions are given in Table 16.4.

TABLE 16.4

freezing points of antifreeze

Weight (%) $C_2H_6O_2$	Molality $C_2H_6O_2$	Observed freezing point (°C)	Observed freezing point (°F)
10	1.8	−3.6	23.7
20	4.0	−7.9	18.7
30	6.9	−14.0	6.8
40	10.8	−22.3	−8.1
50	16.1	−33.8	−28.8

Osmotic pressure of solutions

If a solution such as sugar in water is separated from pure water by **a semipermeable membrane, a membrane through which water can pass but not other large molecules such as sugar** (Figure 16.13), the volume of the solution increases at the expense of the pure water. The water remains pure because sugar molecules cannot pass through the membrane. The water molecules can cross the membrane and do so in both directions. **The process of transfer of molecules through a semipermeable membrane is called osmosis.** In the solution, the water has been diluted by solute, so there are fewer water molecules going into the pure water than there are going into the solution. The volume of the solution increases and the pure water volume decreases. Eventually, no further net transfer occurs, and the level of the solution remains at a specific height above that of the water. The difference in the levels is a measure of the net tendency of water to go through the membrane to dilute the solution. The added height in the solution causes a higher pressure on the solution side of the membrane. This pressure pushes more water

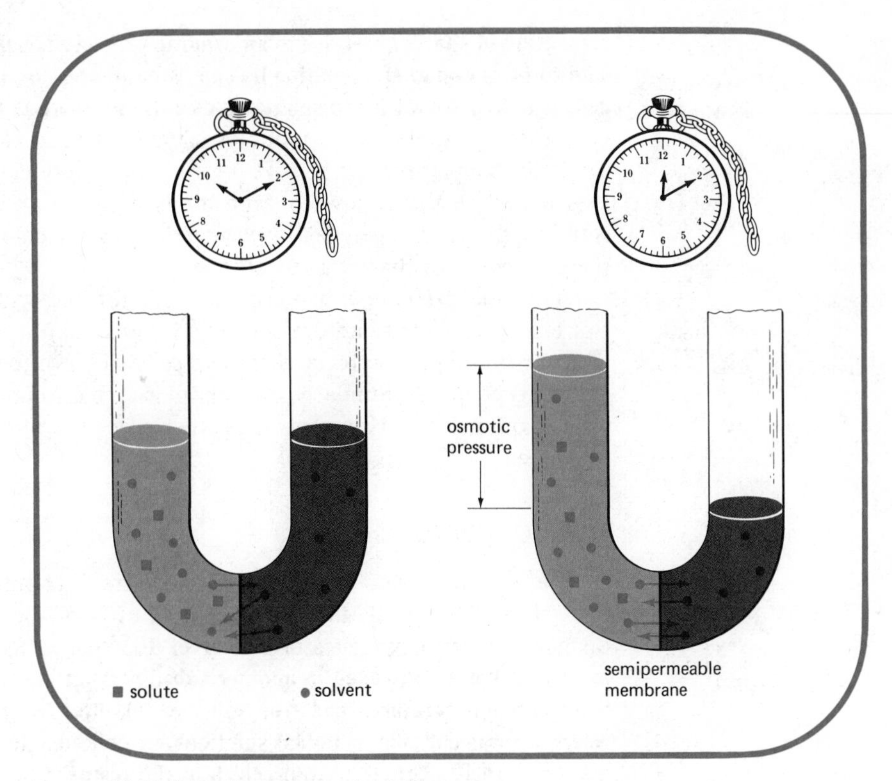

Figure 16.13 Osmotic pressure of solutions.

molecules through to the pure water side. Eventually, the pressure builds up to the point that the number of water molecules going in each direction is equal.

If pressure is applied on the side of the tube containing the solution, the net flow of water molecules can be reversed. **The pressure required to maintain equal levels of the water and the sugar solution is called the osmotic pressure.**

Osmosis of solvent, or the passage through a semipermeable membrane, is a process of great importance in maintaining life processes. Both animals and plants contain membranes through which water passes. If the concentrations of the solutes in the water solution are not properly balanced, water transport will occur so as to impair the function of the cell. Water will transfer from solutions of low osmotic pressure to solutions of high osmotic pressure. **If the osmotic pressure of a solution around a cell is less than within the cell, water will enter the cell and the cell wall will rupture or hemolyze. If the osmotic pressure of a solution is larger than within the cell, the water will leave the cell and cause it to shrivel and shrink (crenate).** Both hemolization and crenation are damaging biological events.

Intravenous administration of dextrose or replacement of ionic constituents of body fluids must be carefully controlled. **A solution whose con-**

centration of solute gives rise to an osmotic pressure equal to that within cells is said to be isotonic. If a solution has an osmotic pressure less than that within a cell, it is hypotonic; if the solution's osmotic pressure is larger than within the cell, it is hypertonic. The osmotic pressure of the red blood cell is 7.7 atm at the body temperature of 98.6°F. This high pressure can be balanced by the presence of relatively low concentrations of solute in surrounding fluid. A solution that is 0.9 percent by weight in sodium chloride has an osmotic pressure equal to that of a red blood cell.

Salt spread on driveways may be injurious to nearby plants and shrubs. Why?

Osmotic pressure differences account for the wilting of plant forms. For example, if the salad dressing of a salad has an osmotic pressure higher than the fluid in the lettuce cells, these cells will lose the water that gives the lettuce its crispness. Similarly, a flower placed in a solution will wilt as it loses water from its cells. However, if it is then placed in pure water, it will regain its shape.

16.9 COLLOIDS

In solutions, individual atoms or molecules are distributed amidst the solvent molecules. The size of these solute particles is of the order of 10^{-7} cm in diameter. **When aggregates of matter of 10^{-7} cm to approximately 10^{-5} cm in diameter are suspended in another substance, an intermediate state between heterogeneous mixtures and true solutions results. Mixtures of this type are referred to as colloids, colloidal solutions, or colloidal suspensions.** Aggregates greater than 10^{-5} cm tend to precipitate and result in heterogeneous mixtures.

The particles suspended in the colloid are called the dispersed phase; the material in which they are suspended is the dispersing phase or medium. A list of some examples of colloids is given in Table 16.5.

TABLE 16.5 types of colloids

Dispersing medium	Dispersed phase	Name	Example
gas	liquid	liquid aerosol	fog
gas	solid	solid aerosol	smoke
liquid	gas	foam	whipped cream
liquid	liquid	emulsion	milk
liquid	solid	sol	latex paints
solid	gas	solid foam	foam rubber
solid	liquid	gel	jello
solid	solid	solid sol	ruby

Types of colloids

A gas can serve as a dispersing phase for either solids or liquids. Small dust particles in air, such as those present in major metropolitan areas, are a well-known example of a solid dispersed in a gas. When water condenses from saturated air and forms tiny drops that remain suspended in the air, the resultant fog is in reality a colloid composed of a liquid dispersed in gas. **Solids and liquids dispersed in a gas are called solid aerosols and liquid aerosols, respectively.**

Many foods that we eat are colloids.

Liquids, too, serve as the dispersing phase for many colloids. Solids such as clay in muddy water or starch in pudding are examples of solids dispersed in a liquid. **Colloids of solids dispersed in liquids are called sols. When a liquid forms a colloid in a liquid, it is called an emulsion.** Mayonnaise, cream, and milk are all emulsions. In the homogenization of milk, the dispersed fat is reduced to such a small size that the emulsion becomes virtually permanent. Examples of gases dispersed in liquids include whipped cream and meringue. **Gases dispersed in liquids are called foams.**

Solids can serve as the dispersing phase for colloids. Foam rubber and styrofoam are examples of a gas dispersed in a solid. **Such colloids are called solid foams. The dispersal of a liquid in a solid is called a gel.** The common examples are jellies and cheeses. Colored glass and gems such as ruby and turquoise are examples of a solid dispersed in a solid. **These colloids are called solid sols.**

The Tyndall effect

Why is a beam of sunlight coming through a window more visible if dust particles are in the air?

When a beam of light is passed through a true solution, there is no evidence of it to an observer at right angles to the path of the light. This fact can be easily observed for a liquid solution, as illustrated in Figure 16.14. Similarly a beam of light passing through dust- and moisture-free air cannot be observed at right angles to the beam. **A colloid, however, contains particles that are larger than those present in true solutions. Colloidal particles are large enough to scatter light. This light scattering is called the Tyndall effect.** The

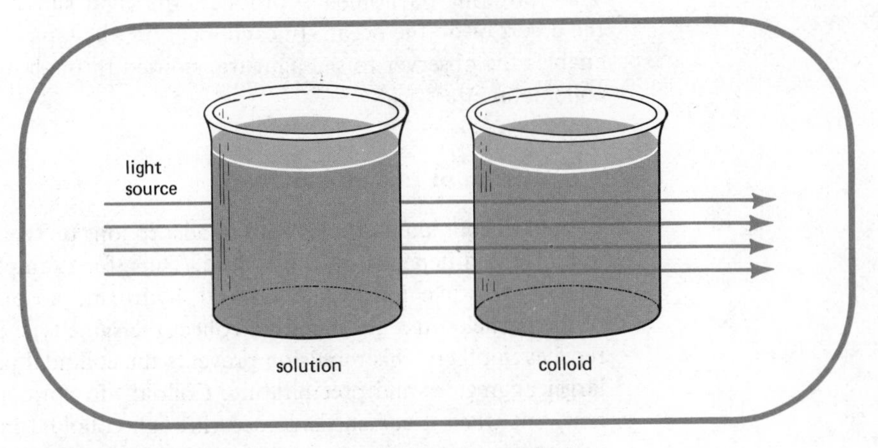

Figure 16.14 Tyndall effect in a liquid dispersing medium.

Figure 16.15 Our atmosphere provides a useful gaseous dispersion medium for spectacular Tyndall effects. Courtesy of Culver Pictures, Inc. Reproduced with permission.

same observation may be made in air containing dust, moisture (fog), or smoke, which are gaseous colloids (Figure 16.15).

The Tyndall effect is controlled by the size and shape of the particles. This scattering phenomenon produces the vivid sunsets that can be seen in the desert or on the ocean. In each location, the long distance to the horizon enables the observer to see light transmitted through miles of air containing colloidal matter.

Adsorption of charged matter

Dispersed colloidal particles tend to adsorb ions on their surface. The type of ion adsorbed depends on the colloid. Thus, for example, arsenic(III) sulfide adsorbs negative ions while iron(III) hydroxide adsorbs positive ions. All of the particles in a given colloid collect the same type of charge and tend to repel each other. This repulsion prevents the colloidal particles from forming larger aggregates and precipitating. Colloids do not conduct an electric current, but when a current is passed through colloidal matter in a liquid, the particles usually move toward one of the charged electrodes in a manner characteristic of the individual colloid. At the electrode the particles are precipitated and no longer migrate.

The Cottrell precipitator

Smoke is a colloidal suspension of solid particles in air. In areas of high industrial concentrations, the effluent from the factories and manufacturing concerns can be a considerable nuisance and in most cases a health hazard. A Cottrell precipitator (Figure 16.16) that discharges the colloidal material by passing the smoke through charged electrodes can be installed in the

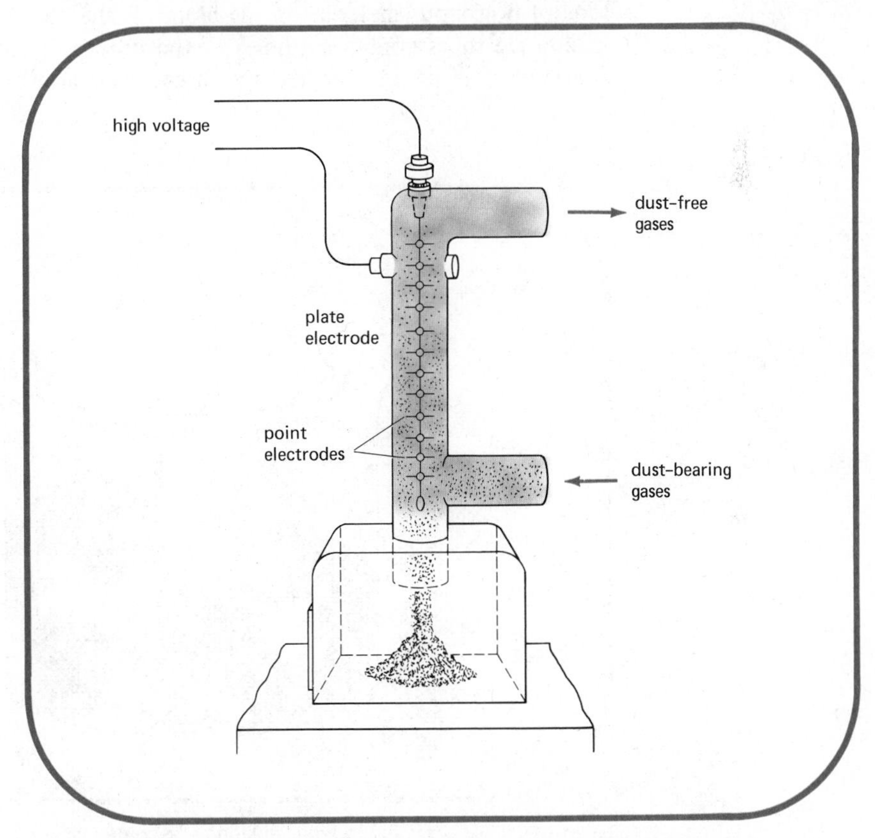

Figure 16.16 The Cottrell precipitator. Adapted from Figure 17.7 of *General Chemistry* by W. A. Neville (New York, McGraw-Hill Book Co., 1968). Reproduced with permission.

chimneys of commercial concerns. It is not uncommon for large industrial plants to recover tons of solids each day through the use of these precipitators. In such cases, the value of the substances recovered may exceed the cost of maintaining the precipitation equipment.

Dialysis of colloids

Ionic and molecular matter of certain limited dimensions can be separated from a colloid by a process called dialysis. The colloid to be purified is separated from pure water by a semipermeable membrane. Ionic substances and molecules of certain dimensions can pass through this membrane, but the colloidal particles cannot. The noncolloidal substances diffuse through the membrane into the surrounding water and can be washed away by the continuous flow of water. By dialysis, it is possible to purify colloids.

The artificial kidney machine is an important practical application of the dialysis process. Complete or partial kidney failure causes an increase in the level of poisonous materials in the blood. If the blood is circulated through membrane tubes containing holes of the proper dimensions, these harmful substances can pass through to the surrounding aqueous solution. However,

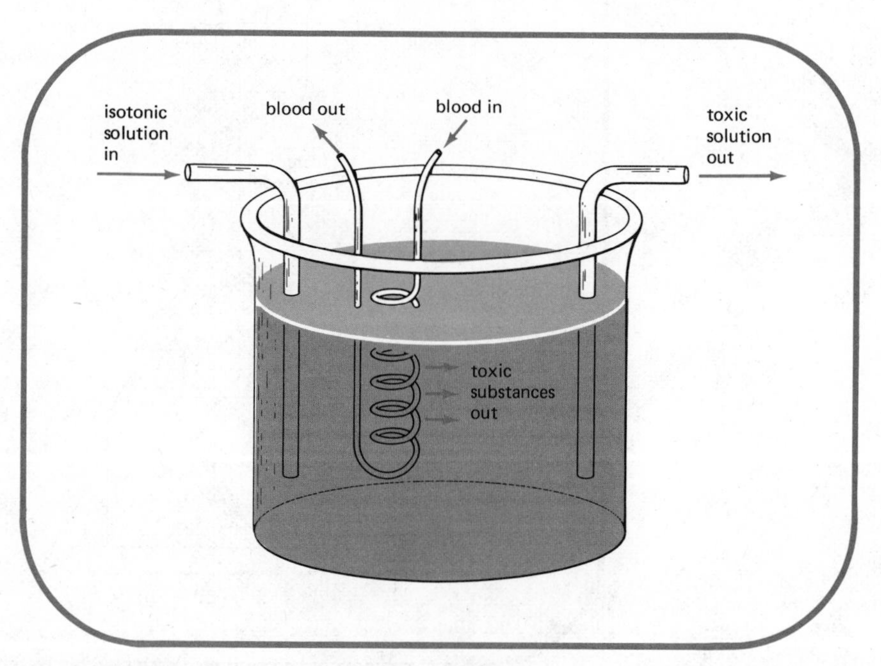

Figure 16.17 Dialysis in the artificial kidney machine.

the colloidal proteins in blood plasma and the cells remain within the tube and are returned to the patient (Figure 16.17). By this means, the life of an individual can be extended until normal kidney functions are established. In the case of complete kidney failure, life can be maintained by treatment approximately once a week with the artificial kidney.

SUMMARY

Solutions *are homogeneous mixtures of substances whose composition may vary.* ***Colloids*** *are an intermediate state of aggregation between heterogeneous mixtures and solutions. Solutions consist of components called the* ***solvent*** *and the* ***solute.*** *Solutions of solid metals in solid metals are called* ***alloys. Amalgams*** *are solutions of liquid metals dissolved in solid metals.*

Concentrations may be expressed as ***percent by mass, molality, molarity,*** *and* ***normality****. The percent by mass of a solute in a solution is equal to the mass of the solute divided by the mass of the total solution with the quotient multiplied by 100. Molality is equal to the number of moles of solute per kilogram of solvent. Molarity is equal to the number of moles of solute per liter of solution. Normality is equal to the number of* ***equivalents*** *of solute per liter of solution.*

A ***saturated*** *solution is one which can exist in equilibrium with undissolved solute. An* ***unsaturated*** *solution contains less dissolved solute than a saturated solution.*

Solubility of gases increases with increasing pressure, while the solubility of liquids and solids is not affected by pressure changes. The solubility of gases in liquids generally decreases with increasing temperature. The solubility of solids or liquids in liquids changes with temperature; some solutes increase in solubility while others decrease.

Solute causes a decrease in vapor pressure, an increase in the boiling point, and a decrease in the freezing point of the solvent. Solutions of high concentration have large ***osmotic pressures.*** *Solvent molecules move from solutions of low osmotic pressure to solutions of high osmotic pressure.*

Colloids consist of a dispersed phase and a dispersing phase. Gases dispersed in liquids and solids are called ***foams*** *and* ***solid foams,*** *respectively. Liquids dispersed in gases, liquids, and solids are called* ***liquid aerosols, emulsions,*** *and* ***gels,*** *respectively. Solids dispersed in gases, liquids, and solids are called* ***solid aerosols, sols,*** *and* ***solid sols*** *respectively. Colloidal particles adsorb ions, scatter light, and cannot pass through membranes.*

QUESTIONS AND PROBLEMS

Definitions

1. Define each of the following terms.

(a) solution — **(b)** colloid
(c) solvent — **(d)** solute
(e) molarity — **(f)** molality
(g) solubility — **(h)** saturated solution
(i) normality — **(j)** equivalent
(k) dispersed particles — **(l)** dispersing phase
(m) osmotic pressure — **(n)** Tyndall effect

(o) Cottrell precipitator
(p) isotonic
(q) hypertonic
(r) hypotonic
(s) hemolyze
(t) crenate
(u) emulsion
(v) gel
(w) dilute
(x) concentrated

2. Distinguish carefully between each of the following.
   (a) a solution and a colloid
   (b) solute and solvent
   (c) molality and molarity
   (d) molarity and normality
   (e) hypertonic and hypotonic
   (f) emulsion and gel

Types of solutions

3. Give one example of a solution of each of the following types.
   (a) gas dissolved in a gas
   (b) gas dissolved in a liquid
   (c) liquid dissolved in a liquid
   (d) solid dissolved in a liquid
   (e) gas dissolved in a solid
   (f) liquid dissolved in a solid
   (g) solid dissolved in a solid

4. Indicate the solute and solvent in each of the following and classify each by phase.
   (a) glucose in blood
   (b) vodka
   (c) oxygen in water
   (d) air
   (e) brass
5. Can a solid metal be a solvent? Explain.
6. Can a gas be a solvent? Explain.
7. What is meant by a dilute solution? How does this differ from a concentrated solution?

Percent by mass

8. Calculate the percent by mass of each of the following solutions.
   (a) 10 g of Na_2SO_4 in 90 g of H_2O
   (b) 5 g of NaCl in 100 g of H_2O
   (c) 50 g of alcohol in 100 g of H_2O
9. How many grams of solute are present in the indicated quantity of solvent?
   (a) sugar in 200 g of a 10% solution in water
   (b) sodium chloride in 100 g of a 5% solution in water
   (c) ammonia in 2000 g of a 1% solution in water
   (d) sodium hypochlorite in 750 g of a 5% solution in water

Molality

10. Calculate the molality of each of the following solutions.
  (a) 124 g of ethylene glycol ($C_2H_6O_2$) in 500 g of water
  (b) 460 g of ethyl alcohol (C_2H_6O) in 750 g of water
  (c) 1.8 g of glucose ($C_6H_{12}O_6$) in 100 g of water

11. How many grams of water are required to dissolve 4.6 g of ethyl alcohol to produce a 2.0*m* solution?

12. How many grams of water are required to dissolve 3.1 g of ethylene glycol ($C_2H_6O_2$) to produce a 0.2*m* solution?

Molarity

13. Calculate the molarity of each of the following.
  (a) 6.2 of ethyl alcohol in 500 ml of solution
  (b) 0.585 g of NaCl in 100 ml of solution
  (c) 49 g of H_2SO_4 in 2000 ml of solution

14. Calculate the number of moles of solute present in each of the following solutions.
  (a) salt in 200 ml of a 0.10 molar solution
  (b) sodium hydroxide in 1500 ml of a 0.50 molar solution
  (c) sugar in 50 ml of a 0.01 molar solution

15. Calculate the number of grams of solute present in each of the following.
  (a) H_2SO_4 in 500 ml of a 0.2*M* solution
  (b) NaOH in 20 ml of a 0.5*M* solution
  (c) ethylene glycol, $C_2H_6O_2$, in 3000 ml of a 2*M* solution

16. Calculate the volume of each of the following solutions necessary to contain the indicated number of moles of solute.
  (a) 0.1 moles of ethyl alcohol from a 2*M* solution
  (b) 0.05 moles of sodium hydroxide from a 0.1*M* solution
  (c) 5.0 moles of alcohol from a 2.0*M* solution

Normality

17. Calculate the normality of each of the following solutions.
  (a) 4.00 g of NaOH in 500 ml of solution
  (b) 49.0 g of H_2SO_4 in 500 ml of solution
  (c) 4.90 g of H_3PO_4 in 50 ml of solution

18. What is the normality of each of the following solutions?
  (a) a 0.2*M* solution of NaOH
  (b) a 0.01*M* solution of $Ca(OH)_2$
  (c) a 0.1*M* solution of H_2SO_4
  (d) a 0.05*M* solution of H_3PO_4

19. Calculate the number of grams of solute necessary to prepare each of the following.
  (a) 150 ml of a 0.10*N* solution of NaOH
  (b) 200 ml of a 0.10*N* solution of H_2SO_4
  (c) 300 ml of a 0.1*N* solution of H_3PO_4

20. How many equivalents are present in each of the following?
  (a) 500 ml of a 0.20*M* H_2SO_4 solution
  (b) 100 ml of a 0.10*M* NaOH solution
  (c) 250 ml of a 0.30*M* H_3PO_4 solution

Kinetic molecular theory

21. Explain the dissolution process of an ionic solid into water in terms of the kinetic molecular theory.

22. In a saturated solution containing excess undissolved solid, an equilibrium exists. Explain what is meant by equilibrium.

Solubility

23. State Henry's law in your own words.

24. Why do bubbles escape from a beer bottle after the cap is removed?

25. What causes the bends? How does replacement of nitrogen by helium prevent the bends?

26. Why is thermal pollution of streams a matter of ecological concern?

27. How is the solubility of ionic solids in water affected by temperature?

Properties of solutions

28. How is the vapor pressure of water affected by dissolved sugar?

29. State Raoult's law in your own words.

30. A maple syrup solution does not boil at the same temperature as water. Explain why.

31. Why is a mixture of rock salt and ice used in hand-crank ice cream freezers?

32. What is osmotic pressure?

33. Explain why liquids used in intravenous feedings must be isotonic.

Colloids

34. How does a colloid differ from a solution?

35. Explain why searchlights used to sweep the skies in night advertising are effective in major metropolitan areas but are not used in areas having very clean, dry air.

36. How are electronic air cleaners used in homes effective in removing airborne colloidal dust and pollen particles?

17 REACTION RATES AND EQUILIBRIA

LEARNING OBJECTIVES

When you finish this chapter you should be able to:

(a) relate the rate of reactions to reactants, the concentration of reactants, temperature, and catalysts.

(b) illustrate the utility of catalysts in industrial and biological processes.

(c) identify chemical reactions which are irreversible.

(d) express the quantities of reactants and products present at equilibrium in terms of an equilibrium constant.

(e) demonstrate the use of Le Châtelier's principle in chemical equilibria.

(f) formulate the solubility of ionic compounds in terms of a solubility product expression.

(g) match each of the following terms with their corresponding definition.

catalyst
enzyme
equilibrium
equilibrium constant
kinetics
mass action
solubility product constant

Up to this point we have viewed chemical changes only in terms of their balanced equations. However, our observations of the chemical changes of matter indicate that they occur at a variety of rates. The reaction of oxygen and the gas from a leaky gas pipe when ignited by a spark occurs with explosive violence. Similarly, the flash of a flashbulb involves the rapid reaction of a fine magnesium ribbon with oxygen. Other reactions, such as the spoiling of wine when alcohol is converted to acetic acid and the rusting of iron, involve much slower reactions with oxygen. In this chapter some of the characteristics of chemical reactions and their rates are discussed.

Many chemical reactions do not lead to complete conversion to products. In these reactions, although the reactants are together, even for prolonged periods of time, there are quantities of reactants remaining. In such reactions, an equilibrium exists between the reactants and the products. At this point, the rate at which reactants are converted to products is balanced by the rate at which products are converted to reactants. For such equilibrium reactions, it is possible to obtain only a limited amount of product. Since chemistry in industry involves economic considerations, the yield of the product at equilibrium should be as high as possible to make optimum use of the costly reactants. Thus an understanding of the effect of reaction conditions on the position of an equilibrium is essential. In this chapter the law of mass action, which governs chemical equilibria, is presented and the principle of Le Châtelier as it applies to chemical reactions is discussed.

17.1 KINETICS

Kinetics is a study of the velocity of chemical reactions. The reaction velocity is a measure of the rate of conversion of reactants into products. The factors that influence the reaction velocity are the structure of the reactants, the temperature, the concentration of reactants, and a class of substances called catalysts.

Reactants and reaction rates

The transformation of reactants into products involves the rupture of some bonds and the formation of others. Therefore, the chemical substances involved are the most important feature controlling the reaction.

Reaction velocities vary considerably depending on molecular structure. Many reactions between ions in aqueous solutions occur so quickly (in microseconds) that they can be monitored only by the most sophisticated electronic devices. In sharp contrast are reactions that require centuries and even perhaps millions of years to reach completion. The formation of oil and related geological reactions that occur deep within the earth are examples of reactions whose velocities are so small as to be beyond our ability to detect.

Concentration and reaction rates

Two common reactions known to everyone are the burning of wood and the rusting of iron. Both of these reactions involve the reaction of the oxygen gas in the air with a solid. The reactants in the different phases must come in contact with each other to react. It is commonly known that the reaction velocity increases with an increase in surface area. If wood is chopped into fine kindling or if iron is ground into powder, more of the solid comes in contact with air and the reaction velocity increases.

For reactions occurring in a single phase, a similar necessity for physical contact of reactants is noted. As the concentration of reactants in either a gaseous or liquid system is increased, the reaction velocity increases. In a gaseous system, the reaction velocity can be increased either by increasing the amount of reactants in a constant volume or by decreasing the volume (increasing the pressure) of a system containing a fixed amount of reactants. Similarly, in the liquid phase, reactant concentrations may be increased by either adding reactants or by removing solvent. In each case the result is to bring reacting particles closer together.

The more crowded a dance floor, the more likely it is that collisions will occur.

Temperature and reaction rates

While the majority of chemical reactions are exothermic, there are a number of endothermic conversions. The energy of a reaction may be added to an equation and regarded as a reactant or product. In an exothermic reaction, the energy given off appears to the right of the arrow. In an endothermic process, the energy used appears to the left of the arrow.

$$2H_2 + O_2 \longrightarrow 2H_2O + 115.6 \text{ kcal} \qquad \text{(exothermic)}$$
$$11.0 \text{ kcal} + 2F_2 + O_2 \longrightarrow 2OF_2 \qquad \text{(endothermic)}$$

Regardless of the net energy difference between reactants and products, all reaction velocities increase with a rise in temperature. This occurs because the speed of the reactant particles and hence their kinetic energies increase with increasing temperature. The faster the particles move, the more chance that they will contact other particles and react. Some reaction velocities are very sensitive to temperature changes, whereas others are only slightly affected. However, a general rule that can be used with some caution is that a 10°C rise in temperature usually doubles or triples the reaction rate. The factor is larger at lower temperatures than at higher temperatures for the same reaction.

Collisions on a dance floor are more likely doing a fast polka than doing a slow waltz.

Catalysts and reaction rates

A catalyst is a substance which speeds up a reaction rate when it is present in the reaction mixture but is not consumed in the process. A catalyst is said to catalyze the reaction, and its effect is known as catalysis. At the termination of the reaction, the catalyst, which is usually required only in trace amounts, can be recovered unchanged. Therefore, in a macroscopic sense, the catalyst does not appear to be involved with the reactants. However, the catalyst must have been involved in a microscopic sense or else no change in the velocity of the reaction would have resulted. In those cases that have been examined in detail, it has been shown that the catalyst does interact either physically or chemically with one or several of the reactants. However, any consumption of the catalyst at a given step in the reaction is always balanced by a regeneration step at a later point.

When potassium chlorate ($KClO_3$) is heated, it slowly decomposes into oxygen and potassium chloride, KCl.

$$2KClO_3 \longrightarrow 2KCl + 3O_2$$

If a small amount of manganese dioxide (MnO_2) is added, the decomposition of potassium chlorate is accelerated. At the conclusion of the reaction, the $KClO_3$ is completely consumed, but all the MnO_2 remains. The manganese dioxide has served as a catalyst in the reaction.

In the chemical industry, catalysts are widely used to facilitate the economical conversion of reactants into a desired product. The catalysts chosen are usually very **specific; that is, they accelerate one chemical reaction while not facilitating other possible competitive reactions.** Since catalysts are only needed in small amounts and are not used up, they are economically desirable in industrial processes. Sulfur dioxide can be converted into sulfur trioxide in the presence of finely powdered platinum. The catalyst is usually written over the arrow in chemical equations:

Time is money and, therefore, catalysts are needed for economical operation of chemical industries.

$$2SO_2 + O_2 \xrightarrow{Pt} 2SO_3$$

Depending on the catalyst employed and the experimental conditions chosen, a set of reactants can be made to produce different products. For example, the reaction of carbon monoxide and hydrogen can produce either methane (CH_4) or methanol (CH_4O), depending on the catalyst chosen.

$$CO + 3H_2 \xrightarrow{Ni} CH_4 + H_2O$$

$$CO + 2H_2 \xrightarrow{ZnO + Cr_2O_3} CH_4O$$

It is the use of catalysts that allows the petroleum industry to produce such a versatile product line. From the same source of crude oil, the industry can produce gasoline, heating oil, jet fuel, and a wide variety of other products. Only by using catalysts can the industry cause specific chemical reactions to occur so that each required product can be obtained as the market demands.

17.2 CATALYSTS IN LIVING ORGANISMS

Catalytic acceleration of reaction velocities also makes it possible to achieve a reasonable reaction rate at a lower temperature than without a catalyst. **Catalysts in the plant and animal kingdoms that enable living organisms to**

function at temperatures that are not destructive to their growth are called enzymes. Each catalyst is highly specific and very efficient.

The rates of chemical reactions are accelerated by factors of 10^{10} to 10^{20} by enzymes. In the body, sucrose is metabolized at 98.6°F or 37°C. The reaction is the primary source of energy for the body.

$$C_{12}H_{22}O_{11} + 12O_2 \longrightarrow 12CO_2 + 11H_2O$$

The actual conversion requires numerous steps, each catalyzed by a specific enzyme. Outside the body, the combustion can occur directly only above 600°C. At body temperature the chemical conversion without enzymes would require months. This rate is obviously too slow to provide the energy necessary to support life.

Where would we be without enzymes? What difficulties could arise with an individual who is born without a certain enzyme?

Enzymes are as common in the plant kingdom as in the animal kingdom. One of the very important processes catalyzed by enzymes is nitrogen fixation. The nitrogen gas in the atmosphere ultimately is the principal natural source of ammonia and other nitrogen compounds essential for protein formation. The gas is quite unreactive but small quantities of nitrogen are converted into nitrogen compounds by lightning discharges. However, the bacteria in the roots of leguminous plants such as alfalfa, beans, clover, and peas are readily able to convert nitrogen gas into ammonia. Because these bacteria contain enzymes that catalyze the conversion under mild conditions, they provide the largest source of nitrogen compounds to organisms.

17.3 EQUILIBRIUM

Reversible reactions and chemical equilibrium

So far we have considered unidirectional reactions in which reactants are converted to products. While some reactants do undergo essentially complete transformation into products, there are many in which the conversion is incomplete. No matter how much time is allowed for reaction, the concentrations of both the reactants and products remain at some fixed concentration. Although beyond this point no net change between reactants and products is observed, molecular interconversions are still occurring. For every transformation that produces a molecule termed a product, another molecule of product is reconverted to the molecule termed the reactant. Therefore, **the rate of one reaction is equal to the rate of the reverse of the reaction. When this state is reached, chemical equilibrium is said to be attained.**

Hydrogen and iodine can react in the gaseous phase at 425°C to yield hydrogen iodide. From one mole each of hydrogen and iodine, two moles of hydrogen iodide could result if a 100% yield were obtained. At the conclusion of this complete reaction there would be no hydrogen or iodine left.

H_2	+ I_2	$\longrightarrow$ 2HI	
1.00 mole	1.00 mole	0.00 mole	(initially)
0.00 mole	0.00 mole	2.00 moles	(theoretical yield)

However, the reaction does not yield the theoretical amount of hydrogen iodide, and there is hydrogen and iodine left no matter how long the reaction is allowed to occur. Only 1.58 moles of hydrogen iodide are obtained and there is 0.21 mole each of hydrogen and iodine remaining. The actual yield is 79%.

H_2	+ I_2	$\longrightarrow$ 2HI	
1.00 mole	1.00 mole	0.00 moles	(initially)
0.21 mole	0.21 mole	1.58 moles	(equilibrium amounts)

These equilibrium quantities are the same as would be eventually achieved if 2.00 moles of hydrogen iodide were allowed to react at 425°C for a sufficient amount of time.

2HI $\longrightarrow$	H_2	+ I_2	
2.00 moles	0.00 mole	0.00 mole	(initially)
1.58 moles	0.21 mole	0.21 mole	(equilibrium amounts)

The equations for the reaction can be combined into a single equation in which the arrows are written in opposing directions:

$$H_2 + I_2 \rightleftharpoons 2HI$$

The observed **equilibrium results from the rate of the forward reaction being equal to the rate of the backward reaction at some combination of concentrations of H_2, I_2, and HI.** When only H_2 and I_2 are present, the forward reaction occurs and begins to deplete the supply of H_2 and I_2 molecules, but as the concentrations of H_2 and I_2 decrease, their rate of conversion decreases. The

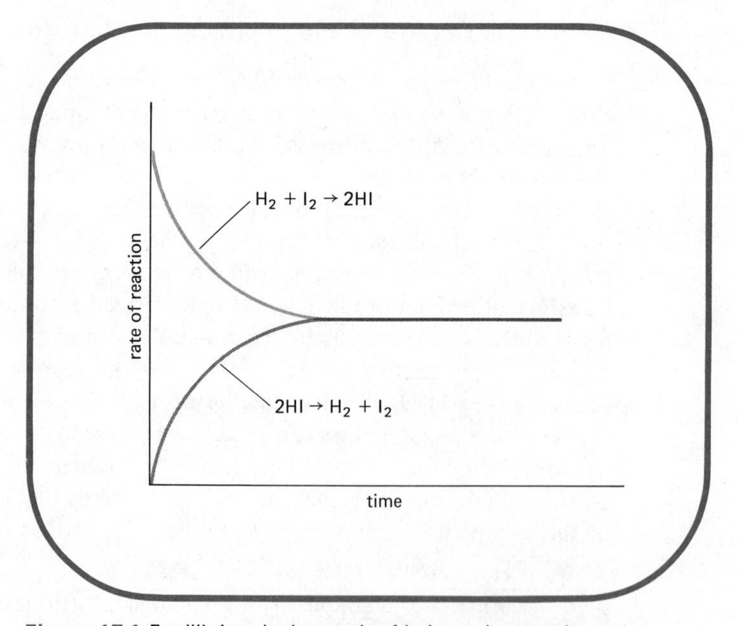

Figure 17.1 Equilibrium is the result of balanced rates of reaction.

rate of the backward reaction of HI is initially zero when only H_2 and I_2 are present. As HI accumulates, the rate of reaction to yield H_2 and I_2 increases. When the rate of reaction of HI eventually equals that of the reaction of H_2 and I_2, equilibrium is achieved (Figure 17.1).

Equilibrium in a chemical sense is a dynamic condition. Chemical activity has not lessened at equilibrium because microscopic processes are occurring. On the macroscopic level, however, nothing appears to be happening.

Irreversible reactions

All chemical reactions in theory are reversible. However, in practice, there are many reactions in which the products are removed or are altered to such a large extent that they cannot react to reform the reactants. In these cases, no equilibrium appears to exist and the yield of the reaction is 100%. Such reactions are said to go to completion or are irreversible.

Is there really such a thing as an irreversible reaction?

Chemical reactions that go to completion usually involve one of the following occurrences:

1 evolution of a gas,
2 formation of a precipitate,
3 formation of a covalent or partially ionized substance.

The first two conditions are the result of removal of a product from the reaction medium so that it cannot partake in the reverse reaction. The third condition involves the formation of a substance from ionic reactants which then has no tendency to reionize in order to form the reactants.

An example of the evolution of a gas which drives a reaction to completion is the formation of carbon dioxide from magnesium carbonate when it reacts with hydrochloric acid. This reaction occurs when certain commercial antacids react with stomach acid, which is dilute HCl.

$$MgCO_3(s) + 2HCl(aq) \longrightarrow MgCl_2(aq) + H_2O + CO_2(g)\uparrow$$

The carbon dioxide escapes so that it is no longer present to react with the other products and reform the reactants. As the gas is removed, more is formed from reaction of the carbonate with the acid.

When a precipitate is produced in a reaction, it causes the reaction forming the insoluble material to form more precipitate. The reaction is reversible since the solid is in equilibrium with the solution. However, if the solubility is low, then very little of the substance is present in solution. Any combination of ions in a metathesis reaction which results in an insoluble substance will cause the reaction to occur. One such reaction is the formation of silver bromide from silver nitrate and hydrobromic acid.

$$AgNO_3(aq) + HBr(aq) \longrightarrow AgBr(s) + HNO_3(aq)$$

The most commonly produced covalent substance in chemical reactions between ionic substances is water. Formation of water which is only slightly ionized is the reason why neutralization reactions occur.

$$KOH(aq) + HNO_3(aq) \longrightarrow KNO_3(aq) + H_2O$$

17.4 LAW OF MASS ACTION

For the general balanced equation at equilibrium:

$$mA + nB \rightleftharpoons pX + qY$$

the following expression is a constant at a specific temperature:

$$\frac{[X]^p[Y]^q}{[A]^m[B]^n}$$

This mathematical expression can be established to be a constant by experiment. **The expression indicates the relationship between the concentration of all substances at equilibrium and is called the mass action expression.**

Since the mass action expression as written is a constant, it follows that the reciprocal of the expression also is a constant. However, by convention, **the expression is always written so that the materials on the right-hand side of the chemical equation appear in the numerator and those on the left-hand side appear in the denominator.**

The mass action expression is valid independent of the source of the molecules *A*, *B*, *X*, and *Y*. An equilibrium state will be reached eventually if sufficient time is allowed. At equilibrium, the concentrations of the four substances always will satisfy the mass action expression.

Mass action expressions for any reaction can be written by inspection of the equation.

$$2N_2O \rightleftharpoons 2N_2 + O_2 \qquad \frac{[N_2]^2[O_2]}{[N_2O]^2}$$

$$2N_2O_5 \rightleftharpoons 4NO_2 + O_2 \qquad \frac{[NO_2]^4[O_2]}{[N_2O_5]^2}$$

$$N_2O_4 \rightleftharpoons 2NO_2 \qquad \frac{[NO_2]^2}{[N_2O_4]}$$

The numerical value of the mass action expression is called the equilibrium constant and is denoted by *K*:

$$K = \frac{[X]^p[Y]^q}{[A]^m[B]^n}$$

For an equilibrium constant less than 1, the numerator of the mass action expression must be smaller than the denominator. This condition means that the concentrations of the products are less than the reactants. Conversely, if the equilibrium constant is greater than one, the reaction system at equilibrium has proceeded toward completion in a left to right direction.

The equilibrium constant for the hydrogen, iodine, and hydrogen iodide equilibrium discussed in Section 17.3 may be calculated from the known equilibrium concentrations given. The mole quantities may be used as the volume of the system cancels in the equilibrium constant expression.

$$\frac{[HI]^2}{[H_2][I_2]} = K = \frac{[1.58/V]^2}{[0.21/V][0.21/V]} = \frac{1.58^2}{0.21^2} = 57$$

When the temperature of a reaction is changed, the value of the equilibrium constant changes in a manner that depends on the energy of the chemical reaction. **If the forward reaction is exothermic, an increase in temperature decreases the equilibrium constant because it favors the reverse reaction. For an endothermic forward reaction, an increase in temperature gives rise to an increase in the equilibrium constant.** These effects are discussed more thoroughly in the next section.

A catalyst does not change the position of a chemical equilibrium and, therefore, the equilibrium constant is independent of catalysts. The catalyst serves to facilitate the achievement of an equilibrium state and functions in allowing both the forward and reverse reactions to proceed at a faster rate. The increase in each rate is exactly equal, and no net positional difference in equilibrium can result.

17.5 LE CHÂTELIER'S PRINCIPLE

The principle of Le Châtelier as stated in Chapter 13 is that if an external force is applied on an equilibrium system, then the system if possible, will readjust to reduce the stress imposed on it. While the principle has physical implications, it also can be applied to chemical systems. The application of the generalization of Le Châtelier is very simple and allows a qualitative evaluation of the alterations that occur in a chemical equilibrium when the conditions are changed.

Can you cite any examples of Le Châtelier's principle in fields other than chemistry?

Three changes in condition will be considered: concentration, pressure, and temperature. **Pressure and concentration changes do not affect the value of the equilibrium constant but rather only the individual concentrations of the substances in equilibrium with each other. Temperature changes alter the value of the equilibrium constant and consequently the concentrations of the substances in equilibrium. Catalysts have no effect on either the value of the equilibrium constant or the concentrations because they affect the rates of the forward and backward reactions to an equal extent.**

Concentration changes

For the reaction of hydrogen and iodine to yield hydrogen iodide, the equilibrium can be upset temporarily by the addition or removal of any component of the system.

$$H_2 + I_2 \rightleftharpoons 2HI$$

If an additional 0.40 mole of I_2 were added to the system at equilibrium, the H_2 would react with it to decrease its concentration. This is predicted by Le Châtelier's principle. In this case, the stress or change in the equilibrium is caused by the addition of I_2, which they must be removed, at least in part, in order to reestablish the equilibrium. The only way in which I_2 can be reduced in concentration is by reaction with H_2. As a consequence of the reactions which occur after the introduction of the I_2, the concentration of H_2 diminishes and that of HI increases. The concentration of I_2 decreases from its value immediately after the addition but only to a value higher than the original one prior to the addition. The equilibrium constant is unchanged because the concentration changes are mathematically balanced.

H_2 +	I_2	$\rightleftharpoons$ 2HI	
0.21 mole	0.21 mole	1.58 moles	(at equilibrium)
0.21 mole	0.61 mole	1.58 moles	(after addition of 0.4 mole of I_2)
0.11 mole	0.51 mole	1.78 moles	(after reestablishing the equilibrium)

$$K = \frac{[\mathrm{HI}]^2}{[\mathrm{H_2}][\mathrm{I_2}]} = \frac{[1.78]^2}{[0.11][0.51]} = 56$$

Within the significant figures used, the value of the equilibrium constant is unchanged.

If the original equilibrium system described in Section 17.3 were disturbed by adding hydrogen iodide, the principle of Le Châtelier indicates that the system should behave so as to decrease the concentration of the substance added. In this case the only way to decrease the concentration of the hydrogen iodide is to produce both hydrogen and iodine. As a consequence, the concentrations of hydrogen and iodine will increase. The concentration of hydrogen iodide will decrease below that which is present immediately after the addition. However, as it decreases, it never quite reaches the value for the hydrogen iodide concentration that was established before the addition.

H_2 +	I_2	$\rightleftharpoons$ 2HI	
0.21 mole	0.21 mole	1.58 moles	(at equilibrium)
0.21 mole	0.21 mole	1.98 moles	(afer addition of 0.4 mole of HI)
0.25 mole	0.25 mole	1.90 moles	(after reestablishing the equilibrium)

$$K = \frac{[\mathrm{HI}]^2}{[\mathrm{H_2}][\mathrm{I_2}]} = \frac{[1.90]^2}{[0.25][0.25]} = 57$$

Again, within the significant figures used, the new concentration values after the reestablishment of the equilibrium yield the same equilibrium constant.

Pressure changes

The concentration of gaseous substances can be changed by altering the pressure and hence the volume of the gas. The position of an equilibrium gaseous reaction *may* be altered by pressure changes. However, this will happen only if the number of gaseous reactant molecules differs from the number of gaseous product molecules. For example, the reaction of H_2 and I_2 in the gaseous phase is not affected by pressure changes. There are two reactant molecules and two product molecules as indicated by the coefficients of the balanced equation. The reaction is independent of pressure because two gas molecules occupy the same volume no matter what their atomic composition.

$$H_2 + I_2 \rightleftharpoons 2HI$$

The reaction of nitrogen and hydrogen to yield ammonia, on the other hand, is very dependent on pressure.

$$N_2 + 3H_2 \rightleftharpoons 2NH_3$$

Here four reactant molecules (three of hydrogen and one of nitrogen) interact to form only two product molecules. When the pressure of the system is increased and as a result the volume decreases, the system shifts to the side with the smallest number of particles to reduce the pressure. In this case, more ammonia is formed. This, of course, reduces the concentrations of the nitrogen and hydrogen.

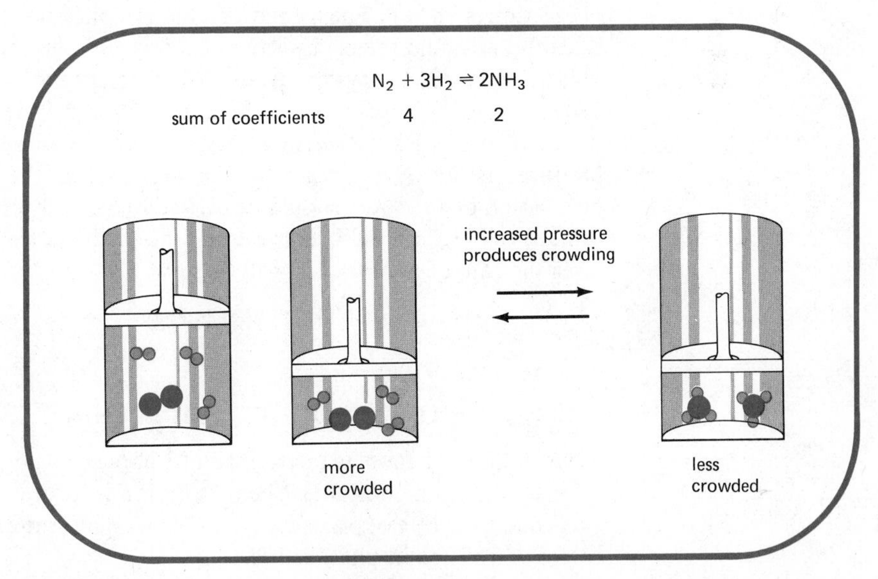

Figure 17.2 How to predict the effect of pressure on the equilibrium concentration of a gaseous reaction.

The effect of pressure on the position of an equilibrium is predicted by Le Châtelier's principle. Under the stress of increased pressure, the molecules become more crowded together. To reduce this crowding, the molecules interact in the way that will produce fewer molecules (Figure 17.2). Therefore, the side of the equation with the lowest sum of the coefficients is favored when the pressure increases. The equilibrium constant does not change with pressure, but the concentrations of reactants and products readjust to alleviate crowding.

While pressure produces changes for liquid and solid components at equilibrium, they are small compared with gaseous components. If an equilibrium involves several phases, only the gaseous components need to be considered to predict the effect of pressure changes:

$$3Fe + 4H_2O \rightleftharpoons Fe_3O_4 + 4H_2$$

For the above reaction at temperatures exceeding 100°C, there are four gas molecules on each side of the equation. The equilibrium is unaffected by pressure changes. The Fe and Fe_3O_4 need not be considered because they are not gaseous at ordinary temperatures.

Temperature changes

It is important to remember that a change in temperature results in a change in the equilibrium constant. We can use Le Châtelier's principle to predict how the temperature change will affect the equilibrium constant if the energy of reaction is known. The formation of ammonia from nitrogen and hydrogen is exothermic, and the backward reaction is endothermic. When the temperature is increased, the system tends to counteract the change by using up the added energy. The reaction to the left is favored. The equilibrium constant therefore decreases with temperature. As the system is cooled, it tends to produce energy by the reaction that proceeds to the right. Therefore, the production of ammonia should be carried out at low temperatures in order to achieve the highest yield. However, it should be noted that at low temperatures the rate of the reaction will be slow.

$$N_2 + 3H_2 \rightleftharpoons 2NH_3 + \text{heat energy}$$

17.6 SOLUBILITY PRODUCT EQUILIBRIA

The solubility phenomena described in Chapter 16 can be expressed in terms of an equilibrium constant. When a solution is saturated, the rate of dissolution equals the rate of precipitation, and an equilibrium between the dissolved solute and the solid phase exists.

Consider the slightly soluble ionic compound A_nB_m in a saturated solution The equilibrium expression is as follows:

$$A_nB_m \rightleftharpoons nA^{m+} + mB^{n-}$$

The equilibrium constant is expressed as:

$$K = \frac{[A^{m+}]^n[B^{n-}]^m}{A_nB_m}$$

Since the solute A_nB_m in the solid phase is a constant, its concentration is incorporated into the constant that is called the solubility product constant, K_{sp}.

$$K_{sp} = K[A_nB_m] = [A^{m+}]^n[B^{n-}]^m$$

Thus **a solubility product constant is equal to the product of the concentrations of the ions each raised to a power which is equal to the coefficients in the equilibrium equation.**

The value of the solubility product constant depends on the nature of the solute and the temperature of the solvent. The change in the solubility product constant with temperature follows those criteria discussed for equilibrium constants in Section 17.6. For an exothermic dissolution process, the solubility product constant will decrease with increasing temperature.

Each solubility product constant has units which depend on the power to which the concentrations of the ions are raised in the equilibrium expression. For the general case cited for A_nB_m, the units are derived as follows:

$$K_{sp} = [\text{mole/liter}]^n[\text{mole/liter}]^m = \frac{\text{mole}^{m+n}}{\text{liter}^{m+n}}$$

Thus the units are mole per liter raised to a power which is the sum of the coefficients in the equilibrium equation.

A comparison of K_{sp} values for compounds of the same type gives the order of solubilities. The compound with the lowest K_{sp} has the lowest solubility. For example, the K_{sp} values for AgCl and AgBr are 1×10^{-10} $\text{mole}^2/\text{liter}^2$ and 7×10^{-13} $\text{mole}^2/\text{liter}^2$. Silver bromide is less soluble than silver chloride. It must be reemphasized that such comparisons are valid only for compounds with the same values of m and n, as only then will the units of K_{sp} be the same. For silver chromate, Ag_2CrO_4, the units of K_{sp} are $\text{mole}^3/\text{liter}^3$ and are not the same as the units for AgCl and AgBr. However, the solubility of Ag_2S could be compared to the solubility of Ag_2CrO_4 by using the K_{sp} values. The solubility products of Ag_2CrO_4 and Ag_2S are 1.9×10^{-12} and 1×10^{-50}, respectively. Thus we see that Ag_2S is less soluble than Ag_2CrO_4.

When the product of the concentrations of the ions present in solution, each raised to their respective powers, exactly equals the solubility product constant, a saturated solution is established. If the product of these concentrations, each raised to their respective powers, is less than the K_{sp} value, the solution is unsaturated.

Problem 17.1 Write the K_{sp} expression for silver bromide.

Solution The formula for silver bromide is AgBr and the equilibrium for the solubility of AgBr is:

$$AgBr(s) \rightleftharpoons Ag^{+}(aq) + Br^{-}(aq)$$

The coefficients for both the silver ion and the bromide ion are one. Therefore, the exponents in the K_{sp} expression are also one.

$$K_{sp} = [Ag^{+}][Br^{-}]$$

Problem 17.2 Write the K_{sp} expression for iron(III) hydroxide.

Solution The formula for iron(III) hydroxide is $Fe(OH)_3$ and the equilibrium for its solubility is:

$$Fe(OH)_3(s) \rightleftharpoons Fe^{3+}(aq) + 3OH^{-}(aq)$$

The coefficient for the hydroxide ion is 3 and therefore the K_{sp} expression must have the same exponent for the concentration of hydroxide ion.

$$K_{sp} = [Fe^{3+}][OH^{-}]^3$$

Problem 17.3 The K_{sp} for AgCl is 1×10^{-10} mole²/liter². What is the molarity of a saturated AgCl solution?

Solution Writing the K_{sp} expression and setting it equal to 1×10^{-10} mole²/liter², we have:

$$K_{sp} = [Ag^{+}][Cl^{-}] = 1 \times 10^{-10} \text{ mole}^2/\text{liter}^2$$

When silver chloride dissolves in water, the concentration of the silver ion is equal to that of the chloride ion. If we let $x = [Ag^{+}]$, then $x = [Cl^{-}]$. Substituting into the K_{sp} expression:

$$[x][x] = 1 \times 10^{-10} \text{ mole}^2/\text{liter}^2$$

$$x^2 = 1 \times 10^{-10} \text{ mole}^2/\text{liter}^2$$

$$\sqrt{x^2} = \sqrt{1 \times 10^{-10} \text{ mole}^2/\text{liter}^2}$$

$$x = 1 \times 10^{-5} \text{ mole/liter}$$

The molarity of dissolved silver chloride is equal to the calculated contration of silver ions.

Problem 17.4 The solubility of $Mn(OH)_2$ is 0.00191 g/liter. What is the K_{sp} of $Mn(OH)_2$?

Solution The formula weight of $Mn(OH)_2$ is 88.9 g/mole. The molarity of a saturated solution of $Mn(OH)_2$ is:

$$\frac{0.00191 \text{ g/liter}}{88.9 \text{ g/mole}} = 2.15 \times 10^{-5} M$$

The concentration of Mn^{2+} is 2.15×10^{-5} mole/liter but the concentration of OH^- is $2(2.15 \times 10^{-5})$ mole/liter $= 4.30 \times 10^{-5}$ mole/liter as two moles of OH^- are produced for every mole of $Mn(OH)_2$ which dissolves.

$$Mn(OH)_2 \rightleftharpoons Mn^{2+} + 2OH^-$$

The K_{sp} expression is now written and the concentrations substituted into it.

$$K_{sp} = [Mn^{2+}][OH^-]^2$$
$$K_{sp} = [2.15 \times 10^{-5}][4.30 \times 10^{-5}]^2$$
$$K_{sp} = 4.0 \times 10^{-14}$$

SUMMARY

***Kinetics** is the study of the velocity of chemical reactions. The velocity of reactions depends on the identity of the reactants, their concentration, the temperature, and the presence of catalysts.*

Catalysts** are substances which speed up a reaction rate but are not consumed in the process. Catalysts which speed up one chemical reaction while not facilitating other possible competitive reactions are called **specific** catalysts. Catalysts in plants and animals are called **enzymes.

*Chemical **equilibrium** is achieved when the rate of one reaction is equal to the rate of the reverse of the reaction. An expression indicating the relationship between the concentrations of all substances at equilibrium is called the **mass action** expression. The value of the mass action expression is the **equilibrium constant.** If the concentrations of any of the substances present at equilibrium change due to addition of material or a pressure change, the system will readjust to try to lessen that change. The concentration values at the new equilibrium will obey the same equilibrium constant. A change in temperature alters the equilibrium constant.*

*The **solubility product constant** for a slightly soluble ionic compound is equal to the product of the concentrations of the ions each raised to a power equal to the coefficients in the equilibrium equation.*

QUESTIONS AND PROBLEMS

Definitions

1. Define each of the following terms.
   (a) reactant
   (b) product
   (c) kinetics
   (d) reaction rate
   (e) equilibrium constant
   (f) reversible
   (g) irreversible
   (h) K
   (i) mass action
   (j) Le Châtelier's principle
   (k) catalyst
   (l) specific
   (m) catalysis
   (n) enzyme
   (o) chemical equilibrium
   (p) K_{sp}
2. Distinguish between each of the following terms.
   (a) enzyme and catalyst
   (b) rate and equilibrium
   (c) K_{sp} and K

Factors controlling rates

3. Describe the effect of the identity of reactants, concentration, temperature, and catalysts on reaction rates.
4. Will an increase in concentration always cause a rate increase?
5. What is a catalyst? Cite examples of uses of catalysts in industry.
6. What are the catalysts of living systems called? Why are they important?
7. Why might an industry carry out a reaction at a high temperature?

Irreversible reactions

8. Outline three types of irreversible reactions and write one equation as an example of each.

Equilibrium constants

9. Write the equilibrium constant expressions for each of the following equations.
   (a) $4NH_3 + 5O_2 \rightleftharpoons 4NO + 6H_2O$
   (b) $CH_4 + Cl_2 \rightleftharpoons CH_3Cl + HCl$
   (c) $N_2 + 3H_2 \rightleftharpoons 2NH_3$
   (d) $3O_2 \rightleftharpoons 2O_3$
   (e) $SO_2 + NO_2 \rightleftharpoons SO_3 + NO$
   (f) $N_2 + O_2 \rightleftharpoons 2NO$
10. Explain what the effect of a catalyst is on the magnitude of the equilibrium constant.
11. How does the magnitude of the equilibrium constant indicate the degree of completeness of a reaction?

Le Châtelier's principle

12. Consider the following equilibrium.

$O_3 + NO \rightleftharpoons O_2 + NO_2$

What effect would each of the following changes have on the reaction?
(a) increasing $[O_3]$
(b) decreasing $[NO]$
(c) increasing $[O_2]$
(d) decreasing $[NO_2]$
(e) increasing the pressure

13. Consider the following equilibrium and indicate the effect each of the changes listed will have on the reaction.

$4HBr(g) + O_2(g) \rightleftharpoons 2H_2O(g) + 2Br_2(g)$

(a) increasing $[O_2]$
(b) decreasing $[HBr]$
(c) increasing $[Br_2]$
(d) decreasing $[H_2O]$
(e) increasing the pressure

14. Predict the effect of temperature on each of the following equilibrium constants.
(a) $3O_2 + 64.8\text{ kcal} \rightleftharpoons 2O_3$
(b) $2F_2 + O_2 + 11.0\text{ kcal} \rightleftharpoons 2OF_2$
(c) $CH_4 + 2O_2 \rightleftharpoons CO_2 + 2H_2O + 213\text{ kcal}$
(d) $2CO + O_2 \rightleftharpoons 2CO_2 + 135\text{ kcal}$
(e) $N_2 + 2O_2 + 16\text{ kcal} \rightleftharpoons 2NO_2$

Solubility product constants

15. Write the K_{sp} expression for each of the following compounds.
(a) $AgCl$
(b) $BaSO_4$
(c) CuS
(d) $Fe(OH)_3$
(e) Ag_2CrO_4
(f) Cu_2S
(g) As_2S_5
(h) Bi_2S_3

16. Calculate the molarity of each of the following saturated solutions using the given K_{sp}.
(a) $AgCl$; $K_{sp} = 1 \times 10^{-10}$ mole^2/liter^2
(b) $PbSO_4$; $K_{sp} = 2 \times 10^{-14}$ mole^2/liter^2
(c) $Fe(OH)_3$; $K_{sp} = 1 \times 10^{-36}$ mole^4/liter^4
(d) Ag_2CrO_4; $K_{sp} = 9 \times 10^{-32}$ mole^3/liter^3

17. Using the indicated solubilities, calculate the K_{sp} for each of the following.
(a) $BaCO_3$; 9.0×10^{-5} mole/liter
(b) FeS; 5.3×10^{-8} mole/liter
(c) $TlBr$; 4.0×10^{-6} mole/liter
(d) AgI; 1.5×10^{-16} mole/liter

Application of kinetics

18. Substances burn more rapidly in pure oxygen than in ordinary air.

Explain why. This phenomenon was responsible for a fire and tragedy aboard an Apollo spacecraft in which three American astronauts died.

19. Explain why persons with an elevated temperature for a prolonged period must either increase food intake or suffer a weight loss.

20. Milk will sour in two days at room temperature but may remain unspoiled for two weeks in a refrigerator. Explain why.

21. Why do potatoes cook in 15 minutes in a pressure cooker at 108°C, whereas 30 minutes are required in boiling water at 100°C?

22. A scout learns to start a fire by first cutting wood into shavings and kindling. After the shavings start to burn, the flames are gently fanned. Explain why these two procedures aid the chemical reaction of combustion.

18 ACID-BASE REACTIONS

LEARNING OBJECTIVES

When you finish this chapter you should be able to:

(a) classify substances as electrolytes and nonelectrolytes.

(b) distinguish between weak and strong electrolytes.

(c) compare acids and bases using the Arrhenius concept.

(d) demonstrate the conjugate acid-base pairs according to the Brønsted-Lowry concept.

(e) relate acid and base strength to the degree of ionization in water.

(f) identify strong and weak acids; identify strong and weak bases.

(g) calculate weak electrolyte equilibria.

(h) calculate the pH and pOH of a solution given either the H_3O^+ or OH^- concentration.

(i) explain the action of a chemical buffer.

(j) calculate the amount of an acid or base present in a sample from titration data.

(k) write balanced ionic equations.

(l) match the following terms with their corresponding definitions.

acid
acid ionization constant
buffer
dissociation
electrolyte
indicator

ionic equation
ionization
neutralization point
pH
POH
spectator ions
strong acid
strong base
strong electrolyte
titration
weak acid
weak base
weak electrolyte

The importance of acids and bases to chemistry should be obvious from the fact that we have encountered them several times already in this book. The nomenclature of acids and bases was presented in Chapter 9 and the neutralization reaction of acids and bases was introduced in Chapter 11. In this chapter, we will examine acids and bases more closely and define them in more explicit terms. Acids and bases play important roles in many diverse areas. Several applications of acid-base reactions will be presented in this chapter.

Acid-base reactions are one of the most important in the maintenance of life. Most of the food which we eat is acidic. Metabolic reactions involve these acids and others produced by chemical breakdown into simpler substances. Although most of the body fluids are acidic, blood is slightly basic. Thus the body has to achieve an acid-base balance constantly. If the acid-base balance changes very slightly in either direction, death will occur.

In agriculture, the acidity of the soil determines what crops may be grown. Fertilizers containing acids or bases are often used to make the soil more suitable for particular crops.

Many household items are either acids or bases. Household ammonia, baking soda, and stomach antacids are bases. Vinegar and aspirin are acids. Toilet bowl cleaners, drain cleaners, and numerous other products for maintaining a home are either acids or bases.

18.1 ELECTROLYTES

Very pure water is essentially a nonconductor. This can be illustrated by the use of the apparatus shown in Figure 18.1. Two electrodes are immersed in water and are not in direct contact with each other. When the switch is closed, electricity would flow and the light bulb would glow if water could conduct the electricity between the two electrodes. However, the light does not glow with pure water in the beaker because pure water is not a conductor.

Any substance which dissolves in water to form a solution that conducts electricity is called an electrolyte. The solution of NaCl in water does conduct

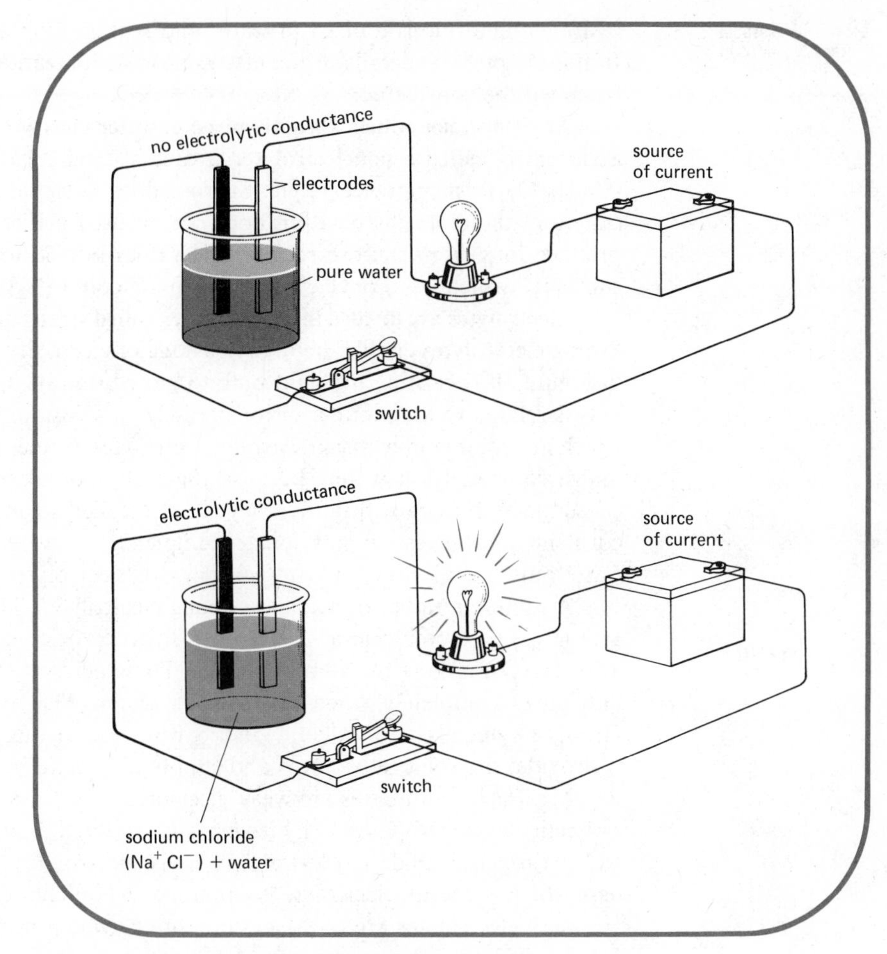

Figure 18.1 Electrical conductance of electrolytes.

electricity. When this solution is used in the apparatus in Figure 18.1, the light glows brightly. The explanation for this phenomenon was provided by the Swedish scientist Svante Arrhenius in 1884. He suggested that substances whose aqueous solutions conduct electricity form ions in the solution. **The cations in the solution migrate to the negative electrode, called the cathode, while the anions migrate to the positive electrode, called the anode.** Such migrations provide the necessary contact between the electrodes and allow for passage of electricity.

When ionic substances are dissolved in water, their ions are separated from the crystal structure by the water molecules. **This separation of ions already present in another phase is called dissociation.** There are covalent substances which produce ions when dissolved in water. **The formation of ions upon dissolution in water is called ionization.** As indicated in Section 9.7,

How does dissociation differ from ionization?

hydrogen chloride is ionized in water and is then termed hydrochloric acid. In this chapter the details of the dissociation or ionization of other acids and bases will be considered.

Any substance which when dissolved in water does not allow the passage of a current is called a nonelectrolyte. Ordinary cane sugar, known as sucrose ($C_{12}H_{22}O_{11}$), is a nonelectrolyte. Sucrose does not contain ions in the solid state and therefore dissociation does not occur. Furthermore, water cannot produce ions from sucrose so ionization does not occur. Another common substance which is a nonelectrolyte is ethyl alcohol (C_2H_6O).

Electrolytes are divided into two classes called strong and weak electrolytes. Strong electrolytes enable the ready passage of electricity in the device shown in Figure 18.1, causing the light bulb to glow brightly. The degree of brightness is related to the number of ions present in solution. All ionic substances which are soluble in water are strong electrolytes. Some covalent compounds ionize completely in water. Because these compounds produce ions which cause the bulb to glow brightly, they are also called strong electrolytes. Other covalent substances do not ionize completely in water. These substances cause only a dull glow and are called weak electrolytes.

A limited number of acids are strong electrolytes. These include sulfuric acid (H_2SO_4), perchloric acid ($HClO_4$), hydrochloric acid (HCl), and nitric acid (HNO_3) among the common acids. These acids are covalent substances but ionize completely when dissolved in water. The hydroxides of all the Group IA metals and calcium and barium are strong electrolytes. These hydroxides are ionic substances and simply dissociate in water.

Most acids and bases are weak electrolytes. For example, the acids such as acetic acid ($HC_2H_3O_2$), hydrogen sulfide (H_2S), nitrous acid (HNO_2), and hydrogen cyanide (HCN) are all weak electrolytes. The most common base which is a weak electrolyte is ammonia, NH_3. The acids and bases which are weak electrolytes are covalent compounds which produce only a limited number of ions in water. Further details of this process will be presented in subsequent sections.

18.2 PROPERTIES OF ACIDS AND BASES

Acids and bases were first mentioned in Chapter 9 but few details of their physical and chemical properties were given. Prior to proceeding with a detailed description of acids and bases, it is now convenient to list some of their properties.

Aqueous solutions of acids have the following properties:

1. Acids turn the color of a dye called litmus from blue to red. The litmus is commonly impregnated in a strip of paper called blue litmus paper.
2. Acids will react with bases to yield salts and water in a process called neutralization.

3 Acids have a sour taste. Examples include the citric acid contained in citrus fruits such as lemons and the acetic acid of vinegar. **(Do not attempt to taste solutions of acids in a laboratory as they are more concentrated than in natural foods and will cause chemical burns.)**

Aqueous solutions of bases have the following properties:

1 Bases turn the color of red litmus to blue. The red litmus is available in strips of red litmus paper.
2 Bases will react with acids to yield salts and water in the neutralization reaction.
3 Bases have a bitter taste. An example is unflavored milk of magnesia, $Mg(OH)_2$.
4 Basic solutions have a soapy or slick feeling when rubbed on the skin. **(Again as in the case of acids, do not touch or taste laboratory bases unless directed to do so with specially prepared solutions by a qualified instructor of chemistry.)**

18.3 THE ARRHENIUS CONCEPT OF ACIDS AND BASES

The Arrhenius definition of acids and bases was based on the ions that are formed upon dissolution of these substances in water. In water, HCl behaves as an acid and a strong electrolyte. This behavior can be explained by the formation of the hydrogen cation, or proton H^+, and the chloride ion Cl^-. According to Arrhenius:

H^+

1 **An acid is a substance that yields hydrogen ions (protons) in water.** For example,

$$HCl \longrightarrow H^+ + Cl^-$$

2 **A base is a substance that produces hydroxide ions, OH^-, in solution,** as in the case for sodium hydroxide:

$$NaOH \longrightarrow Na^+ + OH^-$$

3 **When a base reacts with an acid in water, the reaction involves the proton and the hydroxide ion and is called a neutralization reaction.**

$$H^+ + OH^- \longrightarrow H_2O$$

4 **The proton is responsible for the acidic properties of acids and the hydroxide ion for the basic properties of bases.**

The attraction between the hydrogen ion and water molecules is unusually strong. In fact, this ion, a bare proton, is considered unique among ions. Since it has no electrons, its radius is that of the nucleus, or 10^{-13} cm. For comparison, ions containing electrons are approximately 10^{-8} cm in radius. The concentration of charge in such a small volume makes the attraction between water and the proton much stronger than that between water and other ions. The proton is very tightly bound to water; free protons do not exist in aqueous solutions of acids.

H_3O^+

The hydrated proton, or hydronium ion, H_3O^+, is a stable ion that can be thought of as being derived from the reaction of a free proton with water:

$$H^+ + H\text{-}\ddot{O}\text{-}H \longrightarrow \left[H\text{-}\overset{H}{\underset{|}{O}}{}^{+}\text{-}H\right]$$

When HCl dissolves in water the reaction may be represented by the following equation:

$$HCl + H_2O \longrightarrow H_3O^+ + Cl^-$$

The acid HCl functions by transferring or donating a proton to the water molecule. Water in turn is an acceptor of the proton of the acid HCl.

18.4 THE BRØNSTED-LOWRY CONCEPT OF ACIDS AND BASES

In the Arrhenius concept of acids and bases, basic properties are due to hydroxide ions. However, there are many other substances that react with and neutralize acids which do not form hydroxide ions. For example, ammonia reacts with hydrogen chloride. The ammonia molecule accepts protons in the same way the hydroxide ion does. It seemed necessary, therefore, to provide a broader definition of an acid and a base. **An acid is a substance that can donate protons, and a base is a substance that can accept protons.** This definition was proposed independently by Johannes Brønsted and Thomas Lowry in 1923.

Acids and bases can be either molecules or ions according to the Brønsted-Lowry concept. Indeed, the dependence on water that restricted the Arrhenius concept is completely removed. The base NH_3 accepts a proton from HCl to produce NH_4^+ and Cl^-:

$$\underset{\text{base}}{NH_3} + \underset{\text{acid}}{HCl} \rightleftharpoons \underset{\text{acid}}{NH_4^+} + \underset{\text{base}}{Cl^-}$$

In the reverse reaction, the ammonium ion can behave as an acid by donating a proton to the base chloride ion. **NH_3 and NH_4^+ are related and are called a conjugate pair; NH_3 is the base and NH_4^+ is its conjugate acid.** In like manner, **HCl is an acid and Cl^- is its conjugate base.**

Problem 18.1 What are the conjugate acid and conjugate base of HS^-?

Solution The ion HS^- can act as either an acid or a base. The conjugate acid of HS^- is obtained by addition of a proton to yield H_2S. The conjugate base of HS^-

is derived by removal of a proton to yield S^{2-}.

$$\underset{\text{base}}{HS^-} + \underset{\text{acid}}{HA} \rightleftharpoons \underset{\text{conjugate acid}}{H_2S} + \underset{\text{conjugate base}}{A^-}$$

$$\underset{\text{acid}}{HS^-} + \underset{\text{base}}{B^-} \rightleftharpoons \underset{\text{conjugate base}}{S^{2-}} + \underset{\text{conjugate acid}}{HB}$$

Any substance which can behave either as an acid or a base is called amphoteric. Examples include HS^-, given in Problem 18.1, HSO_4^-, HSO_3^-, $H_2PO_4^-$, HPO_4^{2-}, and HCO_3^-.

18.5 STRENGTHS OF ACIDS AND BASES

Strong acids and bases

When hydrogen chloride is dissolved in water, the resultant solution is called hydrochloric acid and contains virtually no covalent hydrogen chloride. The reaction:

$$HCl + H_2O \rightleftharpoons H_3O^+ + Cl^-$$

Could the relative strengths of strong acids be determined by using a base other than water?

proceeds essentially to completion to the right. **Acids which ionize completely in water are called strong acids.** From the position of the equilibrium, HCl must have a stronger tendency to lose protons than does H_3O^+ and is said to be a stronger acid than H_3O^+. In addition, the direction of the equilibrium must be a reflection of the willingness of H_2O to accept protons as compared to Cl^-. The H_2O molecule is said to be a stronger base than Cl^-.

From consideration of equilibria between acids and bases and their conjugate bases and acids, it can be concluded that **a strong acid, with its great tendency to lose protons, must be conjugate to a weak base that has a low affinity for protons. The stronger the acid, the weaker its related conjugate base. Strong bases attract protons strongly and are conjugate to weak acids that do not readily lose protons.** A list illustrating the relationship between common acid-base pairs is given in Table 18.1.

Notice that hydrochloric, nitric, and perchloric acids are all above water in this table. These three acids are virtually completely ionized in water. Therefore, in water solution, their acid properties, which really only reflect the presence of H_3O^+ in solution, are the same. Their relative strengths cannot be determined in this solvent. This phenomenon is known as the leveling effect of the solvent, which in this case is the base water. The acids below water in Table 18.1 are weaker acids than water and do not completely ionize in water. Their relative strengths can be determined in water.

TABLE 18.1 strengths of acids and related conjugate bases

	Acid	Conjugate base	
strongest ↑	$HClO_4$	ClO_4^-	weakest ↓
	H_2SO_4	HSO_4^-	
	HCl	Cl^-	
	H_3O^+	H_2O	
	HSO_4^-	SO_4^{2-}	
	HF	F^-	
	$HC_2H_3O_2$	$C_2H_3O_2^-$	
	H_2S	HS^-	
	NH_4^+	NH_3	
	HCO_3^-	CO_3^{2-}	
	H_2O	HO^-	
	HS^-	S^{2-}	
weakest	HO^-	O^{2-}	strongest

Weak acids and bases

Most acids are only partially ionized in water and are classified as weak acids. At 25°, a 1*M* solution of acetic acid ($HC_2H_3O_2$) is approximately 0.4 percent ionized and the concentration of ions is very low:

$$HC_2H_3O_2 + H_2O \rightleftharpoons H_3O^+ + C_2H_3O_2^-$$

Acetic acid is a weaker acid than H_3O^+, and $C_2H_3O_2^-$ is a stronger base than H_2O. In this reaction, the equilibrium lies to the side containing the weaker acid and weaker base. This statement is general and quite logical because the proton must reside with the weaker acid or the substance that has the smallest tendency to lose it. Furthermore, the proton remains on the acid because the base on that side of the equation does not have much tendency to remove it.

In order to compare the strengths of acids, it is necessary to measure their tendencies to transfer protons to a common reference base, usually water. The order of acid strengths can be established by measuring the acid ionization constant. For the general acid H*A*, the equilibrium constant for ionization, $HA + H_2O \rightleftharpoons H_3O^+ + A^-$, is

K_i

$$K = \frac{[H_3O^+][A^-]}{[HA][H_2O]}$$

The concentration of water is so large, compared to the other components of the equilibrium, that its value changes very little on a percentage basis when the acid H*A* is added. Therefore, it is included in a constant called the acid ionization constant:

$$K[H_2O] = K_i = \frac{[H_3O^+][A^-]}{[HA]}$$

TABLE 18.2

ionization constants for acids at 25°C

Acid	Reaction	K_i (M)	% ionization in 1M solution
hydrocyanic	$HCN + H_2O \rightleftharpoons H_3O^+ + CN^-$	4.0×10^{-10}	0.002
hypochlorous	$HClO + H_2O \rightleftharpoons H_3O^+ + ClO^-$	3.2×10^{-8}	0.02
acetic	$HC_2H_3O_2 + H_2O \rightleftharpoons H_3O^+ + C_2H_3O_2^-$	1.8×10^{-5}	0.4
formic	$HCO_2H + H_2O \rightleftharpoons H_3O^+ + HCO_2^-$	1.8×10^{-4}	1.4
nitrous	$HNO_2 + H_2O \rightleftharpoons H_3O^+ + NO_2^-$	4.5×10^{-4}	2.1
hydrofluoric	$HF + H_2O \rightleftharpoons H_3O^+ + F^-$	6.7×10^{-4}	2.6

A large K_i reflects a high percent dissociation.

The acid ionization constants of some weak acids and the percent ionization of a 1 molar solution are given in Table 18.2. **The larger the value of K_i, the larger is the percent ionization.**

Ammonia is the most common example of a weak base. When ammonia dissolves in water, a few hydroxide ions are formed as a result of abstraction of a proton from water by ammonia.

$$NH_3 + H_2O \rightleftharpoons NH_4^+ + OH^-$$

Solutions of ammonia in water are often called ammonium hydroxide because of the ions which are present in solution. However, most of the ammonia in solution remains as the molecular compound NH_3 and not the ammonium ion. It is more accurate to refer to the solution as an ammonia solution.

Problem 18.2

Will the following reaction proceed to a significant extent to yield the products on the right-hand side of the equation?

$$H_2O + S^{2-} \rightleftharpoons HS^- + OH^-$$

Solution

H_2O is a stronger acid than HS^- (Table 18.1). Therefore, H_2O should tend to donate protons to S^{2-} with greater facility than HS^- to OH^-. Furthermore, S^{2-} is a stronger base than OH^- and will accept protons with greater facility. The equilibrium should lie to the right.

18.6 CALCULATIONS OF WEAK ELECTROLYTE EQUILIBRIA

The ionization of a weak acid in water or the ionization process of a weak base such as ammonia can be treated using some of the same techniques outlined in Chapter 17. First, express the equilibrium as a properly balanced equation. Second, write the equilibrium constant expression with all of the exponents corresponding to the coefficients in the chemical equation. Then you are prepared to solve for any desired unknown quantities if given enough information about the system.

Problem 18.3 The concentration of the cyanide ion in a $0.010M$ solution of HCN is 2.0×10^{-6} mole/liter. Calculate the K_i for HCN.

Solution The equation for the ionization of HCN and its equilibrium constant for ionization are:

$$HCN + H_2O \rightleftharpoons H_3O^+ + CN^-$$

$$K_i = \frac{[H_3O^+][CN^-]}{[HCN]}$$

From the chemical equation, it can be seen that the concentration of H_3O^+ and CN^- must be equal because they are produced in a 1:1 ratio. Thus the two values required in the numerator of the K_i expression are:

$$[H_3O^+] = [CN^-] = 2.0 \times 10^{-6} \text{ mole/liter}$$

Only the concentration of the un-ionized HCN is needed to solve for K_i. This value can be obtained by subtracting the concentration of the ionized cyanide ion from the molarity of the solution because one cyanide ion is produced for every HCN which ionizes.

$$\begin{array}{ccccc} HCN_{\text{initial}} & - & CN^- & = & HCN_{\text{remaining}} \\ 0.010 \text{ mole/liter} & - & 0.0000020 \text{ mole/liter} & = & 0.010 \text{ mole/liter} \end{array}$$

In this case, the extent of ionization is so small the concentration of un-ionized HCN is essentially 0.010 mole/liter with the two significant figures allowed.

Now substitute into the equilibrium constant expression.

$$K_i = \frac{[2.0 \times 10^{-6}][2.0 \times 10^{-6}]}{[1.0 \times 10^{-2}]} \text{ mole/liter}$$

$$K_i = 4.0 \times 10^{-10} \text{ mole/liter}$$

Problem 18.4 Calculate the hydronium ion concentration of a $0.056M$ solution of acetic acid. The K_i for $HC_2H_3O_2$ is 1.8×10^{-5} mole/liter.

Solution The chemical equation and the equilibrium constant expression are as follows:

$$HC_2H_3O_2 + H_2O \rightleftharpoons H_3O^+ + C_2H_3O_2^-$$

$$K_i = \frac{[H_3O^+][C_2H_3O_2^-]}{[HC_2H_3O_2]}$$

At equilibrium the concentration of H_3O^+ must equal that of $C_2H_3O_2^-$. Both of these quantities may be set equal to X.

$$[H_3O^+] = X = [C_2H_3O_2^-]$$

The amount of un-ionized acetic acid must be equal to $(0.056 - X)$ since one molecule must ionize to yield one H_3O^+ ion.

If the values for H_3O^+, $C_2H_3O_2^-$, and $HC_2H_3O_2$ were substituted into the K_i expression, the equation could be solved for X. However, it would be necessary to solve a quadratic expression. In order to simplify the solution, it is convenient to approximate the concentration of un-ionized acetic acid at 0.056 mole/liter. This approximation will be valid providing the extent of ionization is small.

$$1.8 \times 10^{-5} \text{ mole/liter} = \frac{[X][X]}{[0.056 \text{ mole/liter}]}$$

$$0.10 \times 10^{-5} \text{ mole}^2/\text{liter}^2 = X^2$$

$$\sqrt{1.0 \times 10^{-6} \text{ mole}^2/\text{liter}^2} = X$$

$$1.0 \times 10^{-3} \text{ mole/liter} = X$$

Thus the concentration of H_3O^+ and $C_2H_3O_2^-$ are approximately 1.0×10^{-3} mole/liter. The concentration of $HC_2H_3O_2$ must then be $(0.056 - 0.001)M$ or $0.055M$. The approximation made for the concentration of un-ionized acetic acid is valid within 2%.

18.7 WATER AND THE pH SCALE

Concentrations of hydronium ions produced by the dissolution of acids in water often are very small. As an example, the concentration of hydronium ions in a $1M$ solution of acetic acid is $4.3 \times 10^{-3}M$. It is more convenient to utilize a compact notation called pH, which avoids the use of the exponential notation: **The pH is equal to the negative logarithm of the hydronium ion concentration.**

pH

$$\text{pH} = -\log[H_3O^+]$$

The logarithmic relation was chosen arbitrarily for its ease of use. For example, a $1.0 \times 10^{-4}M$ concentration of hydronium ions has a pH = 4.0:

$$\begin{aligned}\text{pH} &= -\log [H_3O^+] = -\log [1.0 \times 10^{-4}] \\ &= -(\log 1.0 + \log 10^{-4}) \\ &= -(0.0 - 4) \\ &= +4\end{aligned}$$

A solution whose pH is 3.7 has a hydronium ion concentration of $2 \times 10^{-4} M$.

$$\begin{aligned} pH = 3.7 &= -\log [H_3O^+] \\ -3.7 &= \log [H_3O^+] \\ -4 + 0.3 &= \log [H_3O^+] \\ 10^{-4} \times 10^{0.3} &= [H_3O^+] \\ 2 \times 10^{-4} &= [H_3O^+] \end{aligned}$$

In a similar way, the hydroxide ion concentration of solutions can be conveniently expressed by **pOH,** which **is the negative logarithm of the hydroxide ion concentration.**

pOH

$$pOH = -\log [OH^-]$$

The relationships among $[H_3O^+]$, $[OH^-]$, pH, and pOH are illustrated in Table 18.3.

Pure water can act as an acid and a base simultaneously to establish an equilibrium that proceeds as follows to a small but measurable degree.

$$H_2O + H_2O \rightleftharpoons H_3O^+ + OH^-$$

The equilibrium expression that incorporates the concentrations of water in the ion product constant K_w is $1 \times 10^{-14} M^2$ at 25°C:

$$K = \frac{[H_3O^+][OH^-]}{[H_2O]^2}$$

$$K_w = K[H_2O]^2 = [H_3O^+][OH^-] = 1 \times 10^{-14} M^2$$

TABLE 18.3

pH and pOH of aqueous solutions

(H_3O^+)	pH	pOH	(OH^-)	
10^0	0	14	10^{-14}	
10^{-1}	1	13	10^{-13}	
10^{-2}	2	12	10^{-12}	
10^{-3}	3	11	10^{-11}	acidic
10^{-4}	4	10	10^{-10}	
10^{-5}	5	9	10^{-9}	
10^{-6}	6	8	10^{-8}	
10^{-7}	7	7	10^{-7}	neutral
10^{-8}	8	6	10^{-6}	
10^{-9}	9	5	10^{-5}	
10^{-10}	10	4	10^{-4}	
10^{-11}	11	3	10^{-3}	basic
10^{-12}	12	2	10^{-2}	
10^{-13}	13	1	10^{-1}	
10^{-14}	14	0	10^0	

Since the reaction involves the formation of equal quantities of H_3O^+ and OH^-, the concentration of each is $1 \times 10^{-7}M$. Therefore, the pH and the pOH of pure water at 25°C are both 7:

$$[H_3O^+][OH^-] = 10^{-14}M^2 \quad \text{and} \quad [H_3O^+] = [OH^-]$$
$$[H_3O^+]^2 = 10^{-14}M^2$$
$$[H_3O^+] = 10^{-7}M \qquad pH = 7$$
$$[OH^-] = 10^{-7}M \qquad pOH = 7$$

The ionization does not usually complicate our measurement of acid or base strength of substances dissolved in water. For example, when 0.1 mole of HCl is dissolved in sufficient water to make 1 liter of solution, the HCl totally ionizes to produce a $0.1M$ solution of H_3O^+. The H_3O^+ produced from the ionization of water is small compared to the amount produced from HCl. In the $0.1M$ solution formed by adding HCl to water, the H_3O^+ concentration is much greater than that of pure water and the pH is 1.

Why does a basic solution still contain hydronium ions?

Since the K_w value is unchanged in the presence of the added H_3O^+, the concentration of OH^- can be calculated:

$$K_w = [H_3O^+][OH^-] = 10^{-14}M^2$$
$$10^{-1}[OH^-] = 10^{-14}M$$
$$[OH^-] = 10^{-13}M$$

Thus the increase in the concentration of H_3O^+ results in a decrease in the concentration of OH^- as predicted by Le Châtelier's principle. The addition of H_3O^+ causes a disturbance in the equilibrium between the H_3O^+ and OH^- present in water. The system behaves to eliminate some of the added H_3O^+ by using up some of the OH^- to neutralize them.

If 0.001 mole of sodium hydroxide is added to sufficient water to produce 1 liter of solution, the OH^- concentration derived from this completely ionized base is $10^{-3}M$. The OH^- produced from the ionization of pure water is 10^{-7}, which is much less than that obtained by dissolving the sodium hydroxide. The OH^- and H_3O^+ concentrations of the solution can be calculated from the K_w value of water.

$$[OH^-] = 10^{-3}M$$
$$[H_3O^+][OH^-] = 10^{-14}M^2$$
$$[H_3O^+]\,10^{-3} = 10^{-14}M$$
$$[H_3O^+] = 10^{-11}M$$

The solution has pH = 11 and pOH = 3. Note again that Le Châtelier's principle applies in this solution. By increasing the concentration of OH^-, the system tries to eliminate them by reacting with H_3O^+. Thus the concentration of H_3O^+ is decreased.

From the above examples it can be concluded that **the pH of acidic solutions is less than 7 and the pH of basic solutions is greater than 7.** As a general rule, the calculation of pH and pOH and the related hydronium and hydroxide ion concentrations in solutions of strong acids or bases can be done readily because they dissociate completely and are the prime contributors to the concentrations of H_3O^+ and OH^- in solution.

Use of the pH scale

Although the mathematical aspects of pH are not understood by large segments of the general population, there are nevertheless many practical uses of this scale. Because some food crops flourish under certain acid conditions while others require more basic conditions, agriculturalists are dependent on the proper measurement and interpretation of pH. Soil is often tested to determine whether acidic or basic fertilizers are required for a particular crop. A soil containing large amounts of humus is quite acidic and must often be neutralized with powdered limestone. On the other hand, large sections of the agricultural land in California are very basic, requiring acidic phosphate fertilizers.

TABLE 18.4 pH values of body fluids

blood	7.35–7.45
gastric juices	1.6–1.8
bile	7.8–8.6
urine	5.5–7.0
saliva	6.2–7.4
interstitial fluid	7.4
muscle intracellular fluid	6.1
liver intracellular fluid	6.9
pancreatic juice	7.8–8.0

The pH values of several of our body fluids are given in Table 18.4. Except for gastric juices in which the main acid is HCl, the majority of body fluids have pH values near neutrality. None of the body fluids are very basic. This fact will be discussed further in Section 18.8 dealing with buffers.

How do we react to foods of high acidity when we taste them?

In Table 18.5 are listed the pH values of a number of common foods. All of those listed are acidic. Foods with low pH taste tart to us.

TABLE 18.5 pH values of foods

apples	2.9–3.3
cabbage	5.2–5.4
corn	6.0–6.5
grapes	3.5–4.5
lemons	2.2–2.4
milk	6.3–6.6
oranges	3.0–4.0
peas	5.8–6.4
potatoes	5.6–6.0
tomatoes	4.0–4.4

18.8 BUFFERS

The term buffer means, among other things, to lessen the shock of change. **In chemistry, a buffer is something which will lessen the shock of a possible rapid pH change. A chemical buffer is prepared by dissolving both a weak acid and a salt of its conjugate base in water,** for example, acetic acid and sodium acetate. The buffer may react with any added strong acids or bases. Hydronium ions will react with the acetate ions. Hydroxide ions will react with the acetic acid.

$$H_3O^+ + C_2H_3O_2^- \longrightarrow H_2O + HC_2H_3O_2$$
$$OH^- + HC_2H_3O_2 \longrightarrow H_2O + C_2H_3O_2^-$$

In both cases, the substance which could radically change the pH has been eliminated by reaction with a component of the buffer.

Buffers may be prepared from any ratio of concentrations of weak acids and the salt of the weak acid. A 1:1 ratio of acid to salt is the most efficient in handling the addition of either base or acid. If the buffer contains a lot more acid than salt, it will be less efficient in handling acid. Alternatively, a buffer with a lot more salt than acid cannot efficiently counteract the addition of base.

A variety of pH values may be achieved for various ratios of a weak acid and the salt of the weak acid. For a 1:1 ratio of acid to salt, the pH of the acetic acid–acetate buffer is 4.7. For a 10:1 and a 1:10 ratio of acid to salt, the pH values are 3.7 and 5.7, respectively.

Buffers are very important in industrial processes such as the plating of metals and the manufacture of dyes. Many processes proceed at a maximum rate at a specific pH, and the buffer facilitates the maintenance of a rapid and hence economical process.

18.9 BUFFERS IN THE BODY

Human blood is buffered by a combination of substances which maintain the pH at 7.4. If these buffers fail and the pH either increases or decreases by 0.2 pH units, death results. The pH of blood leaving the heart differs only by 0.02 pH units from that returning to the heart after contacting other cells of the body. This is phenomenal considering the many acid-base reactions that take place in the cells of the body and indicates the importance of buffers in human blood.

The buffers of the blood are principally the H_2CO_3–HCO_3^- system and the $H_2PO_4^-$–HPO_4^{2-} system. The H_2CO_3–HCO_3^- buffer occurs by the dissolution of carbon dioxide in the water of blood plasma.

$$CO_2 + H_2O \rightleftharpoons H_2CO_3$$

The dissolved carbon dioxide, which is called carbonic acid, can react with hydroxide ions to prevent an increase in the pH of the blood. The bicarbonate ion can react with hydronium ions to prevent a decrease in the pH of the blood.

$$H_2CO_3 + OH^- \rightleftharpoons HCO_3^- + H_2O$$

$$HCO_3^- + H_3O^+ \rightleftharpoons H_2CO_3 + H_2O$$

If H_2CO_3 and HCO_3^- were present in equal amounts in the blood, the pH would be 6.4. The actual pH of 7.4 is maintained by a high concentration of HCO_3^- relative to H_2CO_3. The ratio $[H_2CO_3]/[HCO_3^-]$ is near 0.05 in blood. Thus the capacity of the buffer to handle base is very limited. However, the buffer has a great capacity to counteract acid.

As indicated earlier, most of the body fluids are acidic. Furthermore, many metabolic reactions form acidic products. These must be neutralized lest the body be overwhelmed. One such acidic compound is lactic acid, which is produced from the metabolism of sugar. Exercise results in the production of lactic acid in muscles which then must be neutralized quickly to prevent acidosis.

Aren't you glad there are buffers in the body?

The phosphate buffer is extremely important in the cell. Many reactions which occur in the cell involve complex compounds which in part contain covalently bonded phosphate groups. In order to maintain the proper pH in cellular fluids, the HPO_4^{2-}–$H_2PO_4^-$ buffer must be very effective. The dihydrogen phosphate ion reacts with the hydroxide ion and prevents an increase in cellular pH. The monohydrogen phosphate ion reacts with the hydronium ion and prevents a decrease in cellular pH.

$$H_2PO_4^- + OH^- \rightleftharpoons HPO_4^{2-} + H_2O$$

$$HPO_4^{2-} + H_3O^+ \rightleftharpoons H_2PO_4^- + H_2O$$

18.10 TITRATION

Titration is an experimental procedure by which the unknown concentration or amount of an acid or base is determined by reacting a sample of it with a known amount of a base or acid to achieve neutralization. The point of exact neutralization is known as the endpoint or neutralization point. The endpoint is indicated by a substance called an indicator.

Chemical indicators are weak acids or bases. Each indicator can exist as a related acid and its conjugate base, or a base and its conjugate acid. Each member of the conjugate pair is a different color so that an indicator changes color as the relative concentrations of the two substances forming the conjugate pair change. For example, phenolphthalein is colorless in acid solution where it exists in its acid form and is pink in basic solution where it exists in its basic form. The colors of a number of other indicators are listed in Table 18.6 along with the pH range over which the color change occurs.

In a titration, a measured volume or mass of an acid or base, whose

TABLE 18.6

acid–base indicators

Indicator	pH range for change	Color at low and high pH
thymol blue	1.2–2.8	red–yellow
methyl orange	3.1–4.4	red–yellow
methyl red	4.2–6.3	pink–yellow
bromocresol purple	5.2–6.8	yellow–red
litmus	4.7–8.2	red–blue
phenolphthalein	8.3–10.0	colorless–pink
thymolphthalein	9.3–10.5	colorless–blue
alizarin yellow	10.1–12.1	yellow–red

quantity is the unknown, is placed in a flask and a couple of drops of indicator is added. The indicator chosen must have a change in color at the pH of the solution of the products. A solution of a base or acid of known concentration is then slowly added in a dropwise fashion. This addition is easily made from a buret, a tube marked for its volume. When the indicator changes color, the addition is terminated and the volume of solution which has been added from the buret is noted (Figure 18.2).

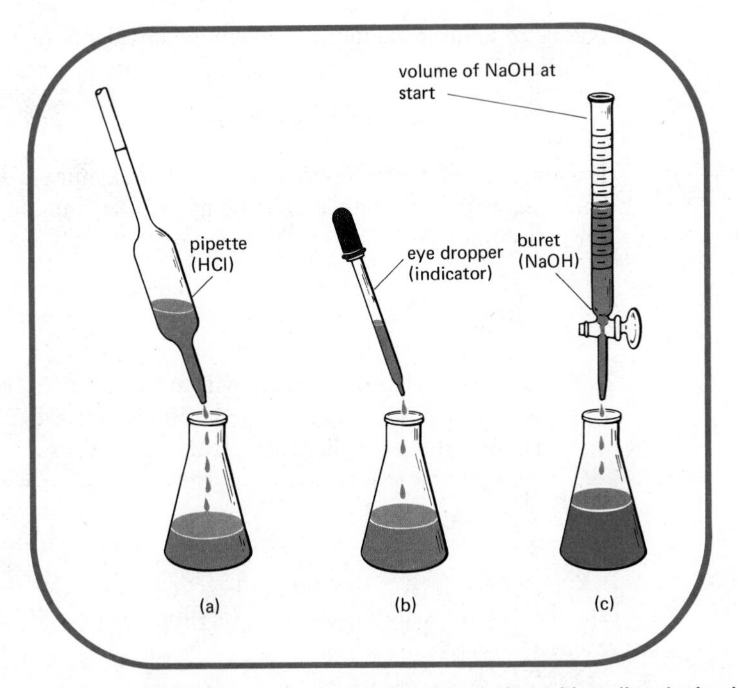

Figure 18.2 Titration of a hydrochloric acid solution with sodium hydroxide: (a) measure an exact amount of a hydrochloric acid solution into a flask; (b) add a couple of drops of phenolphthalein solution as indicated; (c) from the buret, add sodium hydroxide to achieve the neutralization point which occurs when a slight pink color, due to the indicator, is observed.

The number of equivalents of base or acid used from the buret is calculated by multiplying the volume used by its normality. This number of equivalents must be equal to the number of equivalents of the unknown, because one equivalent of an acid is required to neutralize one equivalent of a base. (Refer to Section 16.5 for a review of the use of normality and equivalents.)

Equivalents of base must be equal to the equivalents of acid for neutralization.

Frequently in a laboratory, the volumes of solutions used are not large. Therefore, the units of equivalents and liters are not convenient. For this reason, normality can also be regarded as the number of milliequivalents per milliliter.

$$N = \frac{\text{milliequivalents}}{\text{milliliter}}$$

A milliequivalent is 1/1000 of an equivalent and is abbreviated meq.

Problem 18.5

A 5.00 ml sample of household ammonia is titrated with 48.0 ml of a 0.200N solution of hydrochloric acid to achieve a methyl red endpoint. What is the molarity of the household ammonia?

Solution

The acid solution contains 0.200 milliequivalents of hydrochloric acid per milliliter. Thus the number of milliequivalents of acid used is:

$$0.200\ \text{meq}/\cancel{\text{ml}} \times 48.0\ \cancel{\text{ml}} = 9.60\ \text{meq}$$

The number of milliequivalents of ammonia must also be equal to 9.60. This quantity is contained in 5.00 ml of the ammonia solution, and the normality must be:

$$\frac{9.60\ \text{meq}}{5.00\ \text{ml}} = 1.92N$$

For ammonia, which only reacts with one equivalent of acid per mole, the equivalent weight is the same as the molecular weight. Therefore the molarity is the same as the normality, or 1.92M.

Problem 18.6

Baking soda, which is sodium hydrogen carbonate, may be used to establish the normality of an acid solution such as HCl by the following reaction.

$$NaHCO_3 + HCl \longrightarrow NaCl + CO_2 + H_2O$$

A 0.0420-g sample of sodium bicarbonate is dissolved in water in a flask and 20.0 ml of a hydrochloric acid solution is used to neutralize it. What is the normality of the HCl solution?

Solution

The equivalent weight of $NaHCO_3$ is 84.0. A milliequivalent will have a mass of 0.0840 g. Thus the number of milliequivalents in the sample is:

$$\frac{0.0420\ \cancel{g}}{0.0840\ \cancel{g}/\text{meq}} = 0.500\ \text{meq}$$

The number of milliequivalents in the 20.0 ml of HCl must also be 0.500 meq. Thus the normality is:

$$\frac{0.500\ \text{meq}}{20.0\ \text{ml}} = 0.025N$$

18.11 IONIC EQUATIONS

In Chapter 11 a method of balancing equations as molecular equations was presented. At that time, the balancing of ionic equations was deferred. Now that the nature of electrolytes and nonelectrolytes has been established, it is possible to write balanced ionic equations.

Why is the term spectator ion so descriptive?

Ionic equations state chemical changes in terms of the ions which actually undergo reaction. Many ions do not undergo change during a reaction and they are called spectator ions. The spectator ions are not included in the final balanced ionic equation. Therefore, the ionic equation gives a better representation of the chemical reaction which actually occurs.

The procedure for balancing ionic reactions is essentially the same as given in Chapter 11. In addition to achieving an atomic balance, it **is necessary to insure that an electrical balance is simultaneously achieved.** Some of the rules for representing substances in ionic equations are:

1. **Strong electrolytes are written in terms of the ions present.**
2. **Nonelectrolytes are written in molecular form.**
3. **Weak electrolytes are written in molecular form.**
4. **All insoluble substances and gases are written in molecular form.**
5. **Only those substances undergoing changes are included.**

Problem 18.7

What is the ionic equation for the reaction of aqueous solutions of silver nitrate and sodium chloride?

Solution

The reaction produces insoluble AgCl. The molecular equation is:

$$AgNO_3(aq) + NaCl(aq) \longrightarrow AgCl\downarrow + NaNO_3(aq)$$

Using the rules for expressing ionic equations, the chemicals are written as:

$$Ag^+ + NO_3^- + Na^+ + Cl^- \longrightarrow AgCl\downarrow + Na^+ + NO_3^-$$

The Na^+ and NO_3^- ions are not involved in the reaction and are eliminated.

$$Ag^+ + Cl^- \longrightarrow AgCl\downarrow$$

In this case the equation is balanced as it stands. Both the numbers of atoms and the net charge on each side of the equation are the same.

Problem 18.8 Write the balanced ionic equation for the reaction of sodium bicarbonate with sulfuric acid.

Solution The unbalanced molecular equation is:

$$NaHCO_3 + H_2SO_4 \longrightarrow Na_2SO_4 + H_2O + CO_2$$

In aqueous solution, sodium bicarbonate exists as Na^+ and HCO_3^-. Sulfuric acid exists as hydronium ions and sulfate ions.

$$Na^+ + HCO_3^- + 2H_3O^+ + SO_4^{2-} \longrightarrow 2Na^+ + SO_4^{2-} + H_2O + CO_2$$

The sulfate ion is a spectator ion and should not appear in the final balanced equation. Similarly, the sodium ion is not involved.

$$HCO_3^- + 2H_3O^+ \longrightarrow H_2O + CO_2$$

In order to balance the electrically neutral product, the coefficients of HCO_3^- and H_3O^+ must be identical. If the coefficient of HCO_3^- is made to be 2, then the coefficient of CO_2 would also have to be set at 2.

$$2HCO_3^- + 2H_3O^+ \longrightarrow H_2O + 2CO_2$$

In order to balance the 8 hydrogens on the left in the $2HCO_3^-$ and $2H_3O^+$ a coefficient of 4 is assigned to H_2O. The total number of oxygen atoms simultaneously come into balance.

$$2HCO_3^- + 2H_3O^+ \longrightarrow 4H_2O + 2CO_2$$

This equation is not acceptable as the coefficients are not in the ratio of the simplest numbers. Division by two yields the correctly balanced equation.

$$HCO_3^- + H_3O^+ \longrightarrow 2H_2O + CO_2$$

know terminology

SUMMARY

Electrolytes *are substances which dissolve in water to form solutions containing ions and which conduct electricity. The separation of ions already present in a substance, when dissolved in water, is called* ***dissociation.*** *The formation of ions, upon dissolution in water, from a covalent substance is called* ***ionization.*** *Electrolytes are divided into* *two classes called **strong and weak** electrolytes depending on the number of ions produced from the substance.*

***Acids** yield protons in solution while **bases** yield hydroxide ions according to the **Arrhenius** concept. Acids are **proton donors** and bases are **proton acceptors** according to the **Brønsted-Lowry** concept.*

Strong acids** ionize completely in water. **Strong bases** attract protons strongly. **Weak acids** are only partially ionized in water. The extent of ionization of acids is given by the acid **ionization constant.

***pH** is equal to the negative logarithm of the hydronium ion concentration. Acidic solutions have a pH less than 7. Basic solutions have a pH greater than 7. **pOH** is equal to the negative logarithm of the hydroxide ion concentration.*

***Buffers** consist of a solution of a weak acid and its conjugate base. They function by preventing rapid pH changes as they can react with either base or acid.*

Titration** is an experimental procedure used to determine the unknown concentration of an acid or base. The **neutralization point** in a titration is detected by using a chemical **indicator.

Ionic equations** state chemical changes in terms of the ions which actually undergo reaction. Ions which are not involved in the reaction are called **spectator ions.

QUESTIONS AND PROBLEMS

Definitions

1. Define each of the following terms.

(a) electrolyte	**(b)** nonelectrolyte
(c) strong electrolyte	**(d)** weak electrolyte
(e) hydronium ion	**(f)** hydroxide ion
(g) dissociation	**(h)** ionization
(i) amphoteric	**(j)** indicator
(k) pH	**(l)** pOH
(m) K_w	**(n)** buffer

2. Distinguish between each of the following.

(a) pH and pOH
(b) dissociation and ionization
(c) hydronium ion and hydroxide ion
(d) an Arrhenius acid and a Brønsted-Lowry acid
(e) an Arrhenius base and a Brønsted-Lowry base
(f) molecular equation and ionic equation

Electrolytes

3. Explain how an electrolyte can be experimentally distinguished from a nonelectrolyte.

4. How can a strong electrolyte be distinguished from a weak electrolyte?

5. Give two examples each of strong electrolytes, weak electrolytes, and nonelectrolytes.

Acid-base strength

6. Classify each of the following as weak or strong acids.
(a) HNO_2 **(b)** HNO_3
(c) $HClO_4$ **(d)** H_2SO_4
(e) $HC_2H_3O_2$ **(f)** HCl
(g) HCN **(h)** H_2O

7. Classify each of the following as weak or strong bases.
(a) $NaOH$ **(b)** $Ca(OH)_2$
(c) NH_3 **(d)** H_2O

Conjugate pairs

8. What is the conjugate acid of each of the following?
(a) H_2O **(b)** NO_2^-
(c) HSO_4^- **(d)** NH_3
(e) Cl^- **(f)** ClO_4^-

9. What is the conjugate base of each of the following?
(a) H_2O **(b)** NH_3
(c) HCO_3^- **(d)** HCl
(e) H_2SO_4 **(f)** HNO_3

Acid ionization constants

10. The K_i for hypochlorous acid is 3.2×10^{-8} mole/liter. What is the hypochlorite ion concentration of a $0.31M$ solution?

11. Calculate the hydronium ion concentration of a $0.01M$ solution of HCN. The K_i is 4×10^{-10} mole/liter.

12. A $0.040M$ solution of HF is 13% ionized. Calculate K_i.

13. A $0.10M$ ammonia solution has an equilibrium concentration of NH_4^+ equal to 0.0013 mole/liter. Calculate K for the equation.

$$NH_3 + H_2O \rightleftharpoons NH_4^+ + OH^-$$

pH and pOH

14. Calculate the pH of the following solutions.
(a) a $10^{-3}M$ solution of HCl
(b) a $2 \times 10^{-2}M$ solution of $HClO_4$
(c) a solution with $[H_3O^+] = 10^{-9}$

15. Calculate the pH of the following items.
(**a**) a soft drink with $[H_3O^+] = 2 \times 10^{-4}$ mole/liter
(**b**) milk with $[H_3O^+] = 2 \times 10^{-7}$ mole/liter
(**c**) ammonia with $[H_3O^+] = 2 \times 10^{-12}$ mole/liter
(**d**) vinegar with $[H_3O^+] = 8 \times 10^{-4}$ mole/liter

16. Calculate the pOH of the following solutions.
(**a**) a $10^{-3}M$ NaOH solution
(**b**) a $4 \times 10^{-5}M$ KOH solution
(**c**) a solution with $[OH^-] = 10^{-10}$ mole/liter

Buffers

17. Explain how a buffer works.

18. What buffers are present in the body?

19. Could a solution of HCl and NaCl serve as a buffer? Explain your answer.

Titration

20. What quantity of $0.10N$ NaOH is required to titrate 25.0 ml of a $0.060N$ HCl solution?

21. A 50.0-ml solution of $Ca(OH)_2$ is titrated with 36.0 ml of a $0.05N$ HCl solution to an endpoint. What is the molarity of the $Ca(OH)_2$ solution?

22. A 0.1060-g sample of Na_2CO_3 is neutralized with 48.0 ml of a $HClO_4$ solution. What is the normality of the $HClO_4$ solution?

23. A 0.100-g sample of NaOH is titrated to an endpoint with 12.5 ml of HCl. What is the normality of the HCl?

Ionic equations

24. Write balanced ionic equations for each of the following combinations of reactants.
(**a**) $HCl(aq) + NaOH(aq) \rightarrow$
(**b**) $KCl(aq) + AgNO_3(aq) \rightarrow$
(**c**) $Fe(s) + CuSO_4(aq) \rightarrow$
(**d**) $BaO(s) + HCl\ (aq) \rightarrow$
(**e**) $Cl_2(g) + NaBr(aq) \rightarrow$
(**f**) $MgCl_2(aq) + Na_2CO_3(aq) \rightarrow$
(**g**) $Zn(s) + H_2SO_4(aq) \rightarrow$
(**h**) $HNO_3(aq) + K_2CO_3(aq) \rightarrow$
(**i**) $ZnCl_2(aq) + H_2S(g) \rightarrow$

19 OXIDATION-REDUCTION REACTIONS

LEARNING OBJECTIVES

When you finish this chapter you should be able to:

(a) relate the terms oxidation and reduction to electron loss and gain.
(b) express oxidation-reduction reactions in terms of oxidizing and reducing agents.
(c) balance oxidation-reduction equations using oxidation numbers.
(d) balance oxidation-reduction equations using half reaction.
(e) diagram the galvanic cell and distinguish between the reactions which occur at the anode and at the cathode.
(f) demonstrate how standard electrode potentials are obtained.
(g) calculate the voltage of a galvanic cell.
(h) determine whether an oxidation-reduction reaction will occur in the direction written.
(i) demonstrate how electrolysis is used to prepare, purify, and electroplate metals.
(j) match the following terms with their corresponding definitions.

anode	oxidation
cathode	oxidizing agent
electrolysis	reducing agent
galvanic cell	reduction
half reaction	standard electrode potential

In Chapter 11 you learned that equations representing chemical change can be classified according to five simple types: combination, decomposition, substitution, metathesis, and neutralization. At that time it was emphasized that as you learned more chemistry you would encounter other types of reactions and other ways to classify them.

In Chapter 7 it was noted that atoms can gain or lose electrons to form ions both simple and complex. In order to distinguish between the different ways in which atoms may be combined in compounds, the oxidation number was defined. The oxidation number indicates the number of electrons lost or gained relative to the free element.

There are chemical reactions in which the oxidation number of two or more atoms change. These reactions, called oxidation-reduction reactions, are somewhat more difficult to balance and often require special techniques which will be described in this chapter. Many of the combination, decomposition, and substitution reactions described in Chapter 11 are actually oxidation-reduction reactions. However, the ones which were chosen for discussion could all be balanced by inspection. Now more general techniques will be presented which will enable you to balance all oxidation-reduction equations.

A second feature of this chapter is the establishment of the relationship between chemical energy and electrical energy. Electrical energy can be produced from the stored chemical energy of substances as they undergo oxidation-reduction reactions. Furthermore, electrical energy can be used to cause chemical reactions that would not otherwise occur.

19.1 OXIDATION AND REDUCTION

At one time, the combination of an element with oxygen was called an oxidation reaction. The reaction of magnesium or hydrogen with oxygen involves an oxidation of the substance.

$$2Mg + O_2 \longrightarrow 2MgO$$
$$2H_2 + O_2 \longrightarrow 2H_2O$$

Oxidation is the loss of electrons.

Reduction is the gain of electrons.

In each case, the oxidation number of the element has changed. The oxidation number for magnesium has changed from 0 to +2 while that of hydrogen has changed from 0 to +1. In each case the oxidation number change is the result of loss of electrons. Currently, **oxidation is defined as the loss of electrons by a substance or an increase in its oxidation number.** The reaction does not necessarily involve oxygen. Thus the reaction of sodium with chlorine involves the oxidation of sodium because its oxidation number increases from 0 to +1.

$$2Na + Cl_2 \longrightarrow 2NaCl$$

Early chemists called a reaction in which oxygen was lost a reduction. This frequently occurred when an oxygen-containing substance reacted with hydrogen to produce water. Therefore, any reaction which involves combination with hydrogen was also regarded as a reduction reaction. Examples of reduction of mercury(II) oxide, copper(II) oxide, and fluorine are given below.

$$2HgO \longrightarrow 2Hg + O_2 \qquad \text{(loss of oxygen)}$$
$$CuO + H_2 \longrightarrow Cu + H_2O \qquad \text{(loss of oxygen due to hydrogen)}$$
$$H_2 + F_2 \longrightarrow 2HF \qquad \text{(combination with hydrogen)}$$

In each case, the oxidation number of an element has decreased. This occurs because electrons are gained. In the case of mercury and copper, the changes are from +2 to 0. For fluorine the change is from 0 to −1. Consequently, **reduction is defined as the gain of electrons by a substance or a decrease in its oxidation number.**

When a substance is oxidized, it loses its electrons to another substance which then becomes reduced. In each of the equations cited to this point, this statement can be verified. For example, the electrons lost by magnesium in its reaction with oxygen are gained by the oxygen. The oxidation number of oxygen changes from 0 to −2. Therefore, from a slightly different point of view, it follows that **when a substance becomes reduced, the electrons which it gains are obtained from a substance which then becomes oxidized.** This is why reactions in which an electron transfer occurs are termed oxidation-reduction reactions.

An oxidizing agent becomes reduced.

A reducing agent becomes oxidized.

The close and necessary relationship between oxidation and reduction is emphasized further by use of the terms oxidizing agent and reducing agent. In an oxidation-reduction reaction, **the substance which is reduced is called the oxidizing agent** because it causes oxidation in another substance by gaining electrons. **The substance which is oxidized is called the reducing agent** because it causes the reduction of another substance by losing its electrons. In terms of oxidation number changes, an oxidizing agent is a substance whose oxidation number decreases; a reducing agent is a substance whose oxidation number increases. A summary of the terms used in this section is given in Table 19.1.

TABLE 19.1
oxidation-reduction terminology

Term	Electron change	Oxidation number change
oxidation	loss of electrons	increase
reduction	gain of electrons	decrease
substance oxidized	loses electrons	increase
substance reduced	gains electrons	decrease
oxidizing agent	accepts electrons	decrease
reducing agent	donates electrons	increase

For the equation giving the reaction of hydrogen and fluorine, the oxidation numbers of the elements are given in parentheses. From this information, the terms oxidation, reduction, oxidizing agent, and reducing agent can be assigned.

$$\overset{(0)}{H_2} \quad + \quad \overset{(0)}{F_2} \longrightarrow 2\overset{(+1)}{H}\,\overset{(-1)}{F}$$

H_2	F_2
becomes oxidized reducing agent	becomes reduced oxidizing agent

One way to remember the terms introduced in this section is to draw an analogy between oxidation number and the temperature scale (Figure 19.1). Consider yourself an atom and your clothes electrons. As the temperature

Figure 19.1 An analogy between the relationship of the clothing of an individual and the temperature scale, and the electrons of an atom and the oxidation scale.

increases, you remove clothing to be more comfortable. As the oxidation number increases, that atom loses electrons. Conversely, if the temperature decreases, you add clothing. As the oxidation number decreases, you have reduction occurring and electrons added.

The oxidizing and reducing agents may be compared to a personal valet (Figure 19.2). As you remove clothing, your valet will accept it. When you need clothing, your valet will give it to you.

Figure 19.2 An analogy for the oxidizing and reducing agents as servants of the substance being reduced or oxidized.

Problem 19.1 What becomes oxidized and what is reduced in the following equation?

$$3P + 5HNO_3 + 2H_2O \longrightarrow 5NO + 3H_3PO_4$$

Solution The oxidation numbers of hydrogen and oxygen are +1 and −2, respectively, and are not changed in the reaction as these elements exist only in compounds. The oxidation number of nitrogen in HNO_3 is +5 while it is +2 in NO. Thus nitrogen is reduced. The oxidation number of phosphorus changes from 0 in elemental phosphorus to +5 in H_3PO_4. The phosphorus is oxidized.

Problem 19.2 What is the oxidizing agent and what is the reducing agent in the following equation?

$$4NH_3 + 5O_2 \longrightarrow 4NO + 6H_2O$$

Solution Oxygen is present as a free element as a reactant but is combined in compounds in the products. Its oxidation number changes from 0 to −2. Oxygen becomes reduced and serves as an oxidizing agent. The oxidation number of nitrogen changes from −3 to +2. Nitrogen is oxidized and serves as the reducing agent.

19.2 BALANCING EQUATIONS USING OXIDATION NUMBERS

Review the rules of calculating oxidation numbers now!

The oxidation-reduction equations cited in Section 19.1 are simple enough to balance by inspection. However, for more complex equations, the oxidation number method provides a convenient route to achieving balanced equations. There is no hard and fast procedure or set of rules using oxidation numbers which will work in all circumstances. However, a few guidelines can be given which, if used with some common sense, will produce a balanced equation. These guidelines are:

1. **Determine the oxidation number of each element** using the rules given in Section 7.10.
2. **Establish which elements undergo changes in their oxidation numbers** as they are converted from reactant to product.
3. **Write the oxidation number of each element oxidized or reduced above the elemental symbol.**
4. **Draw an arrow from the element in the reactant to the product for each one that changes oxidation number.**
5. **Calculate the change in the oxidation number required for each element undergoing oxidation and reduction.** Record this change (and its magnitude) as an increase or a decrease over the proper arrow.
6. **Place a factor in front of the change in oxidation numbers so that the total increase is balanced by a total decrease.**
7. **Put this factor as a coefficient in front of the corresponding formula for the reactant or product if the element only is present as a single atom per molecular or formula unit.**
8. **If more than one atom is contained per molecular or formula unit, divide the factor by that quantity to obtain the proper coefficient. If a fraction results, multiply by a number to produce an integer coefficient. Then multiply all other coefficients by the same number.** This must be done to keep the total increase in oxidation number equal to the total decrease in oxidation number.
9. **Balance the remaining parts of the equation by inspection. Do not alter coefficients which affect the oxidation-reduction balance.**

The best way to learn these guidelines is to use them in balancing equations. Several examples follow and additional equations are given at the end of this chapter.

Problem 19.3 Balance the following substitution equation.

$$Zn + AgNO_3 \longrightarrow Zn(NO_3)_2 + Ag$$

Solution According to guideline 1, the oxidation number of each element should be determined. However, note that both nitrogen and oxygen are present only as the nitrate ion in both reactants and products. Clearly these elements do not undergo changes in oxidation number (guideline 2). Thus only zinc and silver change oxidation number. The oxidation numbers are written above the elemental symbol (guideline 3).

$$\overset{(0)}{Zn} + \overset{(+1)}{Ag}NO_3 \longrightarrow \overset{(+2)}{Zn}(NO_3)_2 + \overset{(0)}{Ag}$$

Draw an arrow between elements as reactants and products (guideline 4) and indicate the change in oxidation number (guideline 5).

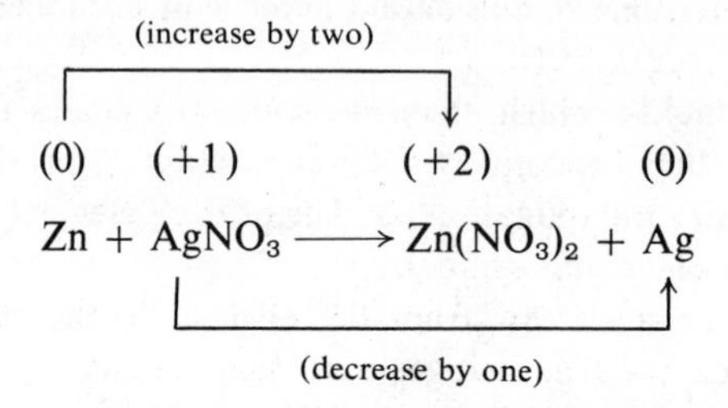

The factors necessary to balance the total decrease by the total increase in oxidation number are 2 and 1, respectively.

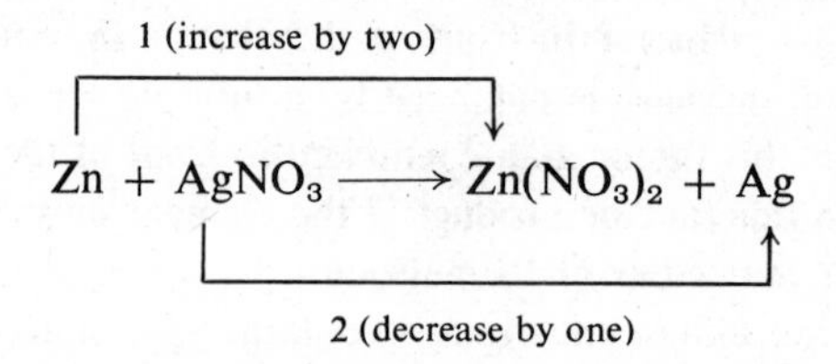

The coefficients are therefore 2 and 1 (guideline 7) because each element is present as a single atom per formula unit.

$$Zn + 2AgNO_3 \longrightarrow Zn(NO_3)_2 + 2Ag$$

The total equation is balanced as it stands. No further changes are required.

Problem 19.4 Balance the following oxidation-reduction reaction.

$$HNO_3 + I_2 \longrightarrow NO_2 + HIO_3 + H_2O$$

Solution In this reaction, hydrogen and oxygen appear only in compounds and their oxidation numbers are $+1$ and -2, respectively. The elements nitrogen and iodine do change oxidation numbers and these values are written above the elemental symbol.

$$H\overset{(+5)}{N}O_3 + \overset{(0)}{I_2} \longrightarrow \overset{(+4)}{N}O_2 + H\overset{(+5)}{I}O_3 + H_2O$$

Next an arrow is drawn between the elements in the reactants and products, and the change in oxidation number is indicated.

(decrease by one)

$$\overset{(+5)}{HNO_3} + \overset{(0)}{I_2} \longrightarrow \overset{(+4)}{NO_2} + \overset{(+5)}{HIO_3} + H_2O$$

(increase by five)

The factors necessary for the total increase in oxidation number of iodine and the total decrease in oxidation number of nitrogen are 1 and 5, respectively.

5 (decrease by one)

$$\overset{(+5)}{HNO_3} + \overset{(0)}{I_2} \longrightarrow \overset{(+4)}{NO_2} + \overset{(+5)}{HIO_3} + H_2O$$

1 (increase by five)

These factors may not be placed as coefficients in front of the formulas for the reactants and products because iodine is present as a diatomic molecule in one of the reactants. Division of the factor by 2 yields a coefficient of $\frac{1}{2}$. In order to obtain an integer coefficient, we multiply by 2. The factor of 5 for nitrogen compounds must then also be multiplied by 2. The coefficients may now be placed in the equation.

$$10HNO_3 + I_2 \longrightarrow 10NO_2 + 2HIO_3 + H_2O$$

The remaining coefficient for water is easily established as 4 by noting that there are 10 hydrogens in the reactant but only 2 accounted for in the $2HIO_3$. Similarly, one could achieve a balance by noting that the 30 oxygens in the reactant produce only 26 oxygens in two of the products whose coefficients are fixed.

$$10HNO_3 + I_2 \longrightarrow 10NO_2 + 2HIO_3 + 4H_2O$$

Problem 19.5 Balance the following oxidation-reduction equation.

$$Na_2CrO_4 + FeCl_2 + HCl \longrightarrow CrCl_3 + FeCl_3 + NaCl + H_2O$$

Solution The elements changing oxidation states are chromium and iron. The oxidation numbers, arrows, and changes in oxidation number are written first.

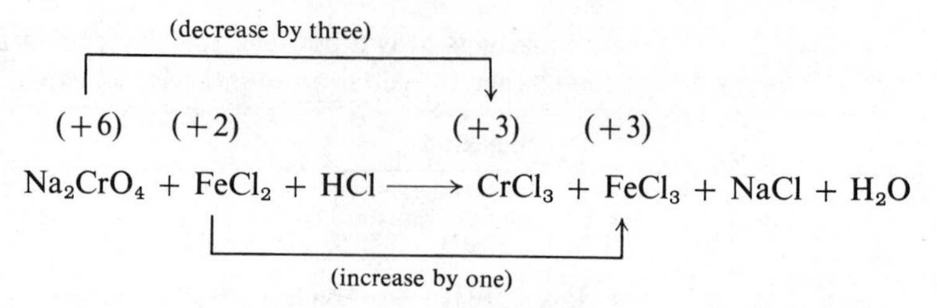

In order to achieve a balance of the total increase in oxidation number with a total decrease in oxidation number, the factors for chromium and iron are 1 and 3, respectively. Since each element appears only as a single atom in the reactants and products, the factors may be used directly as coefficients.

$$Na_2CrO_4 + 3FeCl_2 + HCl \longrightarrow CrCl_3 + 3FeCl_3 + NaCl + H_2O$$

The coefficient for NaCl is next set as 2 because there are two sodium atoms in Na_2CrO_4. Similarly, the four oxygens in Na_2CrO_4 are balanced by placing a coefficient of 4 in front of H_2O.

$$Na_2CrO_4 + 3FeCl_2 + HCl \longrightarrow CrCl_3 + 3FeCl_3 + 2NaCl + 4H_2O$$

The eight hydrogens contained in $4H_2O$ are balanced by placing a coefficient of 8 in front of HCl.

$$Na_2CrO_4 + 3FeCl_2 + 8HCl \longrightarrow CrCl_3 + 3FeCl_3 + 2NaCl + 4H_2O$$

A final check reveals that there are 14 chlorine atoms in both the reactants and products.

19.3 BALANCING EQUATIONS USING THE HALF-REACTION METHOD

A second method of balancing oxidation-reduction equations involves considering the reaction as two half reactions. **A half reaction is an artificial reaction which cannot occur alone.** However, an oxidation half reaction and a reduction half reaction can be regarded as occurring simultaneously within the total reaction. The method of achieving the final balanced equation involves balancing each half reaction and then adding them together.

In Chapter 18 you learned that reactions can be expressed in ionic equations in which only the ions or molecules that are directly involved are written. Ions which do not undergo any change are called spectator ions. While half reactions can be written using all of the components of the total equation, it is much simpler to use only the substances which undergo change. Thus the half reactions when added together will yield a net ionic equation.

As illustrated in several previous outlines of how to balance equations, there are a set of guidelines which, if used with some thought, will enable you to balance equations. In this section you will use the following guidelines to obtain net ionic equations using the half reaction method.

1 **Examine the equation and rewrite it in net ionic form.**
2 **Calculate the oxidation numbers for all of the elements that undergo a change in oxidation number.**
3 **Write the two half reactions:** an oxidation half reaction and a reduction half reaction.
4 **With the exception of hydrogen and oxygen, balance every atom on each side of the equation by using coefficients.**
5 **If hydrogen or oxygen are present in either reactants or products but are not represented on the other side of the equation, then appropriate quantities of H^+, OH^-, or H_2O may be added.** In acidic solutions, H^+ is added as a source for every unit of hydrogen required. In acid solution, water is added for every unit of oxygen required, with $2H^+$ added to the opposite side of the equation. In basic solutions, H_2O is added for every hydrogen required and an OH^- is added to the other side of the equation. For every oxygen atom required, $2OH^-$ are added along with an H_2O on the opposite side of the equation.
6 **Balance each half reaction by adding electrons to one side of the equation to reflect the gain or loss of electrons associated with the reduction or oxidation.**
7 **Check to see that the sums of the charges of the ions and electrons on each side of the equation are equal.**
8 **Multiply each half reaction by a factor so that the total number of electrons gained in one half reaction equals the total number lost in the other half reaction.**
9 **Add the two half reactions and eliminate electrons, water molecules, or ions which are common to each side of the equation and are not needed for the net ionic equation.**

Problem 19.6

Balance the following oxidation-reduction equation using the half reaction method.

$$Ag_2SO_4 + Zn \longrightarrow ZnSO_4 + Ag$$

Solution

The equation in net ionic form is obtained by eliminating the spectator ion, SO_4^{2-} (guideline 1). The oxidation numbers (guideline 2) are calculated and the two half reactions are selected from the unbalanced equation.

$$\overset{(+1)}{Ag^+} + \overset{(0)}{Zn} \longrightarrow \overset{(+2)}{Zn^{2+}} + \overset{(0)}{Ag}$$

$$Ag^+ \longrightarrow Ag \qquad \text{(reduction)}$$
$$Zn \longrightarrow Zn^{2+} \qquad \text{(oxidation)}$$

The equations are balanced with respect to the elements (guideline 4) without changing any coefficients. Furthermore, no hydrogen or oxygen is needed (guideline 5). Thus electrons are added to the equation to indicate the change in oxidation number (guideline 6). Note that the charges also balance (guideline 7).

$$Ag^+ + 1e^- \longrightarrow Ag$$
$$Zn \longrightarrow Zn^{2+} + 2e^-$$

Now the reduction equation is multiplied by 2 so that electrons gained and lost are balanced (guideline 8). Finally the two equations are added (guideline 9).

$$2Ag^+ + 2e^- \longrightarrow 2Ag$$
$$Zn \longrightarrow Zn^{2+} + 2e^-$$
$$\overline{2Ag^+ + Zn + 2e^- \longrightarrow 2Ag + Zn^{2+} + 2e^-}$$
$$2Ag^+ + Zn \longrightarrow 2Ag + Zn^{2+}$$

Problem 19.7 Balance the following equation for a reaction which occurs in acidic solution.

$$H_2S + MnO_4^- + H^+ \longrightarrow Mn^{2+} + S + H_2O$$

Solution

Guideline 1 and 2

$$\overset{(-2)}{H_2S} + \overset{(+7)}{MnO_4^-} + H^+ \longrightarrow \overset{(+2)}{Mn^{2+}} + \overset{(0)}{S} + H_2O$$

Guideline 3 and 4

$$H_2S \longrightarrow S$$
$$MnO_4^- \longrightarrow Mn^{2+}$$

Guideline 5

$$H_2S \longrightarrow S + 2H^+$$
$$8H^+ + MnO_4^- \longrightarrow Mn^{2+} + 4H_2O$$

Guideline 6 and 7

$$H_2S \longrightarrow S + 2H^+ + 2e^-$$
$$5e^- + 8H^+ + MnO_4^- \longrightarrow Mn^{2+} + 4H_2O$$

Guideline 8

$$5(H_2S \longrightarrow S + 2H^+ + 2e^-)$$
$$2(5e^- + 8H^+ + MnO_4^- \longrightarrow Mn^{2+} + 4H_2O)$$

Guideline 9

$$5H_2S + 16H^+ + 2MnO_4^- + 10e^- \longrightarrow 5S + 10H^+ + 2Mn^{2+} + 8H_2O + 10e^-$$

$$5H_2S + 6H^+ + 2MnO_4^- \longrightarrow 5S + 2Mn^{2+} + 8H_2O$$

19.4 GALVANIC CELLS

In Chapter 11 we discussed an order of reactivity of metals called the electromotive series. It was stated that each metal in the series will displace any of those below it in an aqueous solution of the salt. Thus zinc will react with an aqueous solution of copper(II) sulfate to yield copper and zinc sulfate.

$$Zn + CuSO_4 \longrightarrow ZnSO_4 + Cu$$

Now we realize that the replacement reaction is in reality an oxidation-reduction reaction. In terms of a net ionic reaction the equation is:

$$Zn + Cu^{2+} \longrightarrow Zn^{2+} + Cu$$

Although the equation is balanced as it stands, the process of using half reactions or establishing the balance involve the two half reactions.

$$Zn \longrightarrow Zn^{2+} + 2e^-$$
$$2e^- + Cu^{2+} \longrightarrow Cu$$

If the reaction is carried out by immersing the zinc in the copper(II) sulfate, then electron transfer occurs but the electrical energy has not been made to perform useful work. However, this reaction, as well as most oxidation-reduction reactions, can be carried out, with proper experimental modifications, to produce electrical energy. In order to accomplish this, it is necessary to separate the oxidation and reduction half reactions and provide a conductor through which electrons may flow. Transfer of electrons through the conductor allows the oxidation and reduction half reactions to occur and provides a means of utilizing the electric current. **The experimental device that produces electric current from half reactions in solution is called a galvanic cell.**

Galvanic cells produce electric current.

The oxidation of zinc metal by the copper(II) ion, Cu^{2+}, can be carried out in a cell constructed so that the zinc metal and Cu^{2+} are separated (Figure 19.3). A container is partitioned into two compartments by a porous divider that allows solutions placed in the separate compartments to be in contact but retards the rate of mixing. In one compartment a bar of zinc is immersed in a solution of a soluble zinc salt, such as zinc nitrate, $Zn(NO_3)_2$. In the other compartment a bar of copper is immersed in a solution of $Cu(NO_3)_2$. If the two metal bars are then connected to a voltmeter, a potential difference of 1.10 volts is obtained. The significance of this quantity will be discussed in the next section.

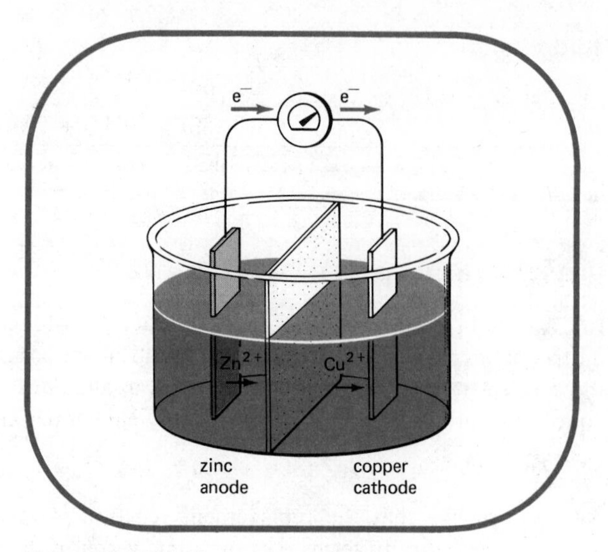

Figure 19.3 The zinc-copper galvanic cell.

The zinc bar slowly dissolves, the copper bar thickens, and the blue copper nitrate solution lightens. Chemical analysis of the zinc nitrate solution indicates that the concentration of zinc ions has increased. The two half reactions are occurring in the two compartments.

Oxidation occurs at the anode.

Reduction occurs at the cathode.

The two metal bars in the galvanic cell are called electrodes. At the zinc electrode, zinc is oxidized and enters the solution as zinc ions: this electrode is called the anode. In the process of being oxidized, the zinc leaves two excess electrons per atom on the zinc bar. Electrons travel to the copper bar where copper ions accept them and are plated out on the copper bar. **The copper bar is the electrode at which reduction occurs and is called the cathode.**

19.5 STANDARD ELECTRODE POTENTIALS

The electromotive series presented in Chapter 11 is based on experimental observations made using galvanic cells. Thus as shown in the preceding section, zinc is more active than copper. However, recall that the reaction in which zinc replaced copper(II) ions in solution occurred with a voltage of 1.10 volts. This voltage is a quantitative measure of the replacement tendency or activity of one metal relative to another.

In Table 19.2 are listed a series of standard electrode potentials ($\mathscr{E}^0$) for the reaction of the oxidized form of a metal. **The standard electrode potentials are the voltages of a galvanic cell in which the metal is compared to a hydrogen electrode. They are assigned positive values if the ion of the metal can gain electrons from hydrogen molecules. Negative values indicate the metal will lose electrons to hydrogen ions in solution to form hydrogen gas.**

Could an electrode other than hydrogen be used to determine standard electrode potentials?

For the measurement of standard electrode potentials, a hydrogen electrode consisting of a strip of platinum (which is resistant to oxidation)

TABLE 19.2 standard electrode potentials of metals

Oxidized form			Reduced form	Volts
Li^+	$+ 1e^-$	$\longrightarrow$	Li	-3.04
K^+	$+ 1e^-$	$\longrightarrow$	K	-2.93
Cs^+	$+ 1e^-$	$\longrightarrow$	Cs	-2.92
Ba^{2+}	$+ 2e^-$	$\longrightarrow$	Ba	-2.90
Sr^{2+}	$+ 2e^-$	$\longrightarrow$	Sr	-2.89
Ca^{2+}	$+ 2e^-$	$\longrightarrow$	Ca	-2.87
Na^+	$+ 1e^-$	$\longrightarrow$	Na	-2.71
Mg^{2+}	$+ 2e^-$	$\longrightarrow$	Mg	-2.36
Al^{3+}	$+ 3e^-$	$\longrightarrow$	Al	-1.66
Mn^{2+}	$+ 2e^-$	$\longrightarrow$	Mn	-1.18
Zn^{2+}	$+ 2e^-$	$\longrightarrow$	Zn	-0.76
Cr^{3+}	$+ 3e^-$	$\longrightarrow$	Cr	-0.74
Fe^{2+}	$+ 2e^-$	$\longrightarrow$	Fe	-0.44
Cd^{2+}	$+ 2e^-$	$\longrightarrow$	Cd	-0.40
Ni^{2+}	$+ 2e^-$	$\longrightarrow$	Ni	-0.25
Sn^{2+}	$+ 2e^-$	$\longrightarrow$	Sn	-0.14
Pb^{2+}	$+ 2e^-$	$\longrightarrow$	Pb	-0.13
$2H_3O^+$	$+ 2e^-$	$\longrightarrow$	$H_2 + 2H_2O$	0.00
Cu^{2+}	$+ 2e^-$	$\longrightarrow$	Cu	$+0.34$
Ag^+	$+ 1e^-$	$\longrightarrow$	Ag	$+0.80$
Hg^{2+}	$+ 2e^-$	$\longrightarrow$	Hg	$+0.85$
Au^{3+}	$+ 3e^-$	$\longrightarrow$	Au	$+1.42$

immersed in a solution of hydronium ions is used. Hydrogen gas is bubbled over the surface of the platinum where the gas and hydronium ion may come in contact. If hydronium ions gained electrons to produce hydrogen, the electrode then would be acting as a cathode. Alternatively, hydrogen could give up electrons and produce hydronium ions, in which case the hydrogen electrode would be acting as an anode. The actual behavior of the hydrogen electrode depends upon the electrode coupled to it. In the presence of zinc and zinc ions, the hydrogen electrode would function as a cathode. The voltage of the hydrogen and zinc galvanic cell is 0.76 V (Figure 19.4). In this reaction, zinc is oxidized.

$$\begin{array}{r} Zn \longrightarrow Zn^{2+} + 2e^- \\ 2H_3O^+ + 2e^- \longrightarrow 2H_2O + H_2 \\ \hline Zn + 2H_3O^+ \longrightarrow 2H_2O + H_2 + Zn^{2+} \quad \mathscr{E}^0 = 0.76 \end{array}$$

Thus the standard electrode potential for the reduction of Zn^{2+} is defined as -0.76 volts. The significance of the negative sign may be understood if you

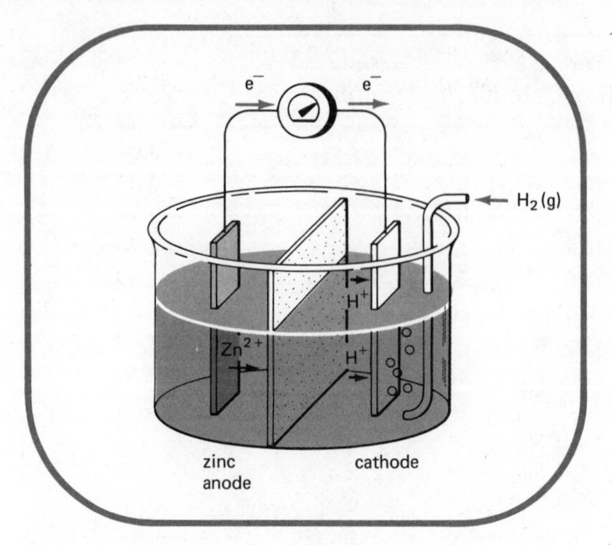

Figure 19.4 The zinc-hydrogen galvanic cell.

consider that the standard electrode potential is for a reaction which is the reverse of that measured in the experiment described.

$$Zn \longrightarrow Zn^{2+} + 2e^- \qquad \text{(observed)}$$
$$2e^- + Zn^{2+} \longrightarrow Zn \qquad \text{(standard electrode potential)}$$

For the observed reaction, the electrons are written on the product side and the reaction is an oxidation. However, all standard electrode potentials are written as reductions. For oxidation reactions, the sign of the related standard electrode potential reaction must be reversed. Therefore, for an oxidation half reaction, the metals above hydrogen in the electromotive series will have a positive oxidation potential while those below hydrogen will have a negative potential.

If copper metal, immersed in a solution containing copper(II) ions, is coupled to the hydrogen electrode, the copper ions will be reduced. This occurs because the electrode potential for the hydrogen electrode is smaller than for the Cu^{2+} ion to copper half reaction. In this case the hydrogen electrode acts as an anode and the voltage for the cell is 0.34 V (Figure 19.5):

$$2e^- + Cu^{2+} \longrightarrow Cu$$
$$2H_2O + H_2 \longrightarrow 2H_3O^+ + 2e^-$$
$$\overline{Cu^{2+} + H_2 + 2H_2O \longrightarrow Cu + 2H_3O^+ \qquad \mathscr{E}^0 = 0.34}$$

Thus the standard electrode potential for the reduction of Cu^{2+} is defined as +0.34 volts.

The reaction of zinc and copper(II) in a galvanic cell was described in the previous section.

$$Zn + Cu^{2+} \longrightarrow Zn^{2+} + Cu$$

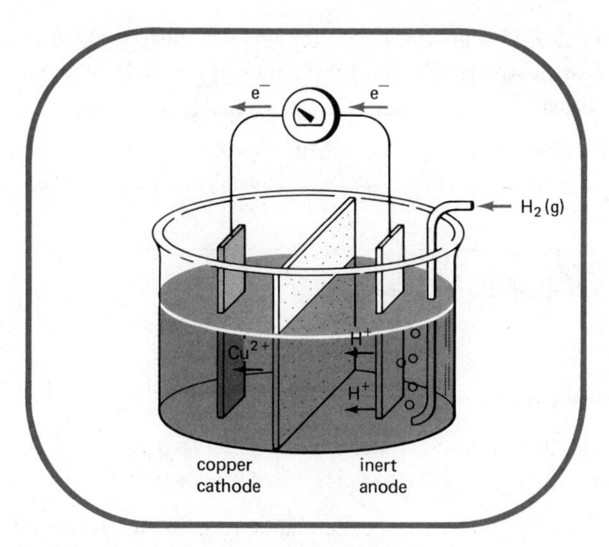

Figure 19.5 The copper-hydrogen galvanic cell.

The voltage observed was 1.10 volts. This voltage is equal to the difference in the abilities of Cu^{2+} and Zn^{2+} to gain electrons in the reactions given as reductions.

$$Cu^{2+} + 2e^- \longrightarrow Cu \qquad \mathscr{E}^0 = +0.34\ V$$
$$Zn^{2+} + 2e^- \longrightarrow Zn \qquad \mathscr{E}^0 = -0.76\ V$$

Copper(II) has a standard electrode potential which is 1.10 volts more positive than zinc(II).

The direction of any oxidation-reduction reaction may be predicted from the list of standard electrode potentials. The reduction half reaction with the largest positive potential will occur in combination with an oxidation reaction of a metal listed above it. The voltage of a cell in which the reaction occurs can be obtained by summing the potentials. Remember that the potential of the oxidation half reaction is obtained by reversing the sign of the standard potential.

$$\begin{array}{rl}
Cu^{2+} + 2e^- \longrightarrow Cu & \mathscr{E}^0 = +0.34\ V \\
Zn \longrightarrow Zn^{2+} + 2e^- & \mathscr{E}^0 = +0.76\ V \\
\hline
Cu^{2+} + Zn \longrightarrow Cu + Zn^{2+} & \mathscr{E}^0 = +1.10\ V
\end{array}$$

Problem 19.8 Will Mg reduce Zn^{2+}?

Solution The electrode potentials to consider are:

$$2e^- + Mg^{2+} \longrightarrow Mg \qquad \mathscr{E}^0 = -2.37\ V$$
$$2e^- + Zn^{2+} \longrightarrow Zn \qquad \mathscr{E}^0 = -0.76\ V$$

Thus Zn^{2+} has the larger tendency to accept electrons and become reduced. The electrons released by magnesium, which acts as the reducing agent, will reduce Zn^{2+}.

$$Mg + Zn^{2+} \longrightarrow Mg^{2+} + Zn$$

Problem 19.9 Will the following reaction occur?

$$Cu^{2+} + Hg \longrightarrow Cu + Hg^{2+}$$

Solution The electrode potentials to consider are:

$$2e^- + Cu^{2+} \longrightarrow Cu \qquad \mathscr{E}^0 = +0.34\ V$$
$$2e^- + Hg^{2+} \longrightarrow Hg \qquad \mathscr{E}^0 = +0.85\ V$$

Thus Hg^{2+} has a larger tendency to gain electrons and become reduced than does Cu^{2+}. The reaction given involves the reduction of Cu^{2+} and therefore will not occur.

Problem 19.10 What is the voltage of a cell so constructed that Mg reduces Pb^{2+}?

Solution The half reactions for the reduction and the oxidation are:

$$Mg \longrightarrow Mg^{2+} + 2e^- \qquad \mathscr{E}^0 = +2.37\ V$$
$$2e^- + Pb^{2+} \longrightarrow Pb \qquad \mathscr{E}^0 = -0.13\ V$$
$$\mathscr{E}^0 = +2.24\ V$$

The voltage is +2.24 volts. Remember that the potential for the oxidation half reaction is obtained by reversing the sign of the standard electrode potential.

An activity series for the nonmetallic halogens was given in Chapter 11. Fluorine is more active than chlorine and will replace the chloride ion from an aqueous solution.

$$F_2 + 2Cl^- \longrightarrow 2F^- + Cl_2$$

Fluorine will also replace bromide and iodide. Chlorine will replace bromide and iodide, while bromine will replace iodide. A quantitative measure of these tendencies is given by the standard electrode potentials listed in Table 19.3. Fluorine has the highest positive electrode potential followed by $Cl_2 >$ $Br_2 > I_2$. Thus the tendency to gain electrons is $F_2 > Cl_2 > Br_2 > I_2$. The

TABLE 19.3

standard electrode potentials for nonmetals

Oxidized form		Reduced form	Volts
$I_2 + 2e^-$	$\longrightarrow$	$2I^-$	+0.54
$Br_2 + 2e^-$	$\longrightarrow$	$2Br^-$	+1.07
$Cl_2 + 2e^-$	$\longrightarrow$	$2Cl^-$	+1.36
$F_2 + 2e^-$	$\longrightarrow$	$2F^-$	+2.87

voltage derived from the reaction of chlorine and bromide is calculated by summing the proper electrode potentials.

$$\begin{array}{lr} Cl_2 + 2e^- \longrightarrow 2Cl^- & 1.36\ V \\ 2Br^- \longrightarrow Br_2 + 2e^- & -1.07\ V \\ \hline Cl_2 + 2Br^- \longrightarrow 2Cl^- + Br_2 & 0.29\ V \end{array}$$

19.6 ELECTROLYSIS

The application of an external source of voltage and electric power to produce chemical changes is called electrolysis. Such processes are extremely important in the industrial preparation of many metals and nonmetals.

Many ions of metals are difficult to reduce because there are no readily available chemical reducing agents for them. As an example, magnesium ions could be reduced by sodium or potassium metal, which have higher electrode potentials, but such a reduction process involves many difficulties in the handling of the very active metals. The commercial preparation of magnesium involves electrolytic reduction of Mg^{2+} at a cathode and electrolytic oxidation of chloride ion at the anode in the molten salt $MgCl_2$ (Figure 19.6):

$$\begin{array}{ll} \text{cathode} & Mg^{2+} + 2e^- \longrightarrow Mg \\ \text{anode} & 2Cl^- \longrightarrow Cl_2 + 2e^- \end{array}$$

The net reaction decomposes the compound $MgCl_2$ into its constituent elements. It is the reverse of the process of compound formation, which in this case is spontaneous. The electrolytic reaction proceeds because a voltage source forces it to occur.

Electric voltage can be used to cause chemical reactions to occur.

Electrolytic processes are used to obtain pure metals from less pure samples. If two strips of copper are connected to the terminals of an appropriate voltage source and are placed in a solution of copper sulfate ($CuSO_4$), copper can be made to dissolve from one electrode and become deposited on the other. The copper that is removed from the anode is changed into Cu^{2+} ions, which then are reduced at the cathode to elemental copper.

$$Cu \longrightarrow Cu^{2+} + 2e^-$$
$$Cu^{2+} + 2e^- \longrightarrow Cu$$

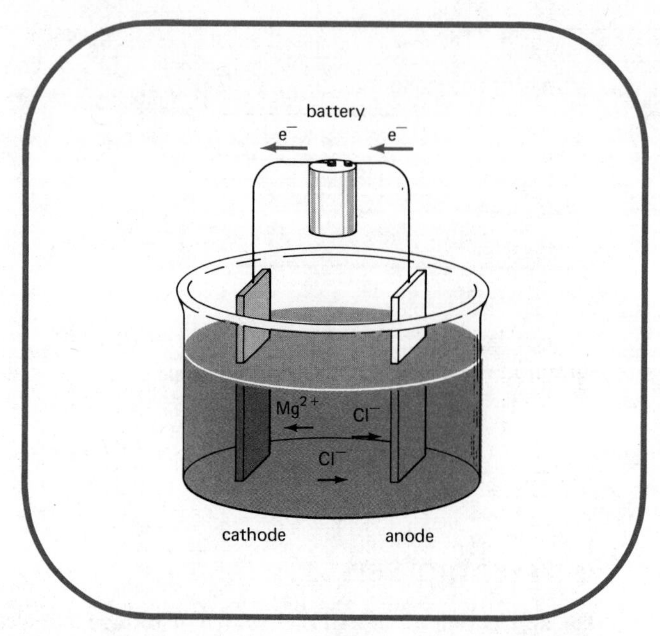

Figure 19.6 The electrolysis of molten magnesium chloride.

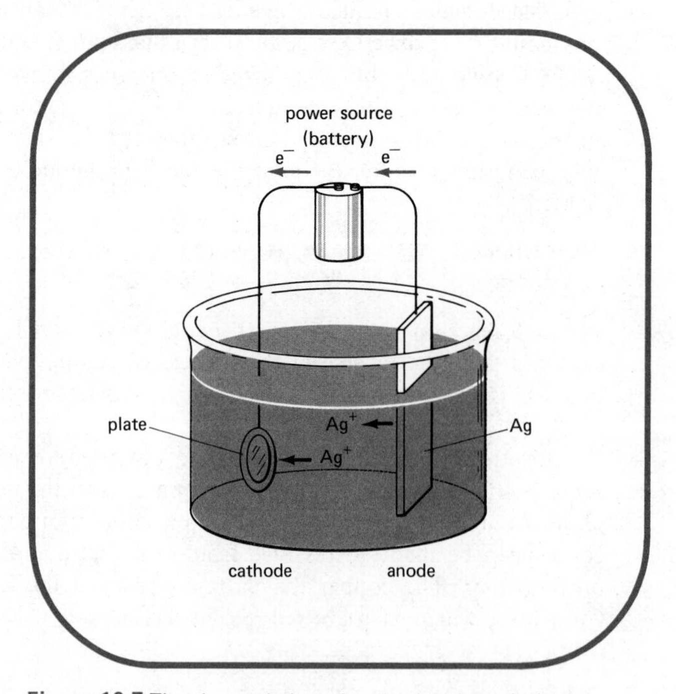

Figure 19.7 The electroplating of a dish by electrolysis.

If the anode is 99.0 percent "pure" copper, the copper deposited at the cathode is approximately 99.98 percent "pure" copper.

Electrolysis is also an important process in industry where metals are coated or plated with a thin layer of a second metal. The second metal may be plated onto the surface for either decorative purposes, as in the case of silver plated utensils, or for protective purposes, as in the case of nickel. In either case, the object is immersed in a solution containing the ions of the metal to be used in plating. Then a bar of the free metal to be used for plating is also immersed in the solution. An electric current is passed through the solution through a wire connecting the object and the pure metal (Figure 19.7). Due to the direction of the current flow the object is the cathode and the metal bar is the anode. When plating an object with silver, the silver at the anode is oxidized and enters solution as Ag^+.

$$Ag \longrightarrow Ag^+ + e^- \qquad \text{(at anode)}$$

The Ag^+ in solution becomes deposited at the cathode as silver and the object is coated with silver.

$$e^- + Ag^+ \longrightarrow Ag \qquad \text{(at cathode)}$$

SUMMARY

__Oxidation__ is the loss of electrons by a substance. __Reduction__ is the gain of electrons by a substance. The substance reduced is also called the __oxidizing agent__. The substance oxidized is called the __reducing agent__.

Oxidation-reduction reactions may be balanced by using oxidation numbers or half reactions. A __half reaction__ is an artificial reaction which cannot occur alone. It may be an oxidation half reaction or a reduction half reaction.

__Galvanic cells__ produce electricity from half reactions in solution. Oxidation occurs at the __anode__ while reduction occurs at the __cathode.__

The __standard electrode potential__ of a metal is the voltage of a galvanic cell in which the metal is compared to a hydrogen electrode. It is positive if the metal ion is reduced to the metal by hydrogen gas. It is negative if the metal is oxidized by the hydrogen ion. Standard electrode potentials and their reverse may be added to obtain the voltage of a galvanic cell.

__Electrolysis__ is the application of an external voltage to produce chemical changes. Electrolysis may be used to reduce metal ions and to purify metals and electroplate metals.

QUESTIONS AND PROBLEMS

Definitions

1. Define each of the following terms.
(a) oxidation **(b)** reduction
(c) oxidizing agent **(d)** reducing agent

(e) anode
(f) cathode
(g) half reaction
(h) standard electrode potential
(i) galvanic cell
(j) electrolysis

2. Why is the standard electrode potential of the hydrogen electrode zero?

3. What occurs at the anode and cathode in a galvanic cell?

Oxidation numbers

4. What is the oxidation number of the indicated element in each of the following substances?
(a) H in $HClO_4$
(b) O in $HClO_4$
(c) Cl in $HClO_4$
(d) S in SO_2
(e) S in H_2S
(f) S in H_2SO_4
(g) N in NO_2
(h) N in HNO_3
(i) Mn in $KMnO_4$
(j) As in H_3AsO_4
(k) Cr in Na_2CrO_4
(l) Mn in MnO_2
(m) Fe in $FeCl_3$
(n) Cr in $CrCl_3$

Oxidizing agents and reducing agents

5. Identify the oxidizing agent and reducing agent in each of the following equations.
(a) $2Al + 3CoCl_2 \rightarrow 2AlCl_3 + 3Co$
(b) $Cu + Br_2 \rightarrow CuBr_2$
(c) $3H_2SO_3 + 2HNO_3 \rightarrow 2NO + H_2O + 3H_2SO_4$
(d) $2S + 3O_2 \rightarrow 2SO_3$
(e) $N_2 + 3H_2 \rightarrow 2NH_3$
(f) $Cl_2 + 2NaBr \rightarrow 2NaCl + Br_2$
(g) $WO_3 + 3H_2 \rightarrow W + 3H_2O$
(h) $SnO_2 + 2C \rightarrow Sn + 2CO$

Balancing equations by oxidation numbers

6. Balance each of the following reactions using the oxidation number method.
(a) $S + O_2 \rightarrow SO_3$
(b) $Fe + Cl_2 \rightarrow FeCl_3$
(c) $ZnS + O_2 \rightarrow ZnO + SO_2$
(d) $CuO + NH_3 \rightarrow Cu + N_2 + H_2O$
(e) $H_2S + Br_2 + H_2O \rightarrow H_2SO_4 + HBr$
(f) $HNO_3 + I_2 \rightarrow NO_2 + H_2O + HIO_3$

7. Balance each of the following reactions using the oxidation number method.
(a) $NaIO_4 + NaI + HCl \rightarrow NaCl + I_2 + H_2O$
(b) $KClO_3 \rightarrow KCl + O_2$
(c) $Cl_2 + KOH \rightarrow KClO_3 + KCl + H_2O$

(d) $KI + H_2SO_4 \rightarrow K_2SO_4 + I_2 + H_2S + H_2O$
(e) $Cu + HNO_3 \rightarrow Cu(NO_3)_2 + NO + H_2O$
(f) $NaOCl + Na_3AsO_3 \rightarrow NaCl + Na_3AsO_4$
(g) $Cu_2S + HNO_3 \rightarrow Cu(NO_3)_2 + NO_2 + S + H_2O$

Balancing equations using half reactions

8. Balance each of the following using the half reaction method.
(a) $Fe^{3+} + Sn^{2+} \rightarrow Fe^{2+} + Sn^{4+}$
(b) $Cl_2 + I^- \rightarrow I_2 + Cl^-$
(c) $Cu + H^+ + NO_3^- \rightarrow Cu^{2+} + NO + H_2O$
(d) $Cr_2O_7^{2-} + Fe^{2+} + H^+ \rightarrow Cr^{3+} + Fe^{3+} + H_2O$
(e) $I^- + MnO_4^- + H^+ \rightarrow Mn^{2+} + I_2 + H_2O$
(f) $H^+Cr_2O_7^{2-} + I_2 \rightarrow Cr^{3+} + IO_3^- H_2O$
(g) $ClO_3^- + H^+ + Fe^{2+} \rightarrow Fe^{3+} + Cl^- + H_2O$

Standard electrode potentials

9. Using the list of standard electrode potentials, indicate which of each of the following has the greater tendency to be reduced.
(a) Zn^{2+} and Cu^{2+} (b) Mg^{2+} and Zn^{2+}
(c) Cd^{2+} and Hg^{2+} (d) Ag^+ and Hg^{2+}
(e) Br_2 and I_2

10. Using the list of standard electrode potentials, indicate which of each of the following has the greater tendency to be oxidized.
(a) Li or K (b) Na or Zn
(c) Zn or Cu (d) Ni or Pb
(e) Cu or Hg (f) I^- or Br^-

Galvanic cells

11. Using the list of standard electrode potentials, predict which of the following reactions will proceed as written.
(a) $Ni + 2Ag^+ \rightarrow Ni^{2+} + 2Ag$
(b) $3Ba + 2Al^{3+} \rightarrow 3Ba^{2+} + 2Al$
(c) $Pb + Hg^{2+} \rightarrow Pb^{2+} + Hg$
(d) $Zn^{2+} + Fe \rightarrow Zn + Fe^{2+}$
(e) $Cl_2 + 2I^- \rightarrow I_2 + 2Cl^-$

12. Calculate the voltage that can be produced for each of the following reactions.
(a) $Mg + Zn^{2+} \rightarrow Mg^{2+} + Zn$
(b) $Mg + Fe^{2+} \rightarrow Mg^{2+} + Fe$
(c) $Zn + Pb^{2+} \rightarrow Zn^{2+} + Pb$
(d) $Cu + 2Ag^+ \rightarrow Cu^{2+} + 2Ag$
(e) $Br_2 + 2I^- \rightarrow 2Br^- + I_2$

13. What process occurs at the cathode of a galvanic cell?

Electrolysis cell

14. In your own words, describe how metals may be purified by electrolysis.

15. Describe the use of an electrolysis cell in electroplating.

16. What is the difference between a galvanic cell and an electrolysis cell?

17. Copper and silver are two examples of metals which can be used in electroplating. Magnesium and aluminum cannot be used in electroplating. What reason can you suggest for this difference?

20 NUCLEAR CHEMISTRY

LEARNING OBJECTIVES

When you finish this chapter you should be able to:

(a) relate the neutron to proton ratio and nuclear stability of isotopes.

(b) distinguish between types of radiation.

(c) describe the units of measurement of radiation.

(d) set up balanced nuclear equations for nuclear reactions.

(e) calculate the fraction of a radioisotope remaining in a decay reaction given its half-life.

(f) demonstrate the processes involved in the transmutation of elements.

(g) interpret nuclear reactions in terms of the nuclear binding energy.

(h) compare nuclear fission and nuclear fusion.

(i) appraise the potential of nuclear energy.

(j) match the following terms with their corresponding definitions.

alpha rays
beta rays
curie
cyclotron
fission
fusion
gamma rays
half-life
linear accelerator
nuclear binding energy
rad
radiation
radioactive decay
radioactivity
radioisotope
rem
roentgen
transmutation
zone of stability

Up to this point, the elements and their properties have been discussed only in terms of the behavior of electrons in atoms and molecules; the nucleus has been mentioned in a purely cursory fashion. Although the mass of an atom is concentrated largely in the nucleus, the chemical reactivity of the electrons does not seem to be significantly altered by the mass of the nucleus. Isotopes of an element undergo the same types of chemical reactions at closely similar rates. Because of their charge, the number of protons present in the nucleus affects ordinary chemical reactions involving the electrons about the nucleus, but the only contribution of the uncharged neutrons is to increase the atomic mass. However, matter undergoes many important transformations that involve the nucleus, and the neutrons play an important role in these processes.

The energy required to make a stable nucleus undergo a change is extremely high and cannot be achieved by ordinary chemical reactions. Furthermore, when unstable nuclei of radioactive isotopes are allowed to react in sufficiently high concentration, there results a rapid chain reaction which is known as an atomic explosion. The energy released in atomic explosions is far beyond the comprehension of anyone who has not actually viewed such an event.

Today's generation of college students has grown up in a nuclear age in which the chemistry of the nucleus is very much a part of modern life. On the one hand we live in fear of a nuclear holocaust and on the other, we enjoy many benefits of radioisotopes.

The atomic devices called atom bombs which were used to devastate two Japanese cities near the end of World War II were miniscule in their power compared to the hydrogen bombs of today. The continuing attempt of the major powers to control their respective arms and to prevent the further spread of nuclear devices to smaller countries is a frequent news item.

Nuclear energy for peaceful uses, such as the generation of power, is a continuing story. The world is limited in its fossil fuels and nuclear energy is regarded by some as the only reasonable alternative in the near future. However, there is much concern about the radioactivity hazards and environmental side effects as the number of nuclear electric generating plants increases.

One major benefit to man from nuclear chemistry is the development of radioisotopes for many diverse purposes other than destruction or energy production. Radioisotopes have found many uses in both medicine and in industrial applications. In this chapter the principles of nuclear chemistry are presented. In addition, the practical aspects of nuclear processes are outlined.

20.1 THE NUCLEUS

The protons and neutrons contained in the nucleus of atoms are collectively called nucleons. The number of protons in a nucleus is its atomic number Z, and the total number of nucleons is its mass number A. The number of neutrons is A − Z. The chemical symbol of a given isotope consists of the elemental letter symbol with a superscript mass number and a subscript atomic number.

$${}^{A}_{Z}X$$

The most striking feature of the nucleus is its ability to maintain an aggregation of positively charged particles in a region that is about 10^{-13} cm in radius. The protons would be expected to fly apart as a result of electrostatic repulsion. Indeed, some nuclei are unstable and do undergo processes

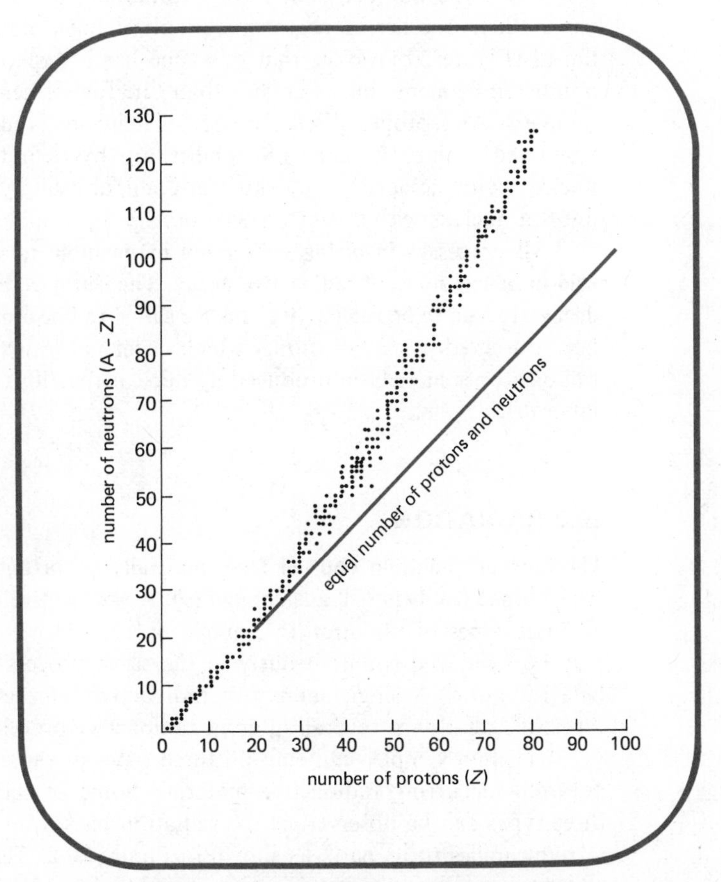

Figure 20.1 The zone of stability. The points represent the nuclear content of stable isotopes. Isotopes which are radioactive are not shown as they lie outside the zone defined by the stable isotopes.

in which other nuclei and atomic particles are produced. The difference between stable and unstable nuclei depends on the number of neutrons present. The neutrons play an important role in binding the nucleus.

For any element, a particular ratio of neutrons-to-protons is required for stability of the nucleus. If the number of neutrons is altered to either increase or decrease that ratio, the nucleus becomes less stable. For the elements of low atomic number, the ratio of neutrons-to-protons in a stable nucleus is approximately 1:1, as in $^{4}_{2}He$, $^{12}_{6}C$, $^{16}_{8}O$, and $^{20}_{10}Ne$. Addition or removal of one neutron from the nucleus of such elements usually leads to unstable isotopes. For elements of higher atomic number, the neutron-to-proton ratio which is required for nuclear stability exceeds one. The required ratio increases with increasing atomic number. Thus the stable isotopes $^{56}_{26}Fe$, $^{79}_{35}Br$, $^{120}_{50}Sn$, $^{179}_{79}Au$ have neutron-to-proton ratios 1.15, 1.26, 1.40, and 1.51, respectively.

When the number of neutrons contained in stable isotopes is plotted versus their number of protons, a graph giving the zone of stability is obtained (Figure 20.1). Note that this zone has a slope of one for low atomic numbered elements but is greater than one for elements with higher atomic numbers. All isotopes whose number of neutrons is more or less than those contained within the zone of stability are by definition unstable. Such a nucleus could achieve a stable nuclear configuration by any process that produces a nucleus with a neutron-to-proton ratio inside the zone of stability.

All processes involving conversion of unstable to more stable nuclei are said to be examples of radioactive decay. The isotopes that undergo radioactive decay are said to be radioactive and are called radioisotopes. Radioactivity has been observed in many isotopes which occur naturally on earth. In addition, radioisotopes have been produced by nuclear reactions carried out in nuclear laboratories.

20.2 RADIATION

$\left.\begin{matrix}\alpha\\ \beta\\ \gamma\end{matrix}\right\}$ radiation

The nuclear radiation emitted from naturally occurring elements is of three types: alpha (α), beta (β), and gamma (γ). A radioactive isotope does not emit all three types of radiation in a single process. However, gamma radiation may be produced simultaneously in the same process with either alpha or beta radiation. A single sample of radioactive material may emit all three types of radiation if several different radioactive processes are occurring.

Uranium samples can emit all three types of radiation characteristic of naturally occurring radioactive material. Some of the differences of these three types can be observed as the radiation passes through an electric field at right angles to its path, as shown in Figure 20.2. This field deflects alpha and beta rays in opposite directions but does not deflect gamma rays. The direction and extent of the deflection of the rays can be determined by their point of collision on a photographic plate. The alpha particles are deflected

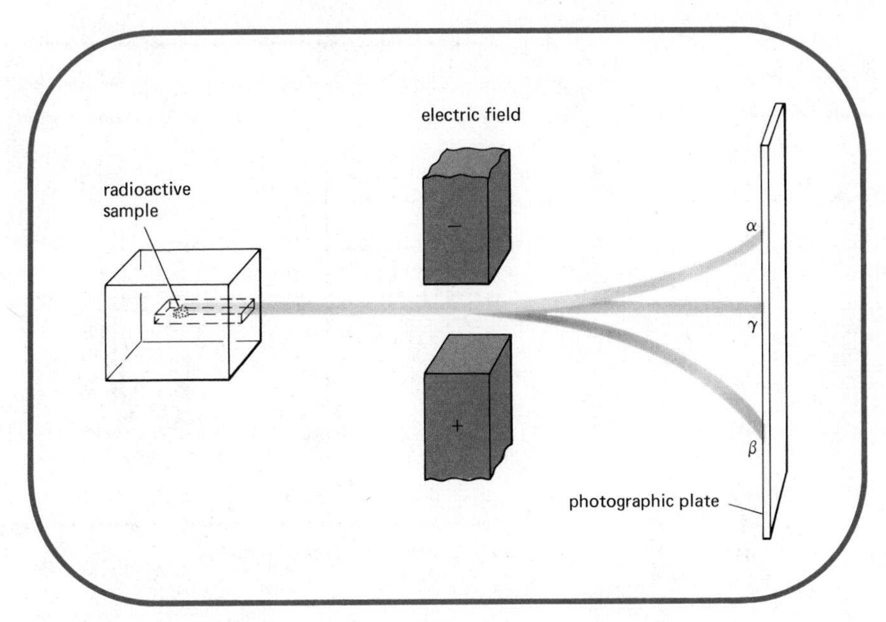

Figure 20.2 Effect of electric field on radiation.

toward the negative pole and are therefore positive, while the beta particles are deflected toward the positive pole and are negative. Note that the beta particles are deflected more than the alpha particles.

Alpha particles

An alpha particle is a helium nucleus, which consists of two protons and two neutrons and may be represented as ${}^{4}_{2}He^{2+}$. However, the 2+ charge is ordinarily omitted because the prime concern of nuclear chemistry is with the nucleus of atoms and not with their electron shell. Alpha particles have no electrons, so they remove electrons from other atoms in the atmosphere and form neutral helium atoms.

When alpha particles are emitted from a nucleus of an atom, they travel at approximately 0.1 times the speed of light, or 18,600 miles per second. These rapidly traveling particles have little penetrating power and can be stopped by this sheet of paper (Figure 20.3). A 0.05-mm layer of dead cells on the surface of skin will stop alpha particles. Thus not even skin burns can result from alpha radiation. However, if radioactive dust particles get inside the body, the alpha particles can affect normal cells and biological damage will occur.

The highly charged alpha particles cause ionization of gases when they accept electrons to form the stable helium atom. This ionization phenomenon is used in the Geiger counter to detect radiation.

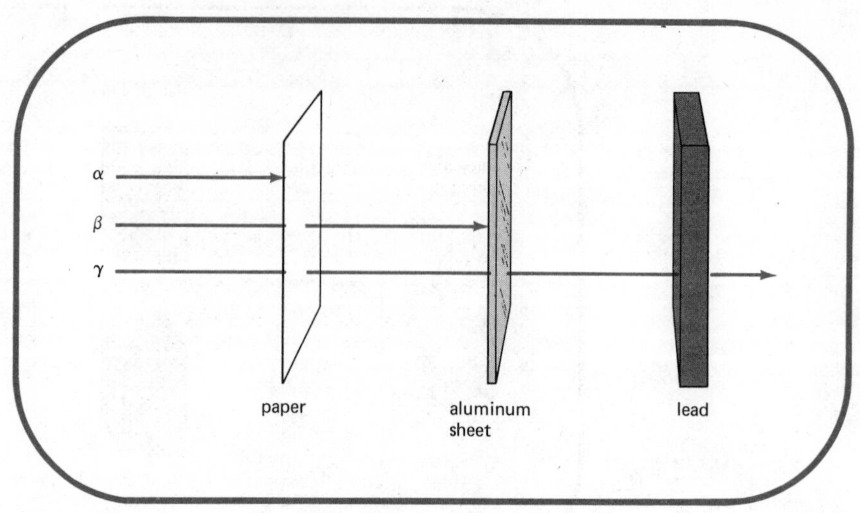

Figure 20.3 Penetrating ability of radiation.

Beta particles

Beta particles are high energy electrons emitted from a radioactive nucleus. Although the electrons do not exist as such in the nucleus, they are produced during a process in which a neutron is transformed into a proton and a beta particle.

neutron $\longrightarrow$ proton + electron

$_{-1}^{0}e$ electron

A beta particle is written as $_{-1}^{0}e$ in balancing nuclear equations. The superscript corresponds to the position of the atomic mass in elements. Since the mass of an electron is very much smaller than any element, it is assigned a value of zero. The -1 subscript occupies the position of the atomic number of an element. Since the atomic number is equal to the charge on the nucleus, the -1 value for the electron emphasizes its charge for electrical balancing purposes.

Beta particles are several thousand times smaller than alpha particles and travel at 0.9 times the speed of light, or about 165,000 miles per second. While beta particles have greater penetrating power than alpha particles, they are stopped by thin sheets of metal (Figure 20.3). Beta particles can easily penetrate into living cells to the depth of about 4 mm. While radiation burns result, the vital internal organs are not affected unless a radioactive source is ingested.

Gamma rays

Gamma rays do not consist of particles but are high energy radiation similar to but of higher frequency than X rays. When gamma rays are emitted by a nucleus, they travel at the speed of light and are so very energetic that they have high penetrating power. Even 50 cm of tissue will only reduce the intensity of gamma radiation by 10 percent. A 5-cm sheet of lead will not stop all gamma rays (Figure 20.3).

Artificial radioactivity

Nuclear chemists have produced many isotopes of naturally occurring elements and 14 new elements. These artificially produced radioactive materials behave similarly to naturally occurring radioactive substances by emitting alpha particles, beta particles, and gamma rays. However, in addition, some emit positrons and neutrons.

Positrons

$^{0}_{1}e$ positron

Positrons might be called positive electrons; they have the same mass as an electron but have a unit positive charge. Positrons are written as $^{0}_{1}e$, where the zero indicates that it has negligible mass and the 1 represents the charge. In the nucleus, the positron is produced when a proton is converted into a neutron.

$$\text{proton} \longrightarrow \text{neutron} + \text{positron}$$

Positrons are of high velocity but have low penetrating power.

Neutrons

Neutrons may be emitted with a variety of velocities ranging from a few thousands miles per second to nearly the speed of light. They have no ionizing effect on gases because of their electrical neutrality. The penetrating power of neutrons is low.

Units of radiation

Radiation has been measured with two different systems. One system measures the intensity of the radiation emitted from the radioactive source. The other measures the radiation absorbed by matter, such as air or tissue.

The most common unit of radiation intensity is the curie. **A curie corresponds to that amount of radioactive material that produces 3.7×10^{10} nuclear disintegrations per second.** This number was chosen because it is the number of disintegrations that 1 g of radium undergoes in 1 second. The mass of other substances that produce 1 curie may be more or less than 1 g. The rating gives no indication of the mass of the radioactive substance.

The concern of society is not with the unit of radioactive intensity but with the absorption dosage. When radiation strikes atoms in any absorbing material, electrons are ejected and a charge imbalance results. In complex molecules, the presence of a charge may result in molecular disruptions or uncharacteristic chemical reactions. Such occurrences in living systems can result in loss of hair, skin burns, cancer, or death. If the absorbing molecule is DNA, then genetic codes can be altered and biological mutations can result. Tissues that reproduce at a rapid rate, such as bone marrow, lymph nodes, and embryonic tissue, are most sensitive to radiation damage.

Three common units of dosage are the roentgen, the rad, and the rem. **The roentgen is that dose of X rays or gamma rays that produces 2.1×10^9 units of electric charge in 1 cm^3 of dry air. The rad is the absorption of 100 ergs of energy per gram of absorbing tissue.** Although the roentgen and the rad are defined very differently, they are numerically quite similar. One roentgen of X ray or gamma ray causes 0.97 rad in muscle tissue. In bone, 1 roentgen delivers 0.92 rad.

The rem is defined as the product of rads and the relative biological effectiveness (RBE). Highly charged and heavy particles such as alpha particles cause more ionization in matter than do the lighter and singly charged beta particles. Accordingly the relative biological effectiveness of beta particles is set at 1 and that of alpha particles at 10. For beta particles, rads are numerically equal to rems. For an alpha particle, 0.1 rad is equal to 1 rem.

20.3 BALANCING NUCLEAR EQUATIONS

All of the symbols of the particles produced in nuclear reactions have been given. These symbols along with those for isotopes are used in balancing nuclear equations.

Why aren't electronic configurations considered in balancing nuclear reactions?

In order to write a balanced nuclear equation, the sum of the mass numbers of the reactants must equal the sum of the mass numbers of the products. Furthermore, the sum of the charges and/or the atomic numbers must be equal for reactants and products.

In every nuclear equation there is a small loss of mass, but it is not indicated in the balanced equation because the quantity is a very small fraction of an atomic mass unit. However, this small mass accounts for the large amounts of energy liberated according to the Einstein equation, $E = mc^2$ (Section 20.6).

Gamma rays do not have any mass or charge and do not appear in balanced nuclear equations. However, they often do accompany the loss of another particle.

Several examples of approaches to balancing nuclear equations follow. The only tools necessary for balancing are simple arithmetic and a list of elements and their atomic numbers.

Problem 20.1 The isotope polonium 212 undergoes alpha decay to yield one alpha particle per atom and a single element. What is the element?

Solution

First, write the symbol for polonium 212 to the left of an arrow using the information from a list of elements that its atomic number is 84. Place ${}^{4}_{2}\mathrm{He}$ on the right of the arrow along with ${}^{A}_{Z}X$ to represent the unknown element.

$$ {}^{212}_{84}\mathrm{Po} \longrightarrow {}^{4}_{2}\mathrm{He} + {}^{A}_{Z}X $$

The sum of the atomic number of the products must equal that of the single reactant, polonium.

$$ 84 = 2 + Z $$
$$ 82 = Z $$

Similarly, the sum of the mass numbers of the products must be equal to the mass of polonium.

$$ 212 = 4 + A $$
$$ 208 = A $$

Now it is established that the element has an atomic number of 82 and a mass number of 208. In the list of elements one finds that lead has an atomic number of 82, and therefore the element is lead 208.

$$ {}^{A}_{Z}X = {}^{208}_{82}\mathrm{Pb} $$

The balanced equation is:

$$ {}^{212}_{84}\mathrm{Po} \longrightarrow {}^{4}_{2}\mathrm{He} + {}^{208}_{82}\mathrm{Pb} $$

Problem 20.2

The isotope carbon 14 is unstable and emits a beta particle per atom. What single element is produced?

Solution

In a manner similar to Problem 20.2, it is first necessary to write the unbalanced equation.

$$ {}^{14}_{6}\mathrm{C} \longrightarrow {}^{0}_{-1}e + {}^{A}_{Z}X $$

The atomic number must be given by the equation:

$$ 6 = -1 + Z $$
$$ 7 = Z $$

The mass number of the element is equal to the mass number of carbon, as the beta particle has a negligible mass.

$$ 14 = 0 + A $$
$$ 14 = A $$

The element is nitrogen as its atomic number is 7.

$$ {}^{14}_{7}X = {}^{14}_{7}\mathrm{N} $$

The balanced equation is:

$$^{14}_{6}C \longrightarrow ^{14}_{7}N + ^{0}_{-1}e$$

Problem 20.3 Phosphorus 30 decays by positron emission. What element is produced?

Solution The unbalanced equation is:

$$^{30}_{15}P \longrightarrow ^{0}_{1}e + ^{A}_{Z}X$$

The atomic number of the element is 14.

$$15 = 1 + Z$$
$$14 = Z$$

The atomic mass is equal to that of phosphorus 30 as the positron has a negligible mass.

$$30 = 0 + A$$
$$30 = A$$

The element is silicon 30.

$$^{30}_{14}X = ^{30}_{14}Si$$

$$^{30}_{15}P = ^{30}_{14}Si + ^{0}_{1}e$$

20.4 HALF-LIVES

It is a characteristic of all radioactive decay reactions that the rate at which nuclear disintegration occurs is governed by a statistical process in which one-half of the nuclei will decay in a given unit of time. **The time required for the decay of one-half of a given number of nuclei is called a half-life.** The half-life of a given nucleus is independent of the number of atoms considered. Thus 100 g of an element might decay at a rate such that its half-life is 5 days. At the end of 5 days, 50 g of the element will remain. In the next 5-day period, one-half of the remaining 50 g will decay. At the end of 50 days, or 10 half-lives, approximately 0.1 g will remain.

$^{14}_{6}C$ 5570 years

One of the best-known uses of the half-lives of radioisotopes is the carbon 14 dating method. The age of ancient materials made of plant or animal matter can be established on the basis of their $^{14}_{6}C$ content. Carbon dioxide in the atmosphere consists mainly of carbon 12 with trace amounts of carbon 14, which is radioactive and decays. However, the concentration of carbon 14 does not decrease because it is constantly being formed in the atmosphere from the action of cosmic rays on nitrogen, $^{14}_{7}N$. All plants absorb carbon dioxide from the atmosphere, and as long as the plant is living, the amount of carbon 14 incorporated into the molecules it produces and uses will be a constant fraction of the amount of carbon present. When the plant dies the carbon compounds in its cells are no longer interacting with the CO_2 of the atmosphere. The amount of carbon 14 in these plants therefore diminishes with time as the carbon 14 decays. Since it is known that the half-life of carbon 14 is 5570 years it is possible to obtain a good measure of the age of an object by determining the amount of radioactive carbon remaining in it. A wooden dish that contains only 25 percent of the carbon 14 that trees have today is therefore approximately 11,000 years old. The carbon 14 dating technique cannot be used accurately for the dating of objects that are older than 50,000 years. After many half-lives have elapsed, only a small amount of carbon 14 remains. It is difficult to measure accurately these amounts.

Because of the time limitation of the carbon 14 dating method, another dating method has been developed involving potassium, $^{40}_{19}K$, and argon, $^{40}_{18}Ar$. The half-life of potassium 40 to produce argon 40 is 1.3×10^9 years. By determining the amount of argon 40 in a potassium-bearing mineral, the approximate data of origin of the mineral can be estimated. Objects as old as 2 million years have been dated by this method, as in the case of moon rocks. The amount of argon 40 increases with the age of the potassium-containing mineral. Hence the measurement becomes more accurate the older the object. However, it is limited to potassium-bearing minerals which are less abundant than carbon-containing materials.

Problem 20.4

The half-life of strontium 90 which was produced by atmospheric nuclear tests is 28 years. How much strontium 90 will remain from a test in 1960 by the year 2016? Assume 1.00 gram was formed.

Solution

After each half-life, the amount remaining will be one-half of the amount remaining from the previous half-life. At 28 years there will be $\frac{1}{2}$ (1.00 g) = 0.50 g. At 56 years, there will be $\frac{1}{2}$ (0.50 g) = 0.25 g. The year 2016 is 56 years after the test. Therefore, 0.25 g will remain.

20.5 TRANSMUTATION OF ELEMENTS

The alchemist could not have succeeded in transmuting elements by ordinary chemical reactions.

Transmutation means the changing of one element into another. This process can be achieved only by nuclear reactions. Elements of atomic number greater than 92 are rare in the earth's crust. **These elements, which are called transuranic since they have atomic numbers greater than uranium, have been synthesized in the laboratory using reactors and accelerators.** The first transuranic element, neptunium 239, $^{239}_{93}Np$, was synthesized in 1940 by an American group of scientists who bombarded the nucleus $^{238}_{92}U$ with deuterons

$^{2}_{1}H$ of very high energy. The first product was an isotope, uranium 239, which then underwent beta emission to produce $^{239}_{93}Np$:

$$^{238}_{92}U + ^{2}_{1}H \longrightarrow ^{239}_{92}U + ^{1}_{1}H$$

$$^{239}_{92}U \longrightarrow ^{239}_{93}Np + ^{0}_{-1}e$$

In a similar fashion, plutonium 239, $^{239}_{94}Pu$, has been produced in atomic reactors by neutron bombardment of $^{238}_{92}U$. The nuclear reactions are given below:

$$^{238}_{92}U + ^{1}_{0}n \longrightarrow ^{239}_{92}U$$

$$^{239}_{92}U \longrightarrow ^{239}_{93}Np + ^{0}_{-1}e$$

$$^{239}_{93}Np \longrightarrow ^{239}_{94}Pu + ^{0}_{-1}e$$

Much of this work has been carried out by the Nobel prize winner Glenn T. Seaborg and other members of the faculty at the University of California at Berkeley, where plutonium, americium, curium, berkelium, and californium have been synthesized and identified.

In order to effectively bombard nuclei with positively charged particles, the electrostatic repulsion between projectile and nuclei must be overcome. The simplest way of accomplishing this is to accelerate the projectiles and thus increase their energy. The positive particles, such as alpha particles, which are emitted in the radioactive decay of nuclei, are usually of insufficient energy to be used for nuclear transformations of other nuclei, so artificial means of accelerating atomic particles have been devised.

Particle accelerators

Two types of particle accelerators are the cyclotron and the linear accelerator. **In the cyclotron, particles are accelerated in circular paths by proper application of magnetic and electric fields.** The source of particles is at the center of the instrument between two hollow D-shaped plates called dees (Figure 20.4). The dees are separated by a gap and are enclosed in an evacuated chamber, which in turn is located between the poles of a powerful electromagnet. The dees are kept oppositely charged by means of a high frequency generator. The charged particles move in a circular path controlled by the magnetic and electric fields. As the particles reach the gap between the dees, the charge of the dees is reversed so that the positive particles are repulsed out of a positive dee and attracted into a negative dee. Because of the acceleration of the particles as they traverse the gap, they travel in a spiral path of increasing radius. Eventually, after several spirals, the particles leave the instrument at high speed and collide with a target nucleus.

Around and around they go and when they come out they react.

In a linear accelerator, the particles are accelerated through a series of charged tubes in an evacuated chamber (Figure 20.5). In the absence of a

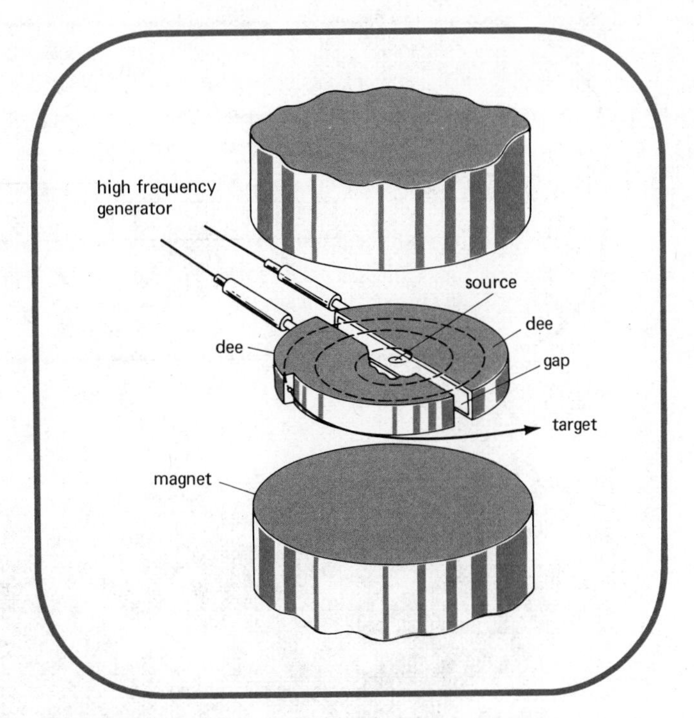

Figure 20.4 Path of a charged particle in a cyclotron.

magnetic field, the particles travel in a straight line, causing the instrument to be very long as compared to the dimensions of the cyclotron. The positive particles are attracted into the first tube, which is at that moment negatively charged. At this time, all odd-numbered tubes are negatively charged and the even-numbered tubes are positively charged. Just as the particles leave tube 1, the polarity of the tubes is reversed. The particles are attracted to tube 2 and repelled by tube 1 and as a result are accelerated. Repetition of this process at constant time intervals leads to increased acceleration. Each tube must be successively larger in order to allow the accelerating particles the same residence time in each tube. At the end of the accelerator, the projectiles meet the target and a nuclear reaction occurs. Linear accelerators require large amounts of land in order to provide the length for sufficient particle acceleration.

Problem 20.5 Bombardment of uranium 238 by nitrogen 14 yields 5 neutrons and an element. What is the element?

Solution The unbalanced equation is:

$$^{238}_{92}\text{U} + ^{14}_{7}\text{N} \longrightarrow 5^{1}_{0}n + ^{A}_{Z}X$$

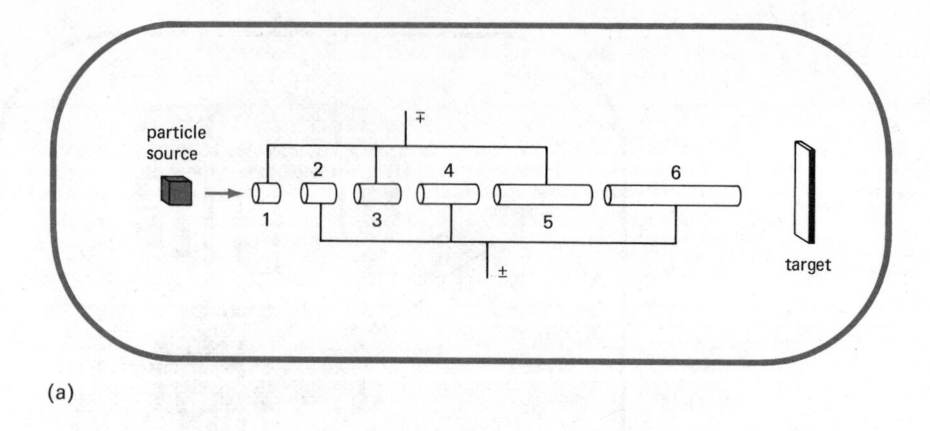

(a)

(b)

Figure 20.5 (a) Representation of a linear accelerator; (b) Aerial view of the 2 mile long linear accelerator at Stanford University. Photo courtesy of Stanford University. Reproduced with permission.

Solving for the atomic number and using the coefficients, we have:

$$92 + 7 = 5(0) + Z$$
$$99 = Z$$

Similarly, the mass number is determined:

$$238 + 14 = 5(1) + A$$
$$247 = A$$

The element is einsteinium 247.

$$^{247}_{99}Es = ^{247}_{99}X$$

20.6 NUCLEAR BINDING ENERGY

The binding energy accounts for the energy liberated in nuclear reactions.

The stability of a nucleus is related to its nuclear binding energy. **The nuclear binding energy is the energy equivalent of the mass difference between the expected sum of the masses of the constituent protons and neutrons of an element and its actual mass.** For example, the masses of the neutron and proton are 1.00867 amu and 1.00728 amu, respectively. On this basis, the mass of the helium nucleus would be expected to be 4.03190 amu. However, experimentally the mass of the helium nucleus has been determined as 4.0026 amu. The mass difference corresponds to an energy that would be liberated in forming the helium nucleus from two neutrons and two protons. Mass and energy are related by the Einstein equation $E = mc^2$, where E represents energy, m is mass, and c is the speed of light. **The 0.0293 amu difference in mass corresponds to 6.3 × 10⁸ kcal/mole of helium atoms and is called the binding energy of the helium nucleus.** In order to disrupt 1 mole of helium atoms and form 2 moles of protons and 2 moles of neutrons, energy equivalent to the binding energy must be supplied.

In a similar manner, the binding energies of other nuclei can be calculated. As the atomic mass increases, the binding energy also increases. However, if the binding energy per nucleon (neutrons and protons) is calculated and plotted as a function of mass number, the intriguing graph in Figure 20.6 is obtained. Elements of intermediate mass have the highest binding energy per nuclear particle and are therefore the most stable. The point of maximum stability occurs in the middle of the transition metals of the fourth period. All other elements are unstable with respect to those at the maximum of the graph. If a heavy element is converted into two or more elements of lower mass, energy would be liberated because of the differences in their nuclear binding energy. **This process of converting heavy elements into two or more lighter elements is called fission. Conversion of two low-mass elements into an element of higher mass can also liberate energy. This process is called fusion.**

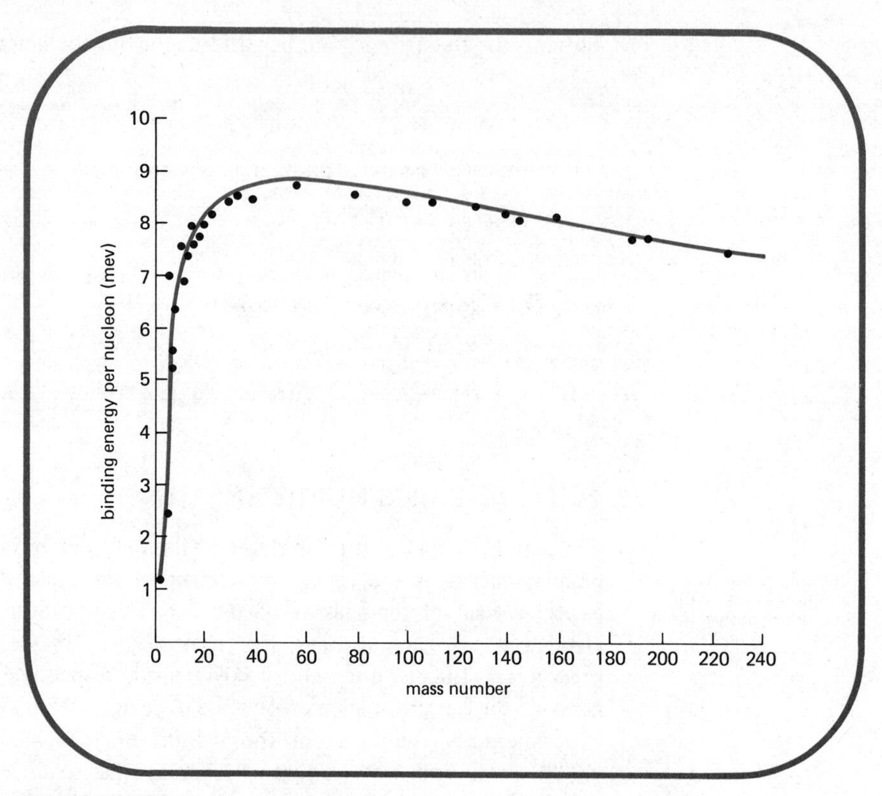

Figure 20.6 Binding energy per nucleon.

20.7 NUCLEAR FISSION

Fission: Heavy element into two lighter elements

Fusion: Two light elements into one heavier element.

An example of nuclear fission is shown by the bombardment of uranium 235 with neutrons. The unstable uranium 236 nucleus is formed, which rapidly splits into nuclei of lower atomic mass. Although many pairs of elements are actually produced, $^{89}_{37}Rb$ and $^{144}_{55}Cs$ are typical.

$$^{235}_{92}U + ^{1}_{0}n \longrightarrow ^{236}_{92}U$$

$$^{236}_{92}U \longrightarrow ^{89}_{37}Rb + ^{144}_{55}Cs + 3^{1}_{0}n$$

The fission process can continue because the neutrons produced by the fission reaction can collide with other uranium atoms causing them to split apart. **Because more neutrons are produced than are used in initiating the reaction, the fission process rapidly becomes a self-sustaining chain reaction** that releases large amounts of energy in a short time. This principle is utilized in the construction of atomic bombs. The chain process is illustrated in Figure 20.7. Other fission processes involving $^{236}_{92}U$ are listed below:

$$^{236}_{92}U \longrightarrow ^{140}_{56}Ba + ^{95}_{36}Kr + ^{1}_{0}n$$

$$^{236}_{92}U \longrightarrow ^{141}_{54}Xe + ^{93}_{38}Sr + 2^{1}_{0}n$$

$$^{236}_{92}U \longrightarrow ^{137}_{52}Te + ^{97}_{40}Zr + 2^{1}_{0}n$$

Atomic energy as contained in the arsenals of numerous countries is potentially destructive and could seriously limit human progress. It should be emphasized that on the other side of the coin, controlled nuclear energy can be of service in a world of limited dimensions and resources. The limited fossil-derived fuels (oil, gas, and coal) are being used at an ever-increasing rate, and in the future this source of fuel will be exhausted. Nuclear fission is and will continue to be an important source of vast amounts of energy.

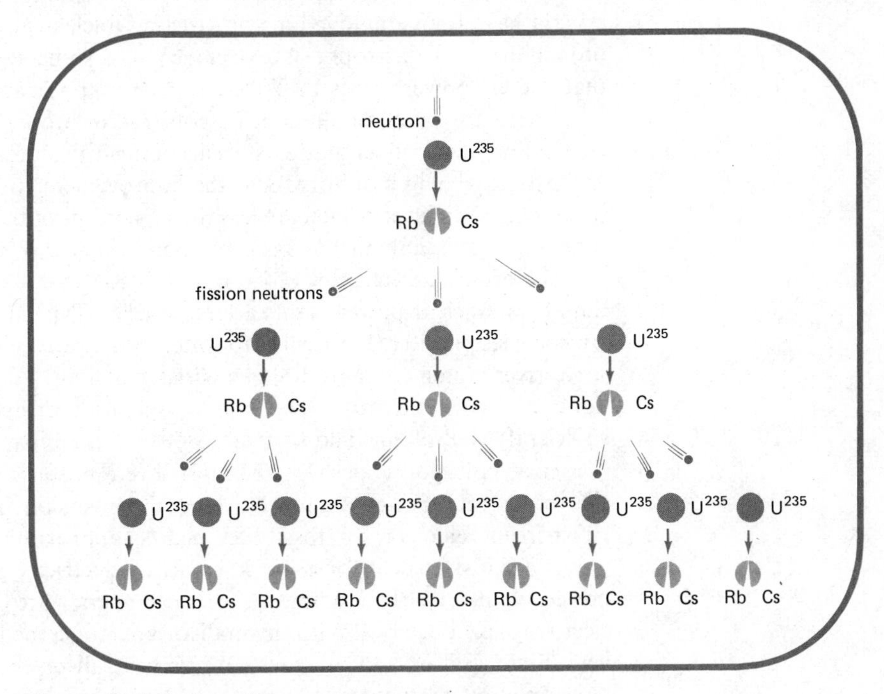

Figure 20.7 Nuclear chain reaction.

The energy equivalent of 1 ton of coal can be produced by only $\frac{1}{2}$ g of uranium 235. To a large degree, the fission process can be controlled in nuclear reactors, in which the rate of the chain reaction is kept in control by the use of materials that can absorb some of the neutrons. The energy produced in atomic reactors is removed as heat and converted into electrical energy.

Nuclear power plants cause concern in the minds of people who do not understand nuclear reactions. They do not realize that an atomic bomb and a nuclear reactor are radically different. In order to produce an atomic bomb, the ^{235}U must be 90 percent pure, whereas in a nuclear reactor the fuel is only 3 to 5 percent ^{235}U. Furthermore, in the atomic bomb, two substantial masses of ^{235}U must be brought together to achieve a critical mass. **The critical mass is that quantity of material which, if suitably shaped, will support a self-sustaining chain reaction.** The shape or geometry of the sample is important as it is necessary that the neutrons have a high probability of encountering ^{235}U nuclei. The geometry of the nuclear fuel in a nuclear reactor will not sustain an explosion.

While nuclear explosions are not possible in a nuclear reactor, there are nevertheless other dangers. If a catastrophe occurred which released radioisotopes into the atmosphere, the radiation hazard could be severe in the vicinity of the power plant. To reduce that possibility, extremely high safety criteria have been employed in constructing nuclear power plants. While the probability of catastrophe is extremely low, some people have suggested that nuclear power plants be placed away from populated centers.

There are small amounts of gaseous radioisotopes such as krypton discharged in the air of ventilation systems of nuclear power plants. This amount is constantly being monitored and the improvements in design have reduced the discharge to levels which are less than 5 percent of the natural background radiation that mankind has been receiving every day of its history.

Waterborne discharges of isotopes of cesium, cobalt, iodine, and strontium from nuclear power plants are monitored. Typically, the water contains 5 picocuries per liter. Domestic tap water averages 20 picocuries per liter and some river water contains 100 picocuries per liter. (A picocurie is one trillionth of a curie.) Clearly, the standard operation of nuclear power plants is sufficiently well planned so that radioactive contamination is minimal.

Heat pollution is perhaps the most severe problem at present in the design of nuclear power plants. No energy conversion process can be made 100 percent efficient. In fossil-fuel plants, approximately 1.3 kilowatts of power are lost as heat for every kilowatt of electricity produced. In nuclear power plants, nearly 2 kilowatts of heat energy are lost per kilowatt of electricity produced. Environmentalists are concerned with the effect of heat discharged into local water sources near all power plants. The nuclear power plant is merely another source of heat which has to be monitored and controlled. The only alternative to the discharge of heat into water is to discharge it into air.

20.8 THE BREEDER REACTOR

Only 0.7 percent of naturally occurring uranium is the fissionable ^{235}U required for nuclear reactors. At the current and projected rate of utilization in present nuclear reactors, the known sources of ^{235}U will be expended near the end of this century. The more abundant ^{238}U (99.28 percent in nature) at the present time cannot be used for generation of nuclear power. However, ^{238}U is a "fertile" isotope that can be converted to fissionable plutonium in a proposed breeder reactor. Unused neutrons in a reactor could convert ^{238}U into ^{239}Pu by a series of reactions.

$$^{238}_{92}U + ^{1}_{0}n \longrightarrow ^{239}_{92}U$$

$$^{239}_{92}U \longrightarrow ^{239}_{93}Np + ^{0}_{-1}e$$

$$^{239}_{93}Np \longrightarrow ^{239}_{94}Pu + ^{0}_{-1}e$$

The fissionable ^{239}Pu emits neutrons which can be used to convert additional "fertile" ^{238}U. **Thus while the fissionable ^{238}Pu is producing energy, it is simultaneously breeding more fuel, hence the term breeder reactor.**

The breeder reactor has the usual hazards of radioactivity associated with it. Plutonium is a long lasting radioisotope. It loses one-half of its radioactivity in 24,360 years. Thus if ^{239}Pu escapes into the environment, its effects will be long-lasting. Another potential hazard of the breeder reactor is that ^{239}Pu is the fissionable substance used in the atomic bomb. If a small part of the plutonium produced in breeder reactors were diverted to military uses, many more nations would have nuclear weapons.

20.9 NUCLEAR FUSION

The second method by which the nuclear binding energy can be liberated is by nuclear fusion. It is a source of even greater energy than nuclear fission because the curve of the binding energy per nuclear particle is very steep in the area of low-mass elements. Unlike nuclear fission, fusion requires extremely high temperatures (1,000,000°C) for initiation.

Fusion occurs on the sun, where the temperature is high enough to convert hydrogen into helium. Only 45 mg of hydrogen are required to produce the energy equivalent of 1 ton of coal. At the present rate, the sun will expend itself in approximately 100 billion years. This problem is not a cause for great concern. The steps proposed to account for the net transformation of $^{1}_{1}H$ into $^{4}_{2}He$ on the sun are as follows:

$$2^{1}_{1}H \longrightarrow ^{2}_{1}H + ^{0}_{1}e$$

$$^{2}_{1}H + ^{1}_{1}H \longrightarrow ^{3}_{2}He$$

$$2^{3}_{2}He \longrightarrow ^{4}_{2}He + 2^{1}_{1}H$$

On earth, nuclear fusion has been achieved by using atomic bomb "triggers" to provide sufficient energy to initiate fusion. This is the principle upon which the hydrogen bomb is based. Since hydrogen is so readily available from water, controlled nuclear fusion for practical purposes would eliminate any concern about the source of energy for future human progress.

Obviously fission atomic bombs cannot be used to produce fusion power in power plants. The fusion reaction will have to be initiated and controlled by technical means not yet devised. The United States has expended a half billion dollars in research toward power from nuclear fusion over the past two decades. Although major technological advances might be made suddenly, it is now estimated that fusion power will not become available until after the year 2000.

Fusion power might be converted directly to electric power and eliminate the problems of thermal pollution associated with fission power plants. Furthermore, there are no radioactive waste products that would have to be stored somewhere in our environment, and the danger of radioactive pollution would be absent. Another advantage to fusion power is that no material is produced which could be used for military purposes, as is the case in the breeder reactor. It appears that fusion power is the answer to our need for clean power.

20.10 USES OF RADIOISOTOPES

Cancer treatment

The use of radium in cancer treatment is well known. The radiation emitted by the decaying radium is focused on a malignant tissue in order to destroy or retard its growth. Great care must be used to limit concomitant destruction of healthy tissue. Radium is being replaced by cobalt 60 for this purpose because it is less expensive and easier to handle.

Hyperthyroidism and cancer of the thyroid gland have been treated with $^{131}_{53}I$. When iodine is introduced into the body, it is concentrated in the thyroid gland, which requires it to function properly. Administration of sodium iodide containing iodine 131 results in transmission of the radioisotope directly to the source of difficulty. Consequently, the danger of damage to healthy tissue is reduced.

Medical diagnosis

The radioisotope sodium 24 is used to monitor and study the circulatory system. Injection of sodium chloride containing the radioisotope into the bloodstream allows a physician to follow the course of the sodium by using the proper radiation detector. In this way, abnormalities of the circulatory system may be examined and diagnosed.

Technetium 99, as the salt sodium pertechnetate, $NaTcO_4$, is used as a scanning agent in the brain. The pertechnate ion is intravenously injected in an isotonic solution of sodium chloride. Cells in brain tumors absorb more pertechnetate than other cells and after a period of time a radiologist can scan the brain to locate and determine the size of brain tumors.

Industry

The petroleum industry uses radioisotopes in pipelines in order to delineate the boundary between batches of different grades of oil being transported through a single pipeline. When the radioisotope is located at a switching station or storage site, the separation can be made. The radioactivity has been used to initiate automatically the switches necessary to separate each batch of oil into its proper pipeline.

The effectiveness of lubricants has been ascertained by exposing the metal object to be lubricated to a nuclear reactor. The radioactive metal, if not properly lubricated, will wear down and the oil becomes radioactive. In a similar manner, the wear of rubber tires has been determined by incorporating radioisotopes in the tread.

SUMMARY

*The stability of a nucleus is controlled by the neutron-to-proton ratio, which graphically represented is called the **zone of stability**. Conversion of unstable to stable nuclei occurs by **radioactive decay**. Isotopes which are **radioactive** are called **radioisotopes**.*

*Nuclear radiation is of three types : **alpha, beta,** and **gamma.** Radiation is measured by a unit of intensity called the **curie**. Absorption of radiation is measured by units of **roentgen, rad,** and **rem.***

*The time required for the decay of one-half of a given number of nuclei is called a **half-life.** The half-life of the radioisotope carbon 14 is used to establish the age of ancient materials containing carbon. Potassium 40 dating is used to determine the age of potassium-bearing minerals.*

***Transmutation** is the changing of one element into another by a nuclear reaction. The **cyclotron** and **linear accelerator** are particle accelerators which are used in transmutation reactions.*

*The **nuclear binding energy** is the energy equivalent of the mass difference between the expected sum of the mass of the constituent protons and neutrons of an element and its actual mass. When a heavy element is converted to two or more lighter elements, the process is called **fission**. The fission reaction occurs in atomic explosions and in atomic power reactors. Conversion of two low-mass elements into a heavier element is called **fusion**. Fusion reactions occur on the sun and in the explosion of hydrogen bombs.*

QUESTIONS AND PROBLEMS

Definitions

1. Define each of the following.

(a) alpha particle
(b) beta particle
(c) gamma ray
(d) radioactivity
(e) radiation
(f) positron
(g) fission
(h) fusion
(i) chain reaction
(j) critical mass
(k) zone of stability
(l) half-life
(m) breeder reactor
(n) cyclotron
(o) linear accelerator
(p) curie

2. Distinguish between each of the following.
   (a) an electron and a positron
   (b) an alpha particle and a beta particle
   (c) natural and artificial reactivity
   (d) a rem and a rad

Nuclear symbols

3. Write the symbols used in balancing nuclear equations for each of the following particles.
   (a) neutron (b) proton
   (c) alpha (d) beta
   (e) electron (f) positron
4. Write the nuclear symbols for each of the following isotopes.
   (a) carbon 14 (b) boron 10
   (c) uranium 238 (d) chlorine 37
   (e) zinc 65 (f) deuterium
   (g) tritium (h) tin 113
5. Indicate the number of protons and neutrons contained in each of the following.
   (a) $^{206}_{82}Pb$ (b) $^{234}_{92}U$
   (c) $^{127}_{53}I$ (d) $^{13}_{7}N$
   (e) $^{25}_{11}Na$ (f) $^{3}_{1}H$

Zone of stability

6. What is the ratio of neutrons-to-protons for stable isotopes of low-atomic-mass elements? How does this ratio change for high-atomic-mass elements?
7. How can beta emission of an element with a high neutron-to-proton ratio lead to a more stable element?
8. How can positron emission of an element with a low neutron-to-proton ratio lead to a more stable element?
9. How can alpha emission of an element with a low neutron-to-proton ratio lead to a more stable element?

Balancing nuclear equations

10. Complete the following nuclear equations.

    (a) $^{234}_{90}Th \longrightarrow ^{234}_{91}Pa + ?$

    (b) $^{226}_{88}Ra \longrightarrow ^{222}_{86}Rn + ?$

    (c) $^{30}_{15}P \longrightarrow ^{30}_{14}Si + ?$

(d) $^{23}_{11}Na + ^{1}_{1}H \longrightarrow ^{23}_{12}Mg + ?$

(e) $^{14}_{6}C \longrightarrow ? + ^{0}_{-1}e$

(f) $^{253}_{99}Es \longrightarrow ? + ^{4}_{2}He$

(g) $^{18}_{10}Ne \longrightarrow ? + ^{0}_{1}e$

11. Write balanced nuclear equations from the following statements.
(a) fluorine 21 undergoes beta emission
(b) bromine 75 undergoes positron emission
(c) silver 104 undergoes positron emission
(d) polonium 212 undergoes alpha emission
(e) curium 240 undergoes alpha emission
(f) silicon 32 undergoes beta emission

Half-lives

12. What percentage of an element will remain after each of the following number of half-lives?
(a) one **(b)** two
(c) three **(d)** four
(e) five **(f)** six

13. A wood object has 12.5 percent of carbon 14 remaining. How old is the object?

14. The half-life of $^{32}_{15}P$ is 14 days. How many grams of the isotope in a 2.00-g sample will remain after 56 days?

15. The half-life of an isotope is 30 minutes. How many hours are required for a 10.00-g sample to decay to 0.62 g of this isotope?

Transmutation of elements

16. Nickel 58 produces an element and an alpha particle when bombarded with a proton. What is the elemental symbol of the product?

17. The bombardment of $^{7}_{3}Li$ by an alpha particle produces a neutron and an element. What is the element?

18. Bombardment of $^{238}_{92}U$ by $^{14}_{7}N$ yields five neutrons and an element. What is the element?

19. Gold 197 can yield mercury 197 and a neutron when bombarded with the proper particle. What is the particle?

20. Lithium 6 can yield tritium and an alpha particle when bombarded with the proper particle. What is the particle?

21. Describe the essential features of the cyclotron.

22. Describe the essential features of the linear accelerator.

Nuclear fission

23. Explain the necessary requirements for an explosion of a fission device.

24. What types of pollution may result from nuclear power plants?

25. What is a breeder reactor?

Nuclear fusion

26. How is a hydrogen bomb detonated?

27. Why haven't nuclear power plants employing nuclear fusion been built?

28. Jupiter has been established as a planet of liquid hydrogen as a result of an American scientific spacecraft fly-by in 1974. Jupiter might be considered a star rather than a planet. Why?

APPENDIX 1
MATHEMATICAL REVIEW

1 MULTIPLICATION

The process of adding a given number a certain number of times is called multiplication. Therefore, 3 times 4 means 3 added four times or 4 added three times to give the product 12.

The various ways of expressing multiplication of 4 and 3 are:

$4 \times 3 \qquad 4 \cdot 3 \qquad 4(3) \qquad (4)(3)$

When a complex quantity contained within parentheses is multiplied, each term must be multiplied.

$$\tfrac{1}{2}(4 + 8) = \tfrac{1}{2}(4) + \tfrac{1}{2}(8) = 2 + 4 = 6$$
$$\tfrac{1}{2}(2x + y) = \tfrac{1}{2}(2x) + \tfrac{1}{2}(y) = x + y/2$$

2 DIVISION

The process of finding out how many times a number is contained in another number is called division. The ways of expressing division are:

$$8 \div 4 \qquad \frac{8}{4} \qquad 8/4$$

The number to the left of the divide symbol, above the horizontal line or to the left of the slant line, is called the numerator, the other number is the denominator. The quotient is the number obtained by dividing one number into another.

3 FRACTIONS

The division of one number by another can be expressed as a fraction. Proper fractions are those in which the numerator is smaller than the denominator; others are improper fractions.

In order to add or subtract two fractions, their denominators must be changed to a common number. The numerators then can be added or subtracted and the resultant placed over the common denominator.

$$\frac{4}{7} + \frac{2}{5} = \frac{4(5)}{7(5)} + \frac{2(7)}{5(7)} = \frac{20}{35} + \frac{14}{35} = \frac{34}{35}$$

Note that conversion of fractions to a common denominator requires that both the numerator and denominator be multiplied by the same number. Of course, identical operations of either multiplication or division on both the numerator and denominator of a fraction leave the fraction mathematically unchanged.

Fractions are multiplied by multiplying both the numerators and the denominators. The product of the numerators is placed over the product of the denominators. The fraction is then reduced to lowest terms by division of numerator and denominator by identical quantities.

$$\frac{4}{7} \times \frac{2}{5} = \frac{4 \times 2}{7 \times 5} = \frac{8}{35}$$

$$\frac{2}{5} \times \frac{3}{8} = \frac{2 \times 3}{5 \times 8} = \frac{6}{40} = \frac{6/2}{40/2} = \frac{3}{20}$$

Division of fractions can be regarded as a multiplication of the numerator by the inverse of the fraction in the denominator.

$$\frac{2/5}{3/7} = \frac{2}{5} \times \frac{7}{3} = \frac{14}{15}$$

The equivalence is achieved by multiplying both numerator and denominator by the inverse of the fraction in the denominator. This changes the denominator to 1.

$$\frac{\frac{2}{5}}{\frac{3}{7}} = \frac{\frac{2}{5} \times \frac{7}{3}}{\frac{3}{7} \times \frac{7}{3}} = \frac{\frac{14}{15}}{\frac{21}{21}} = \frac{\frac{14}{15}}{1} = \frac{14}{15}$$

4 DECIMALS

A decimal can be regarded as a fraction in which the denominator is an unexpressed power of 10. Addition and subtraction of decimals are accomplished in the same manner as similar operations using whole numbers, but the decimals must be placed in the proper column.

add 5.46 + 130.21

$$\begin{array}{r} 5.46 \\ 130.21 \\ \hline 135.67 \end{array}$$

subtract 130.21 − 5.46

$$\begin{array}{r} 130.21 \\ -5.46 \\ \hline 124.75 \end{array}$$

Decimals are multiplied as if they were whole numbers. The location of the decimal point in the product is determined by adding the number of digits to the right of the decimal point in all of the numbers multiplied together. The product contains the same total number of digits to the right of the decimal point.

$$\begin{array}{r} 12.041 \\ \times 0.15 \\ \hline 60205 \\ 12041 \\ \hline 1.80615 \end{array}$$

Division of decimals involves relocation of the decimal points in both the numerator and denominator prior to actually carrying out the division. The relocation is accomplished by multiplying by a power of 10 such that the denominator becomes a whole number. Then the division is carried out and the decimal point is located immediately above its position in the dividend.

$$\frac{84.42}{2.1} = \frac{84.42 \times 10}{2.1 \times 10} = \frac{844.2}{21}$$

$$\begin{array}{r} 40.2 \\ 21\overline{)844.2} \\ 84 \\ \hline 42 \\ \underline{42} \end{array}$$

5 EXPONENTS

Many numbers encountered in science are best expressed in terms of powers of 10. An exponent is a number that is a superscript following another number and indicates how many times the latter number must be multiplied by itself.

$$2^4 = 2 \cdot 2 \cdot 2 \cdot 2 = 16$$
$$10^3 = 10 \cdot 10 \cdot 10 = 1000$$

For both large and small numbers, exponents of the base 10 are employed in order to make the number more compact and easier to handle. A number multiplied by 10^2 is equivalent to another number in which the decimal point is moved two places to the right.

$$3 \times 10^2 = 300$$

If a number is multiplied by 10^{-3}, the decimal is moved three places to the left.

$$3467 \times 10^{-3} = 3.467$$

In expressing a number in exponential form, the decimal point is moved to a new position so that the number value is between 1 and 10. This new number is then multiplied by the proper power of 10 to maintain its original value. The exponent is determined by counting the number of places that the decimal point is moved. If the decimal point is moved to the left, the exponent is a positive number; if moved to the right, the exponent is a negative number:

$$5243 = 5.243 \times 10^{3}$$
$$0.0467 = 4.67 \times 10^{-2}$$

The movement of the decimal point and the introduction of a power of 10 actually involves simultaneous multiplication and division by the same power of 10.

$$5243 = 5243 \times \frac{10^3}{10^3} = \frac{5243}{10^3} \times 10^3 = 5.243 \times 10^3$$

$$0.0467 = 0.0467 \times \frac{10^{-2}}{10^{-2}} = \frac{0.0467}{10^{-2}} \times 10^{-2} = 4.67 \times 10^{-2}$$

In multiplying exponential numbers first multiply the numerical portion of the number and then algebraically add the exponents of the powers of 10.

$$(4 \times 10^2)(2 \times 10^3) = (4 \times 2) \times (10^2 \times 10^3) = 8 \times 10^5$$

Division is carried out in the usual manner with the numerical portion of the number. The powers of 10 then are calculated.

$$\frac{4 \times 10^2}{2 \times 10^3} = \frac{4}{2} \times \frac{10^2}{10^3} = 2 \times \frac{10^2}{10^3} = 2 \times 10^{-1}$$

6 PROPORTIONALITY

It is said that x is proportional to y if x is related to or depends on y. If a certain percentage change in x produces an equal percentage change in y, then x is directly proportional to y. For example, mass and volume are directly proportional to each other because the mass of a substance depends on the volume of the sample under consideration. Mathematically, the direct proportion between mass m and volume V is expressed as follows:

$$m \propto V \qquad m = kV$$

The symbol $\propto$ means "is directly proportional to" and is the proportionality sign. In the second equation, the k is a proportionality constant and allows the replacement of the proportionality sign by an equal sign. In the case of mass and volume, the k is equal to the density of the substance m/V.

The proportionality constant k in a direct proportion indicates the ratio of the two quantities. If one quantity is changed by a factor of 2, then the other quantity also must change by a factor of 2 in order to maintain the equality. For example, the mass and volume of a given substance always must be related such that if a larger mass is considered, the volume of the sample also must be larger. Thus for a given substance:

$$\frac{m_1}{V_1} = \frac{m_2}{V_2} = \frac{m_3}{V_3} = k = \text{density}$$

where the subscripts refer to different quantities of the material.

An inverse proportion indicates that as one variable x is increased by a certain percentage change, the related variable y is decreased by the same percentage change. The inverse proportion may be indicated as:

$$x \propto \frac{1}{y} \qquad x \propto y^{-1}$$

If a proportionality constant k is introduced, the following equations result:

$$x = k\left(\frac{1}{y}\right) \qquad x = ky^{-1}$$

An alternate way of expressing the inverse relationship between the two variable results from rearranging the above equations:

$$xy = k$$

Such an expression indicates that whatever value of x is chosen, y must be such that the product of the two equal a constant k.

At constant temperature the pressure P of a gas and its volume V are inversely related. Therefore, it follows that the equalities listed below are correct. The subscripts refer to different experimental conditions for the same sample of a gas.

$$P_1V_1 = P_2V_2 = P_3V_3 = k$$

7 LOGARITHMS

The power to which the number 10 must be raised to equal a desired number is its logarithm. The log of 100 is 2, because $10^2 = 100$. Similarly, the log of 0.0001 is -4. Since most numbers are not integral powers of 10, the logarithms may be nonintegers. Table 1 gives logarithms of numbers between 1 and 10 to the number of decimal places needed in this text. In order to obtain the logarithm of 6.3, look for 6 in the first vertical column and then across to the column 0.3 to obtain the logarithm 0.799. For numbers greater than 10 or less than 1, write the number in exponential form so it is a number between 1 and 10 multiplied by a power of 10. The number 6300 is 6.3×10^3. Since the logarithm of a product $m \times n$ is equal to the log of m plus the log of n, then:

$$\begin{aligned}\log(6.3 \times 10^3) &= \log 6.3 + \log 10^3 \\ &= 0.799 + 3 \\ &= 3.799\end{aligned}$$

TABLE 1

Table of logarithms

	0.0	0.1	0.2	0.3	0.4	0.5	0.6	0.7	0.8	0.9
1	000	041	079	114	146	176	204	230	255	279
2	301	322	342	362	380	398	415	431	447	462
3	477	491	505	519	532	544	556	568	580	591
4	602	613	623	634	644	653	663	672	681	690
5	699	708	716	724	732	740	748	756	763	771
6	778	785	792	799	806	813	820	826	833	839
7	845	851	857	863	869	875	881	887	892	898
8	903	909	914	919	924	929	935	940	945	949
9	954	959	964	969	973	978	982	987	991	996

The only use of logarithms in this text is in the calculation of pH, which is the negative logarithm of the hydronium ion concentration. For a hydronium ion concentration of 0.00063, the pH is calculated as follows:

$$\begin{aligned}\text{pH} &= -\log(0.00063) \\ &= -\log(6.3 \times 10^{-4}) \\ &= -(\log 6.3 + \log 10^{-4}) \\ &= -(0.799 - 4) \\ &= -(-3.201) \\ &= +3.201\end{aligned}$$

APPENDIX 2
ELECTRONIC CONFIGURATIONS OF THE ELEMENTS

See pages 408–409.

electronic configurations of the elements

		1s	2s	2p	3s	3p	3d	4s	4p	4d	4f	5s	5p	5d	5f
H	1	1													
He	2	2													
Li	3	2	1												
Be	4	2	2												
B	5	2	2	1											
C	6	2	2	2											
N	7	2	2	3											
O	8	2	2	4											
F	9	2	2	5											
Ne	10	2	2	6											
Na	11	2	2	6	1										
Mg	12	2	2	6	2										
Al	13	2	2	6	2	1									
Si	14	2	2	6	2	2									
P	15	2	2	6	2	3									
S	16	2	2	6	2	4									
Cl	17	2	2	6	2	5									
Ar	18	2	2	6	2	6									
K	19	2	2	6	2	6		1							
Ca	20	2	2	6	2	6		2							
Sc	21	2	2	6	2	6	1	2							
Ti	22	2	2	6	2	6	2	2							
V	23	2	2	6	2	6	3	2							
Cr	24	2	2	6	2	6	5	1							
Mn	25	2	2	6	2	6	5	2							
Fe	26	2	2	6	2	6	6	2							
Co	27	2	2	6	2	6	7	2							
Ni	28	2	2	6	2	6	8	2							
Cu	29	2	2	6	2	6	10	1							
Zn	30	2	2	6	2	6	10	2							
Ga	31	2	2	6	2	6	10	2	1						
Ge	32	2	2	6	2	6	10	2	2						
As	33	2	2	6	2	6	10	2	3						
Se	34	2	2	6	2	6	10	2	4						
Br	35	2	2	6	2	6	10	2	5						
Kr	36	2	2	6	2	6	10	2	6						
Rb	37	2	2	6	2	6	10	2	6			1			
Sr	38	2	2	6	2	6	10	2	6			2			
Y	39	2	2	6	2	6	10	2	6	1		2			
Zr	40	2	2	6	2	6	10	2	6	2		2			
Nb	41	2	2	6	2	6	10	2	6	4		1			
Mo	42	2	2	6	2	6	10	2	6	5		1			
Tc	43	2	2	6	2	6	10	2	6	6		1			
Ru	44	2	2	6	2	6	10	2	6	7		1			
Rh	45	2	2	6	2	6	10	2	6	8		1			
Pd	46	2	2	6	2	6	10	2	6	10					
Ag	47	2	2	6	2	6	10	2	6	10		1			
Cd	48	2	2	6	2	6	10	2	6	10		2			
In	49	2	2	6	2	6	10	2	6	10		2	1		
Sn	50	2	2	6	2	6	10	2	6	10		2	2		
Sb	51	2	2	6	2	6	10	2	6	10		2	3		
Te	52	2	2	6	2	6	10	2	6	10		2	4		
I	53	2	2	6	2	6	10	2	6	10		2	5		
Xe	54	2	2	6	2	6	10	2	6	10		2	6		
		2	8		18			18				8			

Electronic Configurations of the Elements (*continued*)

		1	2	3	4*s*	4*p*	4*d*	4*f*	5*s*	5*p*	5*d*	5*f*	6*s*	6*p*	6*d*	7*s*
Cs	55	2	8	18	2	6	10		2	6			1			
Ba	56	2	8	18	2	6	10		2	6			2			
La	57	2	8	18	2	6	10		2	6	1		2			
Ce	58	2	8	18	2	6	10	2	2	6			2			
Pr	59	2	8	18	2	6	10	3	2	6			2			
Nd	60	2	8	18	2	6	10	4	2	6			2			
Pm	61	2	8	18	2	6	10	5	2	6			2			
Sm	62	2	8	18	2	6	10	6	2	6			2			
Eu	63	2	8	18	2	6	10	7	2	6			2			
Gd	64	2	8	18	2	6	10	7	2	6	1		2			
Tb	65	2	8	18	2	6	10	9	2	6			2			
Dy	66	2	8	18	2	6	10	10	2	6			2			
Ho	67	2	8	18	2	6	10	11	2	6			2			
Er	68	2	8	18	2	6	10	12	2	6			2			
Tm	69	2	8	18	2	6	10	13	2	6			2			
Yb	70	2	8	18	2	6	10	14	2	6			2			
Lu	71	2	8	18	2	6	10	14	2	6	1		2			
Hf	72	2	8	18	2	6	10	14	2	6	2		2			
Ta	73	2	8	18	2	6	10	14	2	6	3		2			
W	74	2	8	18	2	6	10	14	2	6	4		2			
Re	75	2	8	18	2	6	10	14	2	6	5		2			
Os	76	2	8	18	2	6	10	14	2	6	6		2			
Ir	77	2	8	18	2	6	10	14	2	6	7		2			
Pt	78	2	8	18	2	6	10	14	2	6	9		1			
Au	79	2	8	18	2	6	10	14	2	6	10		1			
Hg	80	2	8	18	2	6	10	14	2	6	10		2			
Tl	81	2	8	18	2	6	10	14	2	6	10		2	1		
Pb	82	2	8	18	2	6	10	14	2	6	10		2	2		
Bi	83	2	8	18	2	6	10	14	2	6	10		2	3		
Po	84	2	8	18	2	6	10	14	2	6	10		2	4		
At	85	2	8	18	2	6	10	14	2	6	10		2	5		
Rn	86	2	8	18	2	6	10	14	2	6	10		2	6		
Fr	87	2	8	18	2	6	10	14	2	6	10		2	6		1
Ra	88	2	8	18	2	6	10	14	2	6	10		2	6		2
Ac	89	2	8	18	2	6	10	14	2	6	10		2	6	1	2
Th	90	2	8	18	2	6	10	14	2	6	10		2	6	2	2
Pa	91	2	8	18	2	6	10	14	2	6	10	2	2	6	1	2
U	92	2	8	18	2	6	10	14	2	6	10	3	2	6	1	2
Np	93	2	8	18	2	6	10	14	2	6	10	4	2	6	1	2
Pu	94	2	8	18	2	6	10	14	2	6	10	6	2	6		2
Am	95	2	8	18	2	6	10	14	2	6	10	7	2	6		2
Cm	96	2	8	18	2	6	10	14	2	6	10	7	2	6	1	2
Bk	97	2	8	18	2	6	10	14	2	6	10	8	2	6	1	2
Cf	98	2	8	18	2	6	10	14	2	6	10	10	2	6		2
Es	99	2	8	18	2	6	10	14	2	6	10	11	2	6		2
Fm	100	2	8	18	2	6	10	14	2	6	10	12	2	6		2
Md	101	2	8	18	2	6	10	14	2	6	10	13	2	6		2
No	102	2	8	18	2	6	10	14	2	6	10	14	2	6		2
Lw	103	2	8	18	2	6	10	14	2	6	10	14	2	6	1	2
		2	8	18			32			32				9		2

APPENDIX 3
VAPOR PRESSURE OF WATER AT VARIOUS TEMPERATURES

Temperature (°C)	Pressure (torr)	Temperature (°C)	Pressure (torr)
0	4.6	35	42.2
5	6.5	36	44.6
10	9.2	37	47.1
15	12.8	38	49.7
16	13.6	39	52.4
17	14.5	40	55.3
18	15.5	45	71.9
19	16.5	50	92.5
20	17.5	55	118.0
21	18.6	60	149.4
22	19.8	65	187.5
23	21.1	70	233.7
24	22.4	75	289.1
25	23.8	80	355.1
26	25.2	85	433.6
27	26.7	90	525.8
28	28.3	95	633.9
29	30.0	100	760.0
30	31.8		

APPENDIX 4
SOLUBILITY OF SALTS

	F^-	Cl^-	Br^-	I^-	OH^-	NO_3^-	$C_2H_3O_2^-$	O^{2-}	S^{2-}	CO_3^{2-}	SO_4^{2-}
Na^+	*S*	*S*	*S*	*S*	*S*	*S*	*S*	*S*	*S*	*S*	*S*
K^+	*S*	*S*	*S*	*S*	*S*	*S*	*S*	*S*	*S*	*S*	*S*
NH_4^+	*S*	*S*	*S*	*S*	*S*	*S*	*S*	—	*S*	*S*	*S*
Mg^{2+}	*I*	*S*	*S*	*S*	*I*	*S*	*S*	*I*	*d*	*I*	*S*
Ca^{2+}	*I*	*S*	*S*	*S*	*I*	*S*	*S*	*I*	*d*	*I*	*I*
Ba^{2+}	*I*	*S*	*S*	*S*	*s*	*S*	*S*	*s*	*d*	*I*	*I*
Al^{3+}	*I*	*S*	*S*	*S*	*I*	*S*	*S*	*I*	*d*	—	*S*
Fe^{2+}	*s*	*S*	*S*	*S*	*I*	*I*	*S*	*I*	*I*	*s*	*S*
Fe^{3+}	*I*	*S*	*S*	—	*I*	*S*	*I*	*I*	*I*	*I*	*S*
Co^{2+}	*S*	*S*	*S*	*S*	*I*	*S*	*S*	*I*	*I*	*I*	*S*
Ni^{2+}	*s*	*S*	*S*	*S*	*I*	*S*	*S*	*I*	*I*	*I*	*S*
Cu^{2+}	*s*	*S*	*S*	—	*I*	*S*	*S*	*I*	*I*	*I*	*S*
Ag^+	*S*	*I*	*I*	*I*	—	*S*	*I*	*I*	*I*	*I*	*I*
Zn^{2+}	*s*	*S*	*S*	*S*	*I*	*S*	*S*	*I*	*I*	*I*	*S*
Cd^{2+}	*s*	*S*	*S*	*S*	*I*	*S*	*S*	*I*	*I*	*I*	*S*
Hg^{2+}	*d*	*S*	*I*	*I*	*I*	*S*	*S*	*I*	*I*	*I*	*d*
Sn^{2+}	*S*	*S*	*S*	*s*	*I*	*S*	*S*	*I*	*I*	*I*	*S*
Pb^{2+}	*I*	*I*	*I*	*I*	*I*	*S*	*S*	*I*	*I*	*I*	*I*

S = soluble in water
s = slightly soluble in water
I = insoluble in water (less than 1 g/100 g H_2O)
d = decomposes in water

APPENDIX 5
NATURALLY OCCURRING ISOTOPES

Atomic number	Isotope	Percent abundance
1	$^{1}_{1}H$	99.98
	H	0.02
2	$^{3}_{1}He$	Trace
	$^{4}_{2}He$	100.00
3	$^{6}_{2}Li$	7.42
	$^{7}_{3}Li$	92.58
4	$^{9}_{3}Be$	100.00
5	$^{10}_{4}B$	19.6
	$^{11}_{5}B$	80.4
6	$^{12}_{5}C$	98.89
	$^{13}_{6}C$	1.11
7	$^{14}_{7}N$	99.63
	$^{15}_{7}N$	0.37
8	$^{16}_{8}O$	99.76
	$^{17}_{8}O$	0.04
	$^{18}_{8}O$	0.20
9	$^{19}_{9}F$	100.00
10	$^{20}_{10}Ne$	90.92
	$^{21}_{10}Ne$	0.26
	$^{22}_{10}Ne$	8.82
11	$^{23}_{11}Na$	100.00
12	$^{24}_{12}Mg$	78.70
	$^{25}_{12}Mg$	10.13
	$^{26}_{12}Mg$	11.17
13	$^{27}_{13}Al$	100.00
14	$^{28}_{14}Si$	92.21
	$^{29}_{14}Si$	4.70
	$^{30}_{14}Si$	3.09
15	$^{31}_{15}P$	100.00
16	$^{32}_{16}S$	95.00
	$^{33}_{16}S$	0.76
	$^{34}_{16}S$	4.22
	$^{36}_{16}S$	0.01
17	$^{35}_{17}Cl$	75.53
	$^{37}_{17}Cl$	24.47
18	$^{36}_{18}Ar$	0.34
	$^{38}_{18}Ar$	0.06
	$^{40}_{18}Ar$	99.60
19	$^{39}_{19}K$	93.10
	$^{40}_{19}K$	0.01
	$^{41}_{19}K$	6.88
20	$^{40}_{20}Ca$	96.97
	$^{42}_{20}Ca$	0.64
	$^{43}_{20}Ca$	0.14
	$^{44}_{20}Ca$	2.06
	$^{46}_{20}Ca$	Trace
	$^{48}_{20}Ca$	0.18
21	$^{45}_{21}Sc$	100.00
22	$^{46}_{22}Ti$	7.93
	$^{47}_{22}Ti$	7.28

Naturally Occurring Isotopes (*continued*)

Atomic number	Isotope	Percent abundance	Atomic number	Isotope	Percent abundance
	$^{48}_{22}Ti$	73.94		$^{82}_{34}Se$	9.19
	$^{49}_{22}Ti$	5.51	35	$^{79}_{35}Br$	50.54
	$^{50}_{22}Ti$	5.34		$^{81}_{35}Br$	49.46
23	$^{50}_{23}V$	0.24	36	$^{78}_{36}Kr$	0.35
	$^{51}_{23}V$	99.76		$^{80}_{36}Kr$	2.27
24	$^{50}_{24}Cr$	4.31		$^{82}_{36}Kr$	11.56
	$^{52}_{24}Cr$	83.76		$^{83}_{36}Kr$	11.55
	$^{53}_{24}Cr$	9.55		$^{84}_{36}Kr$	56.90
	$^{54}_{24}Cr$	2.38		$^{86}_{36}Kr$	17.37
25	$^{55}_{25}Mn$	100.00	37	$^{85}_{37}Rb$	72.15
26	$^{54}_{26}Fe$	5.82		$^{87}_{37}Rb$	27.85
	$^{56}_{26}Fe$	91.66	38	$^{84}_{38}Sr$	0.56
	$^{57}_{26}Fe$	2.19		$^{86}_{38}Sr$	9.86
	$^{58}_{26}Fe$	0.33		$^{87}_{38}Sr$	7.02
27	$^{59}_{27}Co$	100.00		$^{88}_{38}Sr$	82.56
28	$^{58}_{28}Ni$	67.88	39	$^{89}_{39}Y$	100.00
	$^{60}_{28}Ni$	26.23	40	$^{90}_{40}Zr$	51.46
	$^{61}_{28}Ni$	1.19		$^{91}_{40}Zr$	11.23
	$^{62}_{28}Ni$	3.66		$^{92}_{40}Zr$	17.11
	$^{64}_{28}Ni$	1.08		$^{94}_{40}Zr$	17.40
29	$^{63}_{29}Cu$	69.09		$^{96}_{40}Zr$	2.80
	$^{65}_{29}Cu$	30.91	41	$^{93}_{41}Nb$	100.00
30	$^{64}_{30}Zn$	48.89	42	$^{92}_{42}Mo$	15.84
	$^{66}_{30}Zn$	27.81		$^{94}_{42}Mo$	9.04
	$^{67}_{30}Zn$	4.11		$^{95}_{42}Mo$	15.72
	$^{68}_{30}Zn$	18.57		$^{96}_{42}Mo$	16.53
	$^{70}_{30}Zn$	0.62		$^{97}_{42}Mo$	9.46
31	$^{69}_{31}Ga$	60.4		$^{98}_{42}Mo$	23.78
	$^{71}_{31}Ga$	39.6		$^{100}_{42}Mo$	9.13
32	$^{70}_{32}Ge$	20.52	44	$^{96}_{44}Ru$	5.51
	$^{72}_{32}Ge$	27.43		$^{98}_{44}Ru$	1.87
	$^{73}_{32}Ge$	7.76		$^{99}_{44}Ru$	12.72
	$^{74}_{32}Ge$	36.54		$^{100}_{44}Ru$	12.62
	$^{76}_{32}Ge$	7.76		$^{101}_{44}Ru$	17.07
33	$^{75}_{33}As$	100.00		$^{102}_{44}Ru$	31.61
34	$^{74}_{34}Se$	0.87		$^{104}_{44}Ru$	18.58
	$^{76}_{34}Se$	9.02	45	$^{103}_{45}Rh$	100.00
	$^{77}_{34}Se$	7.58	46	$^{102}_{46}Pd$	0.96
	$^{78}_{34}Se$	23.52		$^{104}_{46}Pd$	10.97
	$^{80}_{34}Se$	49.82		$^{105}_{46}Pd$	22.23

Naturally Occurring Isotopes (*continued*)

Atomic number	Isotope	Percent abundance	Atomic number	Isotope	Percent abundance
	$^{106}_{46}Pd$	27.33		$^{130}_{54}Xe$	4.08
	$^{108}_{46}Pd$	26.71		$^{131}_{54}Xe$	21.18
	$^{110}_{46}Pd$	11.81		$^{132}_{54}Xe$	26.89
47	$^{107}_{47}Ag$	51.82		$^{134}_{54}Xe$	10.44
	$^{109}_{47}Ag$	48.18		$^{136}_{54}Xe$	8.87
48	$^{106}_{48}Cd$	1.22	55	$^{133}_{55}Cs$	100.00
	$^{108}_{48}Cd$	0.88	56	$^{130}_{56}Ba$	0.10
	$^{110}_{48}Cd$	12.39		$^{132}_{56}Ba$	0.10
	$^{111}_{48}Cd$	12.75		$^{134}_{56}Ba$	2.42
	$^{112}_{48}Cd$	24.07		$^{135}_{56}Ba$	6.59
	$^{113}_{48}Cd$	12.26		$^{136}_{56}Ba$	7.81
	$^{114}_{48}Cd$	28.86		$^{137}_{56}Ba$	11.32
	$^{116}_{48}Cd$	7.58		$^{138}_{56}Ba$	71.66
49	$^{113}_{49}In$	4.28	57	$^{138}_{57}La$	0.09
	$^{115}_{49}In$	95.72		$^{139}_{57}La$	99.91
50	$^{112}_{50}Sn$	0.96	58	$^{136}_{58}Ce$	0.19
	$^{114}_{50}Sn$	0.66		$^{138}_{58}Ce$	0.25
	$^{115}_{50}Sn$	0.35		$^{140}_{58}Ce$	88.48
	$^{116}_{50}Sn$	14.30		$^{142}_{58}Ce$	11.07
	$^{117}_{50}Sn$	7.61	59	$^{141}_{59}Pr$	100.00
	$^{118}_{50}Sn$	24.03	60	$^{142}_{60}Nd$	27.11
	$^{119}_{50}Sn$	8.58		$^{143}_{60}Nd$	12.17
	$^{120}_{50}Sn$	32.85		$^{144}_{60}Nd$	23.85
	$^{122}_{50}Sn$	4.92		$^{145}_{60}Nd$	8.30
	$^{124}_{50}Sn$	5.94		$^{146}_{60}Nd$	17.22
51	$^{121}_{51}Sb$	57.25		$^{148}_{60}Nd$	5.73
	$^{123}_{51}Sb$	42.75		$^{150}_{60}Nd$	5.62
52	$^{120}_{52}Te$	0.09	62	$^{144}_{62}Sm$	3.09
	$^{122}_{52}Te$	2.46		$^{147}_{62}Sm$	14.97
	$^{123}_{52}Te$	0.87		$^{148}_{62}Sm$	11.24
	$^{124}_{52}Te$	4.61		$^{149}_{62}Sm$	13.83
	$^{125}_{52}Te$	6.99		$^{150}_{62}Sm$	7.44
	$^{126}_{52}Te$	18.71		$^{152}_{62}Sm$	26.72
	$^{128}_{52}Te$	31.79		$^{154}_{62}Sm$	22.71
	$^{130}_{52}Te$	34.48	63	$^{151}_{63}Eu$	47.82
53	$^{127}_{53}I$	100.00		$^{153}_{63}Eu$	52.18
54	$^{124}_{54}Xe$	0.10	64	$^{152}_{64}Gd$	0.20
	$^{126}_{54}Xe$	0.09		$^{154}_{64}Gd$	2.15
	$^{128}_{54}Xe$	1.92		$^{155}_{64}Gd$	14.73
	$^{129}_{54}Xe$	26.44		$^{156}_{64}Gd$	20.47

Naturally Occurring Isotopes (*continued*)

Atomic number	Isotope	Percent abundance
	$^{157}_{64}Gd$	15.68
	$^{158}_{64}Gd$	24.87
	$^{160}_{64}Gd$	21.90
65	$^{159}_{65}Tb$	100.00
66	$^{156}_{66}Dy$	0.05
	$^{158}_{66}Dy$	0.09
	$^{160}_{66}Dy$	2.29
	$^{161}_{66}Dy$	18.88
	$^{162}_{66}Dy$	25.53
	$^{163}_{66}Dy$	24.97
	$^{164}_{66}Dy$	28.18
67	$^{165}_{67}Ho$	100.00
68	$^{162}_{68}Er$	0.14
	$^{164}_{68}Er$	1.56
	$^{166}_{68}Er$	33.41
	$^{167}_{68}Er$	22.94
	$^{168}_{68}Er$	27.07
	$^{170}_{68}Er$	14.88
69	$^{169}_{69}Tm$	100.00
70	$^{168}_{70}Yb$	0.14
	$^{170}_{70}Yb$	3.03
	$^{171}_{70}Yb$	14.31
	$^{172}_{70}Yb$	21.82
	$^{173}_{70}Yb$	16.13
	$^{174}_{70}Yb$	31.84
	$^{176}_{70}Yb$	12.73
71	$^{175}_{71}Lu$	97.41
	$^{176}_{71}Lu$	2.59
72	$^{174}_{72}Hf$	0.18
	$^{176}_{72}Hf$	5.20
	$^{177}_{72}Hf$	18.50
	$^{178}_{72}Hf$	27.14
	$^{179}_{72}Hf$	13.75
	$^{180}_{72}Hf$	35.24
73	$^{180}_{73}Ta$	0.01
	$^{181}_{73}Ta$	99.99
74	$^{180}_{74}W$	0.14
	$^{182}_{74}W$	26.41
	$^{183}_{74}W$	14.40
	$^{184}_{74}W$	30.64
	$^{186}_{74}W$	28.41
75	$^{185}_{75}Re$	37.07
	$^{187}_{75}Re$	62.93
76	$^{184}_{76}Os$	0.02
	$^{186}_{76}Os$	1.59
	$^{187}_{76}Os$	1.64
	$^{188}_{76}Os$	13.30
	$^{189}_{76}Os$	16.10
	$^{190}_{76}Os$	26.40
	$^{192}_{76}Os$	41.00
77	$^{191}_{77}Ir$	37.3
	$^{193}_{77}Ir$	62.7
78	$^{190}_{78}Pt$	0.01
	$^{192}_{78}Pt$	0.78
	$^{194}_{78}Pt$	32.90
	$^{195}_{78}Pt$	33.80
	$^{196}_{78}Pt$	25.30
	$^{198}_{78}Pt$	7.21
79	$^{197}_{79}Au$	100.00
80	$^{196}_{80}Hg$	0.15
	$^{198}_{80}Hg$	10.02
	$^{199}_{80}Hg$	16.84
	$^{200}_{80}Hg$	23.13
	$^{102}_{80}Hg$	13.22
	$^{202}_{80}Hg$	29.80
	$^{204}_{80}Hg$	6.85
81	$^{203}_{81}Tl$	29.50
	$^{205}_{81}Tl$	70.50
82	$^{204}_{82}Pb$	1.48
	$^{206}_{82}Pb$	23.60
	$^{207}_{82}Pb$	22.60
	$^{208}_{82}Pb$	52.30
83	$^{209}_{83}Bi$	100.00
92	$^{234}_{92}U$	0.01
	$^{235}_{92}U$	0.72
	$^{238}_{92}U$	99.27

APPENDIX 6 ANSWERS TO SELECTED PROBLEMS

CHAPTER 2

6. (a) 3, (b) 1, (c) 4, (d) 5, (e) 3, (f) 7, (g) 2, (h) 3
7. (a) 0.05192, (b) 51.05, (c) 139.2, (d) 6.145, (e) 0.7135, (f) 0.005782, (g) 1.485, (h) 189.4
8. (a) 87.2, (b) 18.3, (c) 12, (d) 0.6, (e) 0.050, (f) $4\bar{0}$, (g) 2.26
9. (a) 5.9, (b) 0.00153, (c) 0.325, (d) 5000, (e) 0.000211, (f) 70,000
10. $1\bar{0}$,000,000
11. 453592.4
13. 36,26,36
14. 25 miles per hour
15. 44
16. 1
17. 10 seconds
18. 5
19. 180 pounds
21. (a) 95, (b) −10, (c) −50, (d) −40, (e) 50, (f) 125
22. (a) 122, (b) 257, (c) 41, (d) −4, (e) −148, (f) 392
23. (a) 298, (b) 510, (c) 273, (d) 173, (e) 0, (f) 373
24. 1762
25. −459
27. (a) 50, (b) 25, (c) 1250
28. 5000
29. 11.2
30. 1114
32. less due to lower gravity
34. 102 g

35. 19.3 g/cm^3
37. (a) 2.00, (b) $1\bar{0}$, (c) 2.2, (d) $2\bar{0}0$, (e) 2.00
38. (a) 62, (b) 18 (c) 9, (d) 92
39. 3.300

CHAPTER 3

3. Chemical changes are (b), (e), (f), (h), (j), (l).
5. Water melts at 0°C and boils at 100°C while the gaseous mixture is a gas at both temperatures.

9. (a) solid, (b) gas, (c) liquid, (d) solid, (e) liquid, (f) solid, (g) solid, (h) liquid, (i) gas, (j) solid
12. (a) element, (b) element, (c) mixture, (d) compound, (e) compound, (f) element, (g) compound, (h) mixture, (i) mixture, (j) mixture, (k) compound, (l) element
13. a chemical change
15. Cooking gas and oxygen are the reactants.
17. one
18. 7.8 g
19. 100 kcal
20. 2.5×10^{15} cal

CHAPTER 4

6. americium, californium, berkelium
8. It is a form of shorthand.
9. There are more than 26 elements.
10. They are based on other languages.
12. (a)–(8), (b)–(5), (c)–(7), (d)–(9), (e)–(2), (f)–(10), (g)–(4), (h)–(6), (i)–(1), (j)–(3)
13. hydrogen
14. iron, oxygen, silicon, magnesium, and nickel
15. oxygen and silicon
16. nitrogen and oxygen
17. oxygen, carbon, and hydrogen
18, 106
19. 83
20. 16
21. 7
22. mercury and bromine
23. Carbon, phosphorus, sulfur, selenium, and iodine are solids.
24. nitrogen, oxygen, fluorine, chlorine, hydrogen, helium, neon, argon, krypton, xenon, and radon
25. CO_2, SO_2, H_2O
26. MgO, HgO
27. pyrite and galena (Table 4.4)
29. (a) 7.8, (b) 3.9, (c) 2.2 g of oxygen will remain, (d) 10.0 g of lead will remain

CHAPTER 5

7. It is much smaller in mass than atoms.
8. protons and neutrons in the nucleus; electron about the nucleus
9. 10^{-13} cm, 10^{-8} cm
10. They are equal in number.
11. The number of protons; it is used as a subscript to the left of the symbol.
12. The number of protons and neutrons combined; it is used as a superscript to the left of the symbol.
13. protons or electrons: (a) 3, (b) 15, (c) 47, (d), 27, (e) 31, (f) 80, (g) 80, (h) 58, (i) 11, (j) 8; neutrons: (a) 4, (b) 16, (c) 60, (d) 32, (e) 38, (f) 121, (g) 119, (h) 84, (i) 11, (j) 9

14. (a) $^{19}_{9}F$, (b) $^{30}_{14}Si$, (c) $^{28}_{14}Si$, (d) $^{56}_{26}Fe$, (e) $^{237}_{92}U$, (f) $^{41}_{19}K$
16. equal amounts of both isotopes
17. the 23.99-mass isotope
18. 10.8
19. (b) H_2, (c) P_4, (d) N_2, (e) O_2, (f) S_8, (g) F_2, (h) Cl_2
21. $C_7H_5N_3O_6$ **22.** C_8H_{18}
23. One molecule contains 10 carbon atoms, 14 hydrogen atoms, and 2 nitrogen atoms.
24. It is a negative ion and contains more electrons than protons.
25. It is a positive ion and contains fewer electrons than protons.
26. S^{2-} **27.** Sc^{3+}
28. (a) oxide, (b) chloride, (c) sulfide, (d) lithium, (e) potassium, (f) calcium
29. (a) hydroxide, (b) ammonium, (c) nitrate, (d) sulfite, (e) sulfate, (f) cyanide, (g) perchlorate, (h) hypochlorite
30. (a) OH^-, (b) O^{2-}, (c) S^{2-}, (d) NH_4^+, (e) Cl^-, (f) SO_4^{2-}, (g) ClO_4^-, (h) CN^-, (i) PO_4^{3-}
31. (a) LiF, (b) ZnO, (c) NaCN, (d) MgF_2, (e) $Zn(CN)_2$, (f) $NaNO_3$, (g) Na_2CO_3, (h) K_2S
32. (a) $FeCl_3$, (b) NaOH, (c) $Mg(OH)_2$, (d) CdS, (e) KBr, (f) Li_3N, (g) $Ba(NO_3)_2$, (h) $CsClO_4$
33. (a) calcium hydroxide, (b) lithium perchlorate, (c) sodium phosphate, (d) potassium sulfate, (e) potassium nitrate, (f) ammonium nitrite, (g) magnesium chloride, (h) lithium cyanide
34. (a) 2, (b) 32, (c) 28, (d) 38, (e) 17, (f) 32, (g) 30, (h) 46, (i) 44, (j) 76, (k) 18, (l) 34, (m) 20, (n) 71, (o) 37, (p) 80

CHAPTER 6

3. (a) two electrons in the *s* subshell of the first energy level (b) four electrons in the *p* subshell of the third energy level (c) five electrons in the *d* subshell of the fourth energy level (d) one electron in the *f* subshell of the fifth energy level
5. (a) 8, (b) 18, (c) 92, .(d) 4, (e) 6, (f) 20, (g) 15, (h) 25
6. (a) 10, (b) 10, (c) 18, (d) 18, (e) 10, (f) 18, (g) 2, (h) 10
7. two in the first, eight in the second, eighteen in the third, and thirty-two in the fourth
9. (a) two in first and seven in second (b) two in first, eight in second, eight in third, and one in fourth (c) two in first, eight in second, eight in third, and two in fourth (d) two in first, eight in second, and six in third (e) two in first and five in second
10. (a) 1, (b) 1, (c) 7, (d) 7, (e) 5, (f) 5, (g) 3, (h) 2, (i) 4, (j) 4
11. No. Mass numbers give protons plus neutrons. Isotopes have different masses but the same electronic configurations.

12. (a) $1s^1$ (b) $1s^22s^1$ (c) $1s^22s^22p^6$ (d) $1s^22s^22p^63s^1$
(e) $1s^22s^22p^63s^23p^4$ (f) $1s^22s^22p^63s^23p^5$ (j) $1s^22s^22p^63s^23p^64s^23d^7$
(k) $1s^22s^22p^63s^23p^64s^23d^{10}4p^65s^24d^2$
(m) $1s^22s^22p^63s^23p^64s^23d^{10}4p^65s^24d^{10}5p^2$
13. (a) 1, 1*s* (b) 1, 3*s* (c) 7, 3*s*, and 3*p* (d) 1, 2*s* (e) 7, 2*s*, and 2*p*
(f) 5, 2*s*, and 2*p* (g) 6, 2*s*, and 2*p* (h) 1, 4*s* (i) 6, 3*s*, and 3*p*
(j) 5, 3*s*, and 3*p* (k) 3, 3*s*, and 3*p* (l) 2, 3*s*, and 3*p*
15. They have the identical number of valence electrons.

CHAPTER 7

5. (b) $1s^2$ (d) $1s^22s^22p^6$ (f) $1s^22s^22p^6$ (h) $1s^22s^22p^6$
(j)$1s^22s^22p^63s^23p^64s^23d^{10}4p^6$ (l) $1s^22s^22p^6$ (n) $1s^22s^22p^6$
8. (a) NaCl, (b) $CaCl_2$, (c) MgO, (d) $MgBr_2$, (e) BaH_2, (f) Ba_3P_2,
(g) Li_3N, (h) $AlCl_3$
9. (a) LiF, (b) $MgBr_2$, (c) Li_2O, (d) MgS, (e) AlF_3, (f) Na_2Se, (g) CaI_2,
Na_3N
14. The more negative end is the more electronegative atom.
16. (a) H, (b) Br, (c) H, (d) O, (e) Cl, (f) Si, (g) H, (h) N, (i) P, (j) P
17. (a) +5, (b) +4, (c) +3, (d) +7, (e) +6, (f) +5, (g) +7, (h) −2,
(i) +2, (j) +6

CHAPTER 8

3. They are parts of groups.
4. Not all the elements were known.
5. They are parts of periods.
7. A is used for representative elements and B for transition elements.
9. (a) 2, VA (b) 4, IIA (c) 5, IIB (d) 4, VIIA (e) 2, IA (f) 3, IIIA
(g) 5, IVA (h) 3, VIA (i) 5, O (j) 4, VIII (k) 4, VIB (l) 4, IB
10. (a) Li, (b) Al, (c) Ba, (d) Ge, (e) I, (f) He, (g) Mn, (h) Ag, (i) W,
(i) none
11. (b), (c), (d), (g), (h), and (j) are metals. (a), (e), (f), and (i) are nonmetals
12. (a) Mg, (b) Ge, (c) Se, (d) Sn, (e) Sb, (f) Te, (g) In, (h) Cs
13. (a) +6, (b) +2, (c) +4, (d) +3, (e) +2, (f) +1, (g) +7, (h) +4
14. (a) +6 and −2, (b) +5 and −3, (c) +7 and −1, (d) +4 and −4,
(e) +3 and −5, (f) +5 and −3
15. (a) 4, (b) 6, (c) 5, (d) 0, (e) 1, (f) 5, (g) 2, (h) 3
16. (a) 2, IIIA (b) 3, IA (c) 3, VIIA (d) 4, VIIB
18. (a) $RbNO_3$, (b) $BaSO_4$, (c) Al_2S_3, (d) CaSe, (e) $NaBrO_4$
(f) $MgSeO_4$
19. (a), (f), (h), and (i) are ionic. (b), (c), (d), (e), (g), and (j) are covalent.
21. (a) Cl, (b) Si, (c) K, (d) C, (e) Se, (f) Pb
22. (a) F, (b) S, (c) Li, (d) N, (e) Cl, (f) S

CHAPTER 9

3. (a) chloride, (b) fluoride, (c) bromide, (d) iodide, (e) hydroxide, (f) sulfide, (g) hydrogen, (h) cyanide, (i) ammonium, (j) sodium, (k) lithium, (l) silver, (m) magnesium, (n) calcium, (o) potassium, (p) aluminum

4. (a) hypochlorite, (b) permanganate, (c) bromate, (d) chlorite, (e) sulfate, (f) sulfite, (g) phosphate, (h) chlorate, (i) carbonate, (j) perchlorate, (k) hypoiodite

6. see Table 9.11

7. (a) Na^+, (b) Cu^+, (c) Fe^{3+}, (d) K^+, (e) Sn^{4+}, (f) Fe^{2+}, (g) Cu^{2+} (h) Hg^{2+}, (i) Ca^{2+}, (j) Al^{3+}, (k) Ba^{2+}, (l) Sn^{2+}

8. (a) Br^-, (b) SO_3^{2-}, (c) ClO_3^-, (d) HCO_3^-, (e) NO_3^-, (f) ClO_4^-, (g) BrO_2^-, (h) NO_2^-, (i) SO_4^-, (j) OI^-, (k) HSO_4^-, (l) CN^-, (m) OH^-, (n) HSO_3^-

9. (a) ammonium chloride (c) aluminum sulfate (e) potassium perchlorate (g) mercury(II) cyanide (i) sodium nitrite (k) aluminum hydroxide (m) calcium bisulfite (o) sulfur trioxide (q) lithium hypoiodite (s) iron(III) oxide (u) phosphorus trichloride (w) aluminum hydroxy chloride (y) potassium hydroxide

11. (b) $CuCl$ (d) Na_2SO_4 (f) HNO_2 (h) $SnCO_3$ (j) $Cu(CN)_2$ (l) FeO (n) $NaKSO_4$ (p)$HgBr_2$ (r) Fe_2O_3 (t) $LiOH$ (v) HCl (x) PCl_5 (z) HNO_3

CHAPTER 10

4. (a) C_5H_7N, (b) $C_7H_5N_3O_6$, (c) $C_{12}H_{22}O_{11}$, (d)C_5H_4, (c) $C_3H_6NO_2S$, (f) $C_6H_8O_7$, (g) $C_2H_4N_2O$

5. (a) 80, (b) 142, (c) 46, (d) 76, (e) 44, (f) 32, (g) 16, (h) 119

6. 162, 227, 342, 128, 240, 192

7. (a) 56.1, (b) 100.1, (c) 40, (d) 74.1, (e) 142, (f) 342, (g) 164, (h) 65.1, (i) 262, (j) 158

8. (a) 2, (b) 0.5, (c) 0.01, (d) 1

9. (a) 10, (b) 0.01, (c) 2, (d) 0.1

10. (a) 0.1, (b) 0.1, (c) 0.00089, (d) 0.005

11. (a) 6.02×10^{22}, (b) 6.02×10^{24}, (c) 3.01×10^{23}, (d) 6.02×10^{21}

12. (a) 6.02×10^{22}, (b) 3.0×10^{24}, (c) 6.02×10^{23}, (d) 1.5×10^{21}

14. (a) Ga_2O_3, (b) PbN_2O_6, (c) $ZnCl_2$, (d) $FeBr_2$

15. (a) CS_2, (b) NH_3, (c) CH_4O, (d) SO_2, (e) $C_2H_5NO_2$, (f) C_2H_4Cl

16. (a) 75% C, 25% H (b) 23.1% C, 73.1% F, 3.8% H (c) 5.9% H, 94.1% O (d) 40.0% C, 6.7% H, 53.3% O (e) 43.7% P, 56.3% O (f) 26.5% Ca, 73.5% Cl (g) 63.5% Ag. 8.2% N, 28.3% O (h) 15.8% Al, 28.0% S, 56.2% O

17. 60% C, 4.4% H, 35.6% O

18. 47.4% C, 2.5% H, 50.1% Cl

19. (a) 55.6% metal, 44.4% oxygen (b) 46.7% element, 53.3% oxygen
(c) 63.6% metal, 36.4% sulfur
20. CHCl, $C_6H_6Cl_6$
21. CF_2Cl_2
22. C_6H_6
23. $C_{20}H_{30}O$

CHAPTER 11

5. (a), (d), and (h) are combination; (c), (f), and (l) are decomposition; (b), (g), and (j) are substitution; (e) (i), and (k) are metathesis.
6. (a) $2P + 3I_2 \longrightarrow 2PI_3$ (b) $H_2O + SO_2 \longrightarrow H_2SO_3$
(c) $Na_2O + H_2O \longrightarrow 2NaOH$ (d) $H_2 + Cl_2 \longrightarrow 2HCl$
(e) $3H_2O + P_2O_5 \longrightarrow 2H_3PO_4$ (f) $2Al + 3Cl_2 \longrightarrow 2AlCl_3$
(g) $CaO + CO_2 \longrightarrow CaCO_3$ (h) $BaO + SO_3 \longrightarrow BaSO_4$
(i) $Si + O_2 \longrightarrow SiO_2$ (j) $Si + 2F_2 \longrightarrow SiF_4$
7. (a) $CdCO_3 \longrightarrow CdO + CO_2$
(b) $CaSO_4 \cdot 2H_2O \longrightarrow CaSO_4 + 2H_2O$
(c) $2NaHCO_3 \longrightarrow Na_2CO_3 + H_2O + CO_2$
(d) $2HgO \longrightarrow 2Hg + O_2$ (e) $2KClO_3 \longrightarrow 2KCl + 3O_2$
(g) $2H_2O \longrightarrow 2H_2 + O_2$
8. (a) $Mg + CO \longrightarrow MgO + C$ (b) $Cl_2 + 2KI \longrightarrow I_2 + 2KCl$
(c) $2Na + 2H_2O \longrightarrow 2NaOH + H_2$
(d) $Zn + NiCl_2 \longrightarrow ZnCl_2 + Ni$ (e) $Cd + CuSO_4 \longrightarrow CdSO_4 + Cu$
(f) $2Al + 3H_2SO_4 \longrightarrow Al_2(SO_4)_3 + 3H_2$
(g) $Cd + 2HCl \longrightarrow CdCl_2 + H_2$
9. (a), (b), (c), and (d) will not occur.
10. (a) $TiCl_4 + 2H_2O \longrightarrow TiO_2 + 4HCl$
(b) $Sb_2S_3 + 6HCl \longrightarrow 2SbCl_3 + 3H_2S$
(c) $Pb(NO_3)_2 + 2HCl \longrightarrow PbCl_2 + 2HNO_3$
(d) $2AgNO_3 + H_2S \longrightarrow Ag_2S + 2HNO_3$
(e) $AgNO_3 + HCl \longrightarrow AgCl + HNO_3$
(f) $FeCl_3 + 3NaOH \longrightarrow Fe(OH)_3 + 3NaCl$
(g) $BaCl_2 + Na_2CO_3 \longrightarrow BaCO_3 + 2NaCl$
11. (a) $NaOH + HCl \longrightarrow NaCl + H_2O$
(b) $Ca(OH)_2 + 2HCl \longrightarrow CaCl_2 + 2H_2O$
(c) $BaO + 2HCl \longrightarrow BaCl_2 + H_2O$
(d) $2NaOH + H_2SO_4 \longrightarrow 2H_2O + Na_2SO_4$
(e) $2Al(OH)_3 + 3H_2SO_4 \longrightarrow Al_2(SO_4)_3 + 6H_2O$
(f) $Zn(OH)_2 + 2HNO_3 \longrightarrow Zn(NO_3)_2 + 2H_2O$
(g) $Fe(OH)_3 + 3HNO_3 \longrightarrow Fe(NO_3)_3 + 3H_2O$
(h) $3KOH + H_3PO_4 \longrightarrow K_3PO_4 + 3H_2O$
13. (Coefficients given in order of appearance in equations.)
(a) 2, 13, 8, 10 (b) 1, 3, 1, 3 (c) 1, 1, 2 (d) 1, 3, 1, 1
(e) 1, 8, 5, 6 (f) 2, 1, 1, 1, 1 (g) 1, 1, 1, 2

CHAPTER 12

3. 5
4. 1
5. 3.5
6. 12
7. 56 liters
8. 250 liters
9. 87.5 liters
10. 896 liters
11. 4000 liters
12. 2.24 liters
13. 0.05 mole
14. 179 liters
15. 0.05 mole
16. 1.12 liters
17. 10 moles
18. 1.2 g
19. 0.33
20. 6
21. 2.0
22. 112 liters
23. 0.654
24. 2240 ml
25. 20,000 g
26. 80,000 g
27. 44.8 liters
28. 127 g
29. 1.2 g
30. 12.8 g
31. 38.5 g
32. 1270 g
33. 1.12 liters
34. 9 g
35. 1.34 liters
36. 17 g
37. 7.17
38. 6.15 g

CHAPTER 13

4. humidity of air and wind
5. On a calm day the rate of evaporation and cooling is less.
7. The vapor pressure of water is equal to the pressure of water in the air.
8. No. Mercury has a small vapor pressure.
9. Its vapor pressure is low.
10. B
11. Same. The vapor pressure depends only on temperature.
12. 100°C, 120°C
13. on
16. The heat released by the condensation of water is large.
17. 5400 cal
18. 5400 cal
20. −3°C
21. High pressure can be produced due to the small area of the blade.
22. Mercury becomes a solid.
23. 80,000 cal
24. 8000 cal
25. Their vapor pressure is high.
29. conversion of water at 100°C to steam
30. 7300 cal
31. 7300 cal
32. 620 cal/g

CHAPTER 14

3. 225 ml
4. 2 atmospheres
5. 3 liters
6. 375°K
7. 2 atmospheres
8. 127°C
9. 750 ml
10. 4 atmospheres
11. 327°C
12. 550 torr, 200 torr
13. 288
14. 309
15. 0.1, 0.075, 1
16. 12.04×10^{23}
17. 44 amu
18. 64 g
19. 1.25 g/liter
20. 2.66 g/liter
21. 0.71 g/liter

CHAPTER 15

3. The electropositive hydrogens are attracted to the pairs of electrons on the oxygen of neighboring water molecules.
4. two
5. HF forms hydrogen bonds
6. $NH_3 > AsH_3 > PH_3$ as NH_3 forms hydrogen bonds
8. Water at 4° is the most dense. The less dense colder water is on the surface.
9. Its density is less than liquid water.
10. no, no
13. coordinated to cation, associated with anion or trapped in the crystal
14. The vapor pressure of the water in air equals the equilibrium vapor pressure of the hydrate.
16. There is a very low vapor pressure of water in the air.
19. Al > Cd > Sn
20. to lower the reactivity with water and oxygen and prevent rusting
21. gold
22. They are unreactive.

CHAPTER 16

3. (a) oxygen in nitrogen (b) oxygen in water (c) alcohol in water (d) sugar in water (e) hydrogen in platinum (f) mercury in gold (g) copper in silver
4. (a) glucose is solute in solvent water
(b) alcohol is solute in solvent water
(c) oxygen in solute in solvent water
(d) oxygen is solute in solvent nitrogen
(e) zinc is solute in solvent copper
5. yes, as in alloys
6. yes, as in air
8. (a) 10% H_2SO_4 (b) 4.7% NaCl (c) 33% alcohol

9. (a) 20 g, (b) 5 g, (c) 20 g, (d) 37.5 g
10. (a) 4, (b) 13.3, (c) 0.1
11. 50 g
12. 250 g
13. (a) 0.27 (b) 0.10 (c) 0.25
14. (a) 0.02 (b) 0.75 (c) 0.0005
15. (a) 9.8 g (b) 0.40 (c) 372
16. (a) 50 ml (b) 500 ml (c) 2500 ml
17. (a) 0.2 (b) 2 (c) 3
19. (a) 0.60 (b) 1.0 (c) 1.0
24. The pressure is decreased.
25. Release of nitrogen at lower pressure; helium is less soluble.
28. It decreases as the concentration of sugar increases.
30. Its vapor pressure is lowered by sugar content.
31. A lower temperature is achieved.

CHAPTER 17

4. No, it depends on the rate law.
6. enzymes
7. to speed up the reaction
9. (a) $\frac{[NO]^4[H_2O]^6}{[NH_3]^4[O_2]^5}$ (c) $\frac{[NH_3]^2}{[N_2][H_2]^3}$ (e) $\frac{[SO_3][NO]}{[SO_2][NO_2]}$
10. There is no effect on the magnitude of K.
11. As K increases, so does the degree of completeness of the reaction.
12. (a) and (d) favor product formation while (b) and (c) favor reactant formation. There is no effect of (e).
13. (a), (d), and (e) favor product while (b) and (c) favor reactant
14. (a), (b), and (e) will increase K with increased temperature
15. (a) $K_{sp} = [Ag^+][Cl^-]$ (c) $K_{sp} = [Cu^{2+}][S^{2-}]$
(e) $K_{sp} = [Ag^+]^2[CrO_4^{2-}]$ (g) $K_{sp} = [As^{5+}]^2[S^{2-}]^5$
16. (a) 10^{-5}, (b) 1.4×10^{-7}, (c) 7.2×10^{-10}, (d) 2.8×10^{-11}
17. (a) 8.1×10^{-9}, (b) 2.8×10^{-15}, (c) 1.6×10^{-11}, (d) 2.2×10^{-32}

CHAPTER 18

3. Nonelectrolytes do not conduct electricity.
4. Strong electrolytes are excellent conductors of electricity.
5. HNO_3, and HCl, HCN and $HC_2H_3O_2$, H_2O and sugar
6. (a), (e), (g), and (h) are weak acids. (b), (c), (d), (f) are strong acids.
7. (c) and (d) are weak bases. (a) and (b) are strong bases.
8. (a) H_3O^+, (b) HNO_2, (c) H_2SO_4, (d) NH_4^+, (e) HCl, (f) $HClO_4$
9. (a) OH^-, (b) NH_2^-, (c) CO_3^{2-}, (d) Cl^-, (e) HSO_4^-, (f) NO_3^-
10. 10^{-4}
11. 2×10^{-6}
12. 6.7×10^{-4}
13. 1.7×10^{-5}

14. 3, 1.7, 9
15. (a) 3.7, (b) 6.7, (c) 11.7, (d) 3.1
16. (a) 3, (b) 4.4, (c) 10
19. No, HCl is a strong acid.
20. 15 ml
21. 0.018*M*
22. 0.042*M*
23. 0.2*M*
24. (a) $H^+ + OH^- \longrightarrow H_2O$ (b) $Cl^- + Ag^+ \longrightarrow AgCl$
(c) $Fe + Cu^{2+} \longrightarrow Fe^{2+} + Cu$
(d) $BaO + 2H^+ \longrightarrow Ba^{2+} + H_2O$
(e) $Cl_2 + 2Br^- \longrightarrow 2Cl^- + Br_2$
(f) $Mg^{2+} + CO_3^{2-} \longrightarrow MgCO_3$
(g) $Zn + 2H^+ \longrightarrow Zn^{2+} + H_2$
(h) $2H^+ + CO_3^{2-} \longrightarrow H_2O + CO_2$
(i) $Zn^{2+} + S^{2-} \longrightarrow ZnS$

CHAPTER 19

4. (a) +1 (b) −2 (c) +7 (d) +4 (e) −2 (f) +6 (g) +4
(h) +5 (i) +7 (j) +5 (k) +6 (l) +4 (m) +3 (n) +3
5. (Answers given as oxidizing agent first followed by reducing agent.)
(a) $CoCl_2$, Al (b) Br_2, Cu (c) HNO_3, H_2SO_3 (d) O_2, S
(e) N_2, H_2 (f) Cl_2, NaBr (g) WO_3, H_2 (h) SnO_2, C
6. (Coefficients given in order of corresponding sequence in equation.)
(a) 2, 3, 2 (b) 2, 3, 2 (c) 2, 3, 2, 2 (d) 3, 2, 3, 1, 3
(e) 1, 4, 4, 1, 8 (f) 10, 1, 10, 4, 2
7. (Coefficients given in order of corresponding sequence in equation.)
(a) 1, 7, 8, 8, 4, 4 (b) 2, 2, 3 (c) 3, 6, 1, 5, 3, (d) 8, 5, 4, 4, 1, 4
(e) 3, 8, 3, 2, 4, (f) 1, 1, 1, 1, (g) 1, 8, 2, 4, 1, 4
8. (Coefficients given in order of corresponding sequence in equation.)
(a) 2, 1, 2, 1 (b) 1, 2, 1, 2 (c) 3, 8, 2, 3, 2, 4 (d) 1, 6, 14, 2, 6, 7
(e) 10, 2, 16, 2, 5, 8 (f) 34, 5, 3, 10, 6, 17 (g) 1, 6, 6, 6, 1, 3
9. (a) Cu^{2+}, (b) Zn^{2+}, (c) Hg^{2+}, (d) Hg^{2+}, (e) Br_2
10. (a) Li, (b) Na, (c) Zn, (d) Ni, (e) Cu, (e) I^-
11. (a) yes, (b) yes, (c) yes, (d) no, (e) yes
12. (a) 1.60, (b) 1.92, (c) 0.63, (d) 0.46, (e) 0.53
13. reduction

CHAPTER 20

3. (a) ${}^1_0n$, (b) ${}^1_1H$, (c) ${}^4_2He$, (d) ${}^0_{-1}e$, (e) ${}^0_{-1}e$, (f) ${}^0_1e$
4. (a) ${}^{14}_6C$, (b) ${}^{10}_5B$, (c) ${}^{238}_{92}U$, (d) ${}^{37}_{17}Cl$, (e) ${}^{65}_{30}Zn$, (f) ${}^2_1H$, (g) ${}^3_1H$, (h) ${}^{113}_{50}Sn$
5. (a) 82 and 124, (b) 92 and 142, (c) 53 and 74, (d) 7 and 6, (e) 11 and 14, (f) 1 and 2
6. One; it increases with increasing atomic weight.
7. A neutron is destroyed and a proton is formed.
8. A proton is destroyed and a neutron is formed.

10. (a) $_{-1}^{0}e$, (b) $_{2}^{4}\text{He}$, (c) $_{1}^{0}e$, (d) $_{0}^{1}\text{n}$, (e) $_{7}^{14}\text{N}$, (f) $_{97}^{249}\text{Bk}$, (g) $_{9}^{18}\text{F}$

11. (a) $_{9}^{21}\text{F} \longrightarrow {}_{-1}^{0}e + {}_{10}^{21}\text{Ne}$ (b) $_{35}^{75}\text{Br} \longrightarrow {}_{1}^{0}e + {}_{34}^{75}\text{Se}$

(c) $_{47}^{104}\text{Ag} \longrightarrow {}_{1}^{0}e + {}_{46}^{104}\text{Pd}$ (d) $_{84}^{212}\text{Po} \longrightarrow {}_{2}^{4}\text{He} + {}_{82}^{208}\text{Pb}$

(e) $_{96}^{240}\text{Cm} \longrightarrow {}_{2}^{4}\text{He} + {}_{94}^{236}\text{Pu}$ (f) $_{14}^{32}\text{Si} \longrightarrow {}_{-1}^{0}e + {}_{15}^{32}\text{P}$

12. (a) 50, (b) 25, (c) 12.5, (d) 6.25, (e) 3.12, (f) 1.06

13. 3(5570) years

14. 0.12 g

15. 4(30) minutes

16. $_{27}^{55}\text{Co}$

17. $_{5}^{10}\text{B}$

18. $_{99}^{247}\text{Es}$

19. $_{1}^{1}\text{H}$

20. $_{0}^{1}n$

GLOSSARY

Absolute zero the lowest attainable temperature (−273°C).
Acid a compound in which a hydrogen atom is the most electropositive element.
Acid oxide an oxide of a nonmetal which may react with water to yield an acid.
Acid salt a salt which contains one or more hydrogen atoms in the negative complex ion.
Actinide any element of atomic numbers 90 through 103 in which the 5*f* subshell becomes filled.
Actual yield the amount of product actually obtained from a chemical reaction.
Alkali metal any element of Group IA.
Alkaline earth metal any element of Group IIA.
Alloy a mixture of solid metals.
Alpha particle the helium nucleus consisting of two protons and two neutrons.
Amalgam a solution of a liquid in a solid.
Amphiprotic a substance which can act as either an acid or a base.
Amphoteric the property of behaving both as an acid and a base.
Anion an ion bearing a negative charge.
Anode the electrode to which negative ions migrate.
Artificial radioactivity radioactivity emitted from artificial isotopes.
Atmosphere a unit of pressure. Normal atmospheric pressure is 76 cm of Hg.
Atom the smallest unit of an element.
Atomic mass the mass equaling $\frac{1}{12}$ the mass of ^{12}C.
Atomic number the number of protons in the nucleus of an atom.
Atomic radius the radius of an atom usually expressed in angstrom units.
Atomic weight the average relative weight of a mixture of isotopes.

Avogadro's hypothesis equal volumes of different gases under the same conditions contain the same number of particles.
Avogadro's number the number of units of particles present in one mole, which is 6.02×10^{23}.

Barometer a device used to measure atmospheric pressure.
Base a compound in which the hydroxide ion is the negative ion.
Basic oxide an oxide of a metal which may react with water to yield a base.
Basic salt a salt which contains one or more hydroxide ions in addition to other anions.
Beta particle high energy electrons emitted from a nucleus.
Binary compound a compound consisting of only atoms of two elements.
Boiling point the temperature at which the vapor pressure of a liquid equals the atmospheric pressure.
Bond an interaction between atoms caused by a change in electronic structure of the atoms associated with each other.
Boyle's law the volume of a gas sample is inversely proportional to its pressure at a constant temperature.
Brass an alloy containing copper and a lesser amount of zinc.
Bronze an alloy of zinc and tin.
Buffer a solution of a weak acid and its conjugate base which resists pH changes.

Calorie the amount of heat required to raise the temperature of 1 g of water 1°C. The dietic Calorie is 1000 calories.
Catalyst any substance which causes a rapid rate of a reaction but is unchanged in the process.
Cathode the electrode to which positive ions migrate.
Cation an ion bearing a positive charge, such as Li^+.
Chain reaction a series of nuclear reactions maintained by particles produced in the decay process.
Charles' law the volume of a gas sample is directly proportional to the absolute temperature at constant pressure.
Chemical change a change in the composition of substances as a result of a chemical reaction.
Chemical energy energy stored in chemicals which may be released during a reaction.
Chemical equation a summary of the reactants and products involved in a reaction using coefficients and chemical symbols.
Chemical equilibrium a state in which the rate of one reaction is equal to the rate of the reverse of the reaction.
Chemical properties properties that depend on the composition and reactivity of substances.

Chemical reaction a chemical change characterized by the formation of new substances.
Chemical symbol a form of chemical shorthand identifying an element with one or two letters.
Chemistry a science dealing with composition, structure, and reactions of matter.
Colloid a mixture which is an intermediate state between heterogeneous mixtures and true solutions.
Combination reaction the direct union of two or more substances to produce one substance.
Combining capacity the number of hydrogen or chlorine atoms with which an element may combine.
Common chemical name a name in wide usage which does not indicate chemical composition.
Compound a substance containing combined elements which can be broken down only by chemical reactions.
Concentrated a relative term indicating a high proportion of solute to solvent.
Concentration a measure of the relative amounts of solute to solvent.
Coordinate covalent bond a covalent bond in which both shared electrons are supplied by one atom.
Cottrell precipitator a device to eliminate the charge of colloids and allow aggregation of particles.
Covalent bond a bond resulting from a shared pair of electrons.
Crenation the shrinking of cells.
Critical mass the mass of an isotope necessary to maintain a chain reaction.
Curie a unit of radiation intensity based on 3.7×10^{10} nuclear disintegrations per second.
Cyclotron a particle accelerator used in nuclear transmutations.

Decomposition reaction the reaction of a single substance to produce two or more simpler substances.
Deliquescence the process of absorbing sufficient water to form a solution.
Density mass per unit volume.
Deuterium the isotope of hydrogen having a mass number of 2.
Dialysis the removal of soluble low molecular weight materials from colloidal substances.
Dilute a relative term indicating a low proportion of solute to solvent.
Dilute solution a low concentration of solute in a solvent.
Dipole having two poles within a molecule as a result of unequal sharing of electrons.
Dipole moment a measure of the tendency of dipoles to become oriented in a field.
Dissociation the separation of ions in a crystal into an aqueous solution.
Dobereiner triad a group of three elements having similar chemical properties.
Ductile capable of being drawn into wires.

Efflorescence the process of losing water from a hydrate.
Electrolysis the passage of an electric current through an electrolyte to cause a chemical change.
Electrolyte any substance which when dissolved in water yields a conducting solution.
Electromotive series a list of metals in terms of their reactivity.
Electron a negative subatomic particle with a mass of 1/1840 that of a hydrogen atom.
Electron dot symbol a symbol depicting the valence electrons of an atom.
Electronegativity the measure of the attraction of electrons by an element in the presence of a second element.
Electron shell a group of electrons possessing similar energies and positions with respect to the nucleus.
Element a substance that cannot be constructed from or decomposed into simpler substances.
Empirical formula the simplest formula giving the smallest ratio of atoms present in a compound.
Emulsion a colloidal suspension of two liquids.
Endothermic reaction a chemical reaction in which heat is absorbed.
Energy the ability to do work.
Enzyme a catalyst in biological systems.
Equation a statement of a chemical change in terms of the composition of reactant and products.
Equilibrium a balance or equality between opposing forces or reactions.
Equilibrium constant the numerical value of the mass action expression.
Evaporation the slow transfer of liquid into the gaseous phase at a temperature below the boiling point.
Exothermic reaction a reaction in which chemical energy is released as heat.

Fission conversion of a heavy element into two light elements.
Foam a colloid of a gas dispersed in a liquid.
Formula weight the weight of one formula unit on the atomic weight scale.
Freezing point the temperature at which a liquid and a solid can exist in equilibrium.
Fusion conversion of two light elements into a heavy element.

Gamma ray high energy X rays emitted by radioactive nuclei.
Gay-Lussac's law the pressure of a gas is directly proportional to its absolute temperature at constant volume.
Gel a colloidal dispersion of a liquid within a solid.
Gram atomic weight the atomic weight of an element in grams.
Gram molecular weight the molecular weight of a substance in grams.
Group a column of the periodic table and the elements contained in it.

Half-life the time necessary for one-half of a radioisotope to decay.
Halogen any element of a Group VIIA.
Hard water any naturally occurring water that contains a high concentration of cations, such as calcium, which interferes with the cleansing action of soap.
Heat of fusion the energy required to melt 1 g of a solid at its melting point.
Heat of vaporization the energy required to convert 1 g of a liquid into a gas at the boiling point.
Heavy water water in which deuterium has replaced hydrogen.
Hemolysis the rupture of a cell wall.
Henry's law the solubility of a gas in a liquid is proportional to its partial pressure.
Heterogeneous mixture a mixture containing parts that are unlike.
Homogeneous mixture a mixture whose properties are uniform throughout.
Hydrate a crystalline compound which contains water in definite proportion by weight.
Hydrocarbon an organic compound containing only carbon and hydrogen.
Hydrogen bonding the intermolecular attraction between a positive hydrogen atom and the electron pair of an electronegative element.
Hydrometer the device used to measure the specific gravity of solutions.
Hydronium ion the H_3O^+ ion present in aqueous acid solutions.
Hydroxide ion the OH^- ion present in aqueous basic solutions.
Hygroscopic any substance which absorbs water from the atmosphere.
Hypertonic solution a solution whose salt concentration and osmotic pressure are higher than in blood.
Hypothesis a tentative model.
Hypotonic solution a solution whose osmotic pressure is smaller than within a cell.

Immiscible the phenomena of two liquids that will not dissolve in each other.
Indicator a substance which can change color within a specific pH range.
Ion any electrically charged atom or group of atoms.
Ionic bond a bond existing between ions as a result of electron transfer.
Ionic compound a compound composed of ions.
Ionization the formation of ions from some covalent substance when dissolved in water.
Ionization potential the energy required to remove an electron from an atom.
Isotonic solution a solution whose salt concentration and osmotic pressure are close to that of blood.
Isotope atoms of an element that contain differing numbers of neutrons.
Isotope effect a difference in physical and chemical properties of isotopes and their compounds.
IUPAC International Union of Pure and Applied Chemistry.

Kelvin scale the absolute temperature scale.
Kilo a prefix denoting 1000.

Kinetic energy the energy of motion.
Kinetics a study of the velocity of chemical reactions.
Kinetic theory of gases the theory that gases consist of moving particles that undergo elastic collisions.

Lanthanide any element of atomic numbers 58 through 71 in which the 4f subshell becomes filled.
Law an explicit statement of scientific fact.
Le Châtelier's principle if an external force is applied to an equilibrium, the system will readjust to reduce the stress imposed upon it.
Limiting reagent the reagent present in the smallest mole ratio.
Linear accelerator a device that accelerates electrically charged particles in a straight line.
Liter the unit of volume in the metric system.

Malleable the property of being able to be hammered into sheets.
Mass the amount of matter.
Mass action expression an expression indicating the relationship between the concentration of all substances at equilibrium.
Mass number the sum of the number of protons and neutrons in the nucleus of an atom.
Matter anything that has mass and occupies space.
Measurement a comparison to the standard measuring device.
Melting point the temperature at which a solid and a liquid can exist in equilibrium.
Metals elements which are malleable, ductile, and conduct electricity and heat.
Metathesis reaction the reaction of two or more compounds to produce two different compounds by an exchange of atoms or groups of atoms.
Meter the standard unit of length in the metric system.
Metric system a measuring system based on powers of ten.
Milli a prefix representing 1/1000.
Mixed salt a salt containing two or more different cations.
Mixture two or more substances mixed together.
Model an idea that corresponds to what is responsible for natural phenomena.
Molality the concentration of a solution expressed in moles of solute per 100 g of solvent.
Molar volume the volume of 1 mole of a gas at standard temperature and pressure (22.4 liters).
Mole the amount of material which contains a mass in grams equal to its mass on the atomic weight scale.
Molecule two or more atoms combined to give a neutral substance.
Molecular formula a formula which gives the total number of atoms of each type of element that is present in the compound.
Molecular weight the weight of a molecule on the atomic weight scale.

Natural radioactivity radioactivity of naturally occurring isotopes rather than transmuted elements.
Neutralization reaction the reaction of an acid or an acid oxide with a base or a basic oxide.
Neutron a neutral subatomic particle having the same atomic weight as a proton.
Newlands' octave an arrangement of elements by atomic weight into repeating sequences of seven elements.
Noble gas any of the elements of Group 0.
Nonmetals elements which are nonconductors of electricity.
Normality the concentration of a solute expressed in equivalents per liter of solution.
Nuclear fission the cleavage of a radioactive isotope into two or more lighter elements and other nuclear particles.
Nuclear fusion the combination of two light nuclei to form a heavier nucleus.
Nuclear reactions reactions involving the transformation of the nucleus of atoms.
Nuclear reactor a device that controls the rate of nuclear reactions and produces energy.
Nucleus the center of an atom which contains the protons and neutrons.

Octet rule atoms tend to achieve a noble gas electron configuration.
Orbital a region of space about a nucleus that contains electrons.
Organic chemistry a study of carbon-containing substances.
Osmosis the passage of liquids through a semipermeable membrane.
Osmotic pressure the pressure required to prevent net passage of liquids through a semipermeable membrane.
Oxidation the loss of electrons.
Oxidation number a positive or negative integer indicating the number of electrons lost or gained by an atom.
Oxidizing agent an agent that causes loss of electrons from a substance.

Partial pressure the pressure exerted by a single gas in a mixture of gases.
Percent composition the mass of each element in a compound divided by the total mass of all elements with the quotient multiplied by 100.
Percent yield the actual yield divided by the theoretical yield multiplied by 100.
Period a horizontal row of the periodic table.
Periodic law elements arranged in order of their atomic number exhibit periodic reoccurrence of properties.
Periodic table an arrangement of elements into periods and groups according to their physical and chemical properties.
pH the negative logarithm of the hydronium ion concentration.
Phase a homogeneous part of a system.
Physical change a change in state but not composition of a substance.

Physical property property which is determined without altering the chemical composition of the substance.
pOH the negative logarithm of the hydroxide ion concentration.
Polar covalent bond a covalent bond in which electrons are not shared equally between two atoms.
Potential energy energy of position or storage.
Pressure force per unit area.
Product the substance produced in a chemical reaction.
Protium the most abundant isotope of hydrogen, ${}^{1}_{1}H$.
Proton a positively charged subatomic particle.

Rad the absorption of 100 ergs of energy per gram of absorbing tissue.
Radioactive isotope an isotope that undergoes spontaneous nuclear reactions.
Raoult's law for nonvolatile solutes, the lowering of the vapor pressure of a solvent is directly proportional to the concentration of the solute.
Rare gases see *Noble gases.*
Reactant the starting material of a chemical reaction.
Reaction a process in which the composition of matter is altered.
Reducing agent the agent that causes the gain of electrons by a substance.
Reduction the gain of electrons.
Rem the product of rads and relative biological effectiveness.
Representative element any of the Group A elements and Group 0.
Roentgen a unit of radioactive dosage based on ionization of air.

Salt a compound containing the cation of a base combined with the anion of an acid.
Saturated solution a solution which cannot dissolve any additional solute.
Scientific method a method of observing and using the observations in the solution of problems.
Semimetals elements with some properties of both metals and nonmetals.
Significant figure the number of digits in a number which gives reliable information.
Solubility a description of the extent to which a solute dissolves in a solvent.
Solute the substance dissolved in a solvent.
Solution a homogeneous mixture.
Solvent the dissolving substance of a solution.
Specific gravity the ratio of the density of a substance to the density of water.
Specific heat the energy required to raise the temperature of 1 g of a substance by 1°C.
Spectator ions ions required for electrical neutrality but not involved in a chemical reaction.
State form in which matter may be found: gas, liquid, and solid.
Steel an alloy of iron with small amounts of other metals or nonmetals.
Stock system a method of indicating the oxidation number of metals in compounds.

Stoichiometry mathematical calculation of the quantities of reactants and products of a reaction from an equation.
STP standard temperature and pressure: one atmosphere and 0°C.
Sublimation the conversion of a solid directly into a gas.
Substitution reaction the reaction in which one element substitutes for or replaces another element in a compound.
Symbol a shorthand representation of an element.
Systematic chemical name a name which conveys information about the constitution of a compound and which has been agreed to by the IUPAC.

Ternary three components.
Theoretical yield the amount of product that would be obtained from a complete reaction.
Theory a model which has been tested.
Thermometer a device to measure the temperature of matter.
Titration a quantitative determination of the concentration of an acid or a base.
Transition element any of the Group B elements and Group VIII.
Transmutation the interconversion of elements accomplished by nuclear reactions.
Tritium a radioactive isotope of hydrogen having two neutrons in the nucleus.
Tyndall effect the scattering of light by suspended colloidal particles.

Valence the combining capacity of an element with either hydrogen or chlorine.
Valence electrons the electrons in the highest energy level of an atom.
Vapor pressure the pressure of a vapor in equilibrium with a liquid.

Weight a measure of the gravitational attraction for matter.

Zone of stability a region of isotopes which are not radioactive.

INDEX

76 77 78 79 9 8 7 6 5 4 3 2 1

(Based on the Atomic Mass of $^{12}C = 12$. The values for atomic masses given in the table apply to elements at they exist in nature, without artificial alteration of their isotopic composition.)

element	symbol	atomic number	atomic mass
Mercury	**Hg**	80	200.59
Molybdenum	**Mo**	42	95.94
Neodymium	**Nd**	60	144.24
Neon	**Ne**	10	20.183
Neptunium	**Np**	93	....
Nickel	**Ni**	28	58.71
Niobium	**Nb**	41	92.906
Nitrogen	**N**	7	14.0067
Nobelium	**No**	102	....
Osmium	**Os**	76	190.2
Oxygen	**O**	8	15.9994
Palladium	**Pd**	46	106.4
Phosphorus	**P**	15	30.9738
Platinum	**Pt**	78	195.09
Plutonium	**Pu**	94	....
Polonium	**Po**	84	....
Potassium	**K**	19	39.102
Praseodymium	**Pr**	59	140.907
Promethium	**Pm**	61	....
Protactinium	**Pa**	91	....
Radium	**Ra**	88	....
Radon	**Rn**	86	....
Rhenium	**Re**	75	186.2
Rhodium	**Rh**	45	102.905
Rubidium	**Rb**	37	85.47
Ruthenium	**Ru**	44	101.07
Rutherfordium	**Rf**	104	....
Samarium	**Sm**	62	150.35
Scandium	**Sc**	21	44.956
Selenium	**Se**	34	78.96
Silicon	**Si**	14	28.086
Silver	**Ag**	47	107.868
Sodium	**Na**	11	22.9898
Strontium	**Sr**	38	87.62
Sulfur	**S**	16	32.064
Tantalum	**Ta**	73	180.948
Technetium	**Tc**	43	....
Tellurium	**Te**	52	127.60
Terbium	**Tb**	65	158.924
Thallium	**Tl**	81	204.37
Thorium	**Th**	90	232.038
Thulium	**Tm**	69	168.934
Tin	**Sn**	50	118.69
Titanium	**Ti**	22	47.90
Tungsten	**W**	74	183.85
Uranium	**U**	92	238.03
Vanadium	**V**	23	50.942
Xenon	**Xe**	54	131.30
Ytterbium	**Yb**	70	173.04
Yttrium	**Y**	39	88.905
Zinc	**Zn**	30	65.37
Zirconium	**Zr**	40	91.22